Lecture Notes in Civil Engineering

Volume 371

Lecture Notes in Civil Engineering (LNCE) publishes the latest developments in Civil Engineering—quickly, informally and in top quality. Though original research reported in proceedings and post-proceedings represents the core of LNCE, edited volumes of exceptionally high quality and interest may also be considered for publication. Volumes published in LNCE embrace all aspects and subfields of, as well as new challenges in, Civil Engineering. Topics in the series include:

- Construction and Structural Mechanics
- Building Materials
- Concrete, Steel and Timber Structures
- Geotechnical Engineering
- Earthquake Engineering
- Coastal Engineering
- Ocean and Offshore Engineering; Ships and Floating Structures
- Hydraulics, Hydrology and Water Resources Engineering
- Environmental Engineering and Sustainability
- Structural Health and Monitoring
- Surveying and Geographical Information Systems
- Indoor Environments
- Transportation and Traffic
- Risk Analysis
- Safety and Security

To submit a proposal or request further information, please contact the appropriate Springer Editor:

- Pierpaolo Riva at pierpaolo.riva@springer.com (Europe and Americas);
- Swati Meherishi at swati.meherishi@springer.com (Asia—except China, Australia, and New Zealand);
- Wayne Hu at wayne.hu@springer.com (China).

All books in the series now indexed by Scopus and EI Compendex database!

Eric Strauss
Editor

Proceedings of the 7th International Conference on Civil Engineering

ICOCE 2023, 24–26 March, Singapore

Springer

Editor
Eric Strauss
Michigan State University
Dimondale, MI, USA

ISSN 2366-2557 ISSN 2366-2565 (electronic)
Lecture Notes in Civil Engineering
ISBN 978-981-99-4044-8 ISBN 978-981-99-4045-5 (eBook)
https://doi.org/10.1007/978-981-99-4045-5

This Springer imprint is published by the registered company Springer Nature Singapore Pte Ltd.
The registered company address is: 152 Beach Road, #21-01/04 Gateway East, Singapore 189721, Singapore

Conference Committees

Conference Chair

Prof. Zongjin Li, University of Macau, China

Conference Co-chair

Prof. Shane Snyder, Nanyang Technological University, Singapore

Program Chairs

Prof. Prashant Kumar, University of Surrey, UK
Prof. Pen-Chi Chiang, National Taiwan University, Taiwan

Publication Chair

Prof. Eric Strauss, Michigan State University, USA

Technical Program Committees

Prof. Akmal Abdelfatah, American University of Sharjah, UAE
Dr. Mubarak Al Alawi, Sultan Qaboos University, Oman
Dr. Ana Almerich-Chulia, Universitat Politecnica de Valencia, Spain

Dr. Godwin Akpeimeh, University of Leeds, UK
Dr. J. Amudhavel, Bhopal-Indore Highway, India
Prof. Michele Barbato, University of California, USA
Dr. Irina Benedyk, University at Buffalo, USA
Prof. Klaas van Breugel, Delft University of Technology, The Netherlands
Prof. Ta-Peng Chang, Civil and Construction Engineering, Taiwan Tech, Taiwan
Prof. Fadi Hage Chehade, Lebanese University, Beirut, Lebanon
Assoc. Prof. Chow Ming Fai, Monash University Malaysia, Malaysia
Assoc. Prof. Luigi Coppola, University of Bergamo, Italy
Dr. Yasmin Fathy, University of Cambridge, UK
Dr. Koorosh Gharehbaghi, RMIT University, Australia
Assoc. Prof. Habib Gurbüz, Suleyman Demirel University, Turkey
Dr. Hassan Hemida, University of Birmingham, UK
Prof. Gordon Huang, University of Regina, Canada
Dr. Yuner Huang, The University of Edinburgh, UK
Dr. Farhad Jazaei, The University of Memphis, USA
Dr. Fei Jin, Cardiff University, Wales, UK
Dr. Yongmin Kim, University of Glasgow Singapore, Singapore
Assoc. Prof. Paulo Mendonca, Universidade do Minho, Portugal
Dr. Saber Moradi, Toronto Metropolitan University, Canada
Prof. Hue Thi Nguyen, University of Wisconsin-Madison, Vietnam
Dr. Fernando Pachego Torgal, Universidade do Minho, Portugal
Dr. Kedsarin Pimraksa, Chiang Mai University, Thailand
Dr. Zhao Qin, Syracuse University, USA
Assoc. Prof. Siti Fatin Mohd. Razali, Universiti Kebangsaan Malaysia (UKM), Malaysia
Assoc. Prof. Pier Paolo Rossi, University of Catania, Italy
Assoc. Prof. Wong Wah Sang, University of Hong Kong, Hong Kong
Assoc. Prof. Korb Srinavin, Khon Kaen University, Thailand
Assoc. Prof. Sudharshan N. Raman, Monash University Malaysia, Malaysia
Assoc. Prof. June Tay, Singapore University of Social Sciences, Singapore
Dr. Kong Fah Tee, University of Greenwich, UK
Dr. Linh Truong-hong, Delft University of Technology, The Netherlands
Prof. Chien Ming Wang, The University of Queensland, Australia
Prof. Kejin Wang, Iowa State University, USA
Dr. Yuandong Wang, University of Utah, USA
Dr. Yan Xiao, Dalian University of Technology, China
Dr. Jinhui Yan, University of Illinois, USA
Assoc. Prof. Mijia Yang, North Dakota State University, USA
Assoc. Prof. He Yihong, Singapore University of Social Sciences, Singapore
Prof. Xiong Yu, Case Western Reserve University, Cleveland, USA

Preface

This publication contains peer-reviewed articles presented at the 2023 7th International Conference on Civil Engineering (ICOCE 2023). The meeting was successfully held in hybrid mode in Singapore during the time period of March 24–26. The conference has been held in Hanoi in 2017, Da Nang in 2018, Hue in 2019, virtual mode from 2020 to 2022. ICOCE is held annually to provide an ideal platform for researchers and practitioners to report the most recent innovations and developments, summarize state-of-the-art technologies, and exchange ideas and advances in all aspects of civil engineering. The material contained in this book covers both technical and policy issues present in the built environment.

This year, the event featured with four keynote speeches, two invited speeches, and eight oral sub-sessions as well as one poster session. The meetings had an international focus including both the developed and developing economies. Researchers, engineers, and academicians as well as industrial professionals, coming from various countries including USA, Mexica, Singapore, China, South Korea, Japan, Australia, Italy, Malaysia, Thailand, India, Cyprus, Bangladesh, Nepal, Indonesia, Philippines, United Arab Emirates, Mauritius, Croatia, Peru, Egypt, Sudan, and Saudi Arabia, have presented their research results and development activities during the conference.

ICOCE 2023 proceedings are a collection of outstanding submissions from universities, research institutes, and industries. All of the papers were subjected to peer review by conference committee members and international reviewers prior to publication. The papers selected depended on their quality and their relevancy to the conference. The ideas presented in this volume are all based on research which is designed to answer emerging questions in the field. There is a focus on structures of many types. All parts of the development process are the subject of material in the chapters. This includes design, construction, and impact on the surrounding area. The material has been organized to provide comparisons along similar lines of intellectual inquiry. There are many common themes contained in the book. Modeling is featured in many examples. The focus is on the evaluation of new technologies that solve problems, save energy, and promote sustainable development. In particular, advances in the use of common building materials to improve existing construction techniques are highlighted in the book. The improvement and use of concrete play a role in many chapters. Also discussed are small- and large-scale water-based projects, including dams, hydropower, water conservation, and flood control. Transportation issues are discussed in some chapters. These include the use of new technologies for the maintenance of existing highways and the building of new streets. There are chapters on monitoring existing pavement conditions as well as discussions on railroads.

The proceedings serve as a reference and hopefully inspire new ideas for scientists, engineers, graduate students, equipment manufacturers, and operators, as well as technical managers who are working in the broad field of civil engineering. The proceeding is divided into 6 chapters, with specific chapter themes as follows: Preparation and Properties of Advanced Building Materials; Hydraulic Engineering, Flood Control, and Bridge Engineering; Urban Planning, Construction, and Sustainable Development; Architectural Design and Structural Mechanics; Reinforced Concrete Structural Engineering and Durability of Concrete Structures; and Building Information Technology, Road Condition Monitoring and Construction Management.

We would like to express our sincere gratitude to all the authors who have contributed to this volume and to the organizing committees, reviewers, speakers, chairpersons, and sponsors as well as all the conference participants for their support to ICOCE 2023.

Dimondale, USA Prof. Eric Strauss

Contents

Preparation and Properties of Advanced Building Materials

Assessment of Waste Polyethylene Terephthalate (PET) as Sand in Sustainable Geopolymer Concrete: Non-destructive Tests Investigation 3
Mahmood Fawzi Ahmed

Achievement of Roller Compacted Concrete Incorporating GGBS by Using Soil Compaction Approach 11
Sorabh Saluja, Kulwinder Kaur, and Shweta Goyal

Effect of Paraformaldehyde Fibers on Mechanical and Shrinkage Properties of High-Strength Concrete 25
Ying Hu, Sheng Zhang, Junlong Jin, Bo Chen, Shenlin Hu, Yang Li, and Yixin Hong

Comparing Between Crushed and Fine Aggregate Recycled in Concrete 37
Ebraheem Alrashidi and Abdulaziz Almutairi

An Overview on Utilization of Steel Slag as Road Construction Materials 51
Ashkar Rahman Aquib, Zarrin Tasnim Probha, and Md. Arifin Haque

Study on the Performance of Ultra-Fine Cement Slurry Reinforced Coral Aggregates and Coral Concrete 69
Shuang Li, Zhuolin Xie, Jianmin Hua, Lepeng Huang, Jian Kang, and Xuran Liu

Development of Pavement Condition Index for Philippine Asphalt National Roads 81
Jamie Alea B. Ramos, Lea B. Bronuela-Ambrocio, Hilario Sean O. Palmiano, John Paul T. Dacanay, Lestelle V. Torio-Kaimo, and Jonas Christian R. Quero

Hydraulic Engineering, Flood Control, and Bridge Engineering

Overview of Critical Vortex on Horizontal Jet Fluidization for Sediment Flushing Systems 97
Rudi Azis, Farouk Maricar, Muhammad Arsyad Thaha, and Bambang Bakri

Large-Scale in Situ Direct Shear Test in the Construction of Keureuto Dam, Indonesia 107
Abi Maulana Hakim, Samira Albati Kamaruddin, Andhika Sahadewa, Ramli Nazir, and Haris Eko Setyawan

The Prediction of Lahar Flood Event Impact on the Inundation Areas in Gendol River, Indonesia 119
Jazaul Ikhsan, Elang Afif Hafizh Zhafran, Ani Hairani, and Mohd. Remy Zainol

Captive Use Mini Hydropower Project for Pumping Station 131
Gautam Narula, Vijayinder Kumar Dogra, Rahul Sharma Vaibhav Sapkal, Komal Bharadwaj, and Aradhyesh Sharma

An OSINT-Driven Security Analysis of Intelligent Construction of Water Conservancy Projects in China 139
Yuanbo Qi, Ronghui Yang, and Chenghe Su

Urban Planning, Construction, and Sustainable Development

Research on the Application of Comprehensive Geophysical Methods in Tunnel Investigation 153
Yong Hai He and Hong Qiang Zhang

Experimental Assessment of Leakage in Water Distribution Network 161
S. Atabay, T. A. Ali, Md. M. Mortula, and S. Sharifi

A Multi-task Oriented Optimization Method for Urban Rail Overhaul Workflow Based on Critical Chain Method 169
Shan Huang, Qin Luo, Jingjing Chen, and Tian Lei

The Role of Public–Private Partnership on Preservation-Led Projects in Urban China—A Comparative Perspective 185
Ting Zhang, Fangqian He, Zhouquan Li, and Ran Xu

Gamification to Stimulate Green Behaviors in Cities 195
Joyce Ngo, Emmanuel Fragnière, Blaise Larpin, and Jean-Michel Sahut

Challenges of Municipal Solid Waste Management in Jalandhar, Punjab (India): A Case Study 207
Davinder Singh and Sanjeev Kumar

Architectural Design and Structural Mechanics

Evaluating Energy-Saving Potential of Passive Design Technologies Based on Residential Architectural Prototypes 217
Jiuwei Liu, Yuanli Ma, and Wu Deng

3D Modeling of Folded Footings with Ring Beam on Sand Using Various Folding Angles 231
Ahmed E. Gomaa, Ahmed M. M. Hasan, Yasser M. Mater, and Sherif S. AbdelSalam

Approximate Estimation for Global Buckling Load of Cylindrical Single-Layer Grid Shells: Fitting of Envelope Equations Based on Regression Analysis 241
Baoxin Liu, Pei-Shan Chen, Jialiang Jin, and Xiangdong Yan

Reinforced Concrete Structural Engineering and Durability of Concrete Structures

Predicting the Performance of Shear Wall Structures Using the Confidence Nets Model 257
Nouraldaim F. A. Yagoub and Wang Xuxin

Impacts of Web Stiffener Locations on Capacities of Cold-Formed Steel SupaCee Sections 267
Ngoc Hieu Pham

Study on the Early Shrinkage Behavior of Coral Aggregate Concrete Reinforced with Ultra-Fine Cement 277
Guosong Hu, Zhuolin Xie, Jianmin Hua, Lepeng Huang, Songxiao Huang, and Qiming Luo

Evaluation of Soil-Structure Interaction on RC Framed Irregular Building Under Varying Ground Conditions 291
Arnab Chatterjee and Heleena Sengupta

Electrochemical Technique to Evaluate Carbonation Behavior of Reinforced Concrete 305
Kulwinder Kaur, Sorabh Saluja, and Shweta Goyal

Performance Improvement of Reinforced Concrete Beams Strengthened with GFRP Sheet 317
Achmad Zultan Mansur, Rudy Djamaluddin, Herman Parung, and Rita Irmawaty

Assessment of the Pressure-Impulse Curves of Reinforced Concrete Panels Considering Full Blast Loading History 329
Nasser A. Alarfaj and Omar M. Alawad

Effect of Steel Fiber Volume Ratio on Bending Moment Transfer Coefficient of SFRC Shield Tunnel Under Staggered Assembling 339
Shuo Yu, Huajun Sun, Miaofeng Cao, and Changbao Liu

Building Information Technology, Road Condition Monitoring and Construction Management

An Assessment of Road Condition Monitoring Practice and Technologies in the Philippines 355
John Paul T. Dacanay, Lestelle V. Torio-Kaimo, and Lea B. Bronuela-Ambrocio

BIM Cost Calculator: Contract Costing of Building Information Modeling Services Using Parametric Estimates for BIM-Based Projects in the Philippines 365
Nel Ann Beloso and Dante Silva

Preparation and Properties of Advanced Building Materials

Assessment of Waste Polyethylene Terephthalate (PET) as Sand in Sustainable Geopolymer Concrete: Non-destructive Tests Investigation

Mahmood Fawzi Ahmed

Abstract The utilization of waste plastic and clay brick in geopolymer concrete will protect the environment, reduced carbon dioxide (CO_2) emission and preservation of raw materials. This paper aims to evaluate the substitution of natural fine aggregate with shredded polyethylene terephthalate (PET) waste on some properties of geopolymer concrete based on blended (50:50)wt% metakaolin (Mk) and waste brick powder (WBP). Toward that end, four geopolymer concrete mixtures were prepared; three mixtures containing PET aggregate as sand replacement by volumetric levels (10%, 15% and 20%), and one mixture without PET aggregate as control mixture for comparison. Some non-destructive tests have been made to examine the influence of PET particles on the voids content; ultrasonic pulse velocity (UPV) and dynamic modulus of elasticity (Ed) to all geopolymer concrete mixtures at 7 and 28 days. Results indicated that inclusion of PET particles as fine aggregate could yield smaller voids content, higher pulse velocity and improved dynamic modulus of blended Mk-WBP geopolymer concrete. However, the mixture of 20% PET aggregate accorded the lowest voids content (8.90%) and higher UPV (4.444 km/s), while the 10% PET aggregate provided the optimum dynamic modulus of elasticity (28.41 GPa). This research could support the waste management and upgrade of sustainable geopolymer concrete for wide civil engineering applications.

Keywords Geopolymer · Brick powder · Waste PET · Dynamic modulus · Metakaolin

1 Introduction

Massive expansion in the construction industry has caused a significant environmental global concern. Concrete of Ordinary Portland Cement (OPC) is a primary building material and the most extremely used overall. The excessive exploitation of

M. F. Ahmed (✉)
General Directorate of Education Anbar-Ministry of Education, Hit 31007, Iraq
e-mail: mahmodmfa81@gmail.com

E. Strauss (ed.), *Proceedings of the 7th International Conference on Civil Engineering*, Lecture Notes in Civil Engineering 371, https://doi.org/10.1007/978-981-99-4045-5_1

(OPC) and depletion of natural coarse and fine aggregates have made the concrete production unsustainable with adverse environmental effects [1]. The manufacturing of (OPC)—key ingredient—of concrete has notable environmental drawbacks associated with energy consumption, carbon dioxide (CO_2) gas emission and exhaustion of natural resources. It is reported that the production of one ton of OPC releases one ton of CO_2, which amounts approximately 7–8% of global CO_2 emission [2]. Hence, low carbon concrete, i.e., geopolymer concrete (GPC) appears as a suitable eco-alternative to conventional cement concrete and possesses the same OPC concrete performance in strength and durability [3]. Moreover, sizeable amounts of clay brick waste and plastic waste are generated every year causing an environmental challenge around the world. Most of the brick waste are disposed to landfills or dumped into open area, which cause a serious environmental pollution problem. Similarly, Polyethylene Terephthalate (PET) waste, which mostly comes from single-use plastic bottles, is accumulated on earth surface or ends up in the rivers and oceans, causing a pollution crisis over a long time because of its non-biodegradable character [4]. The best practice in overcoming the aforementioned environmental problems is recycling-reuse of these waste materials as a substitution for concrete ingredients. Due to the significant content of silica oxide SiO_2 and alumina oxide Al_2O_3 in chemical composition, many researchers have focused on using waste clay brick in powder form as a precursor for geopolymer concrete [5]. Tuyan et al. [6] reported the possibility to gain a compressive strength of 36.2 MPa from clay brick powder-based geopolymers by curing the specimens at 90 °C and 40% relative humidity for 5 days. Mahmoodi et al. [7] optimized the compositions of red clay brick waste-based geopolymer binders under ambient temperature curing. Furthermore, experiments by Ahmed et al. [2] displayed that the dynamic modulus of metakaolin-based geopolymer concrete increased from 20.44 GPa to 20.93 GPa for mixtures that contain 20% of brick powder as a substitution for metakaolin. On the other hand, several studies are carried out by many researchers on employed waste PET plastic as a fine aggregate substitution in ordinary concrete [8]. Kangavar et al. [9] observed that the modulus of elasticity reduced gradually with an increase of PET replacement ratio in mixture, especially for the specimens with 30%–50% PET content. Similarly, Al-Hadithi and Al-Ani [10] investigated the effect of substitution fine aggregate with 2.5%, 5% and 7.5% of PET waste on the properties of high performance concrete. They found that the static and dynamic modulus of elasticity have been dropped with an increase of PET replacement level. Some studies discussed the use of PET as fibers in geopolymer concrete. In addition, Faiz Uddin Ahmed Shaikh [11] investigated the flexural and tensile behavior of fly ash-slag-based geopolymer concrete which was included recycled PET fibers. From the previous reviews, it can be seen that there is limited information about using waste PET as fine aggregate to synthesize geopolymer composite based on the precursor of metakaolin and brick powder. Therefore, this study aims to examine the effect of inclusion of PET particles as fine aggregate, in particular; on voids content, ultrasonic-pulse velocity and dynamic modulus of elasticity for blended Mk-WBP-based GPC. Consequently, this paper will suggest a sustainable solution toward minimizing wastes of clay brick and plastic PET and tries to enhance the sustainability of concrete.

2 Experimental Program

2.1 Materials

In this study, the main precursor consisted of blended Iraqi metakaolin and clay brick powder at (50:50)% by weight. The process of manufacturing WBP has been presented in previous research [3], and both Mk and WBP are complied with ASTM C-618 [12] as a natural pozzolan. The alkaline activator was a chemical solution prepared by mixing sodium silicate (Na_2SiO_3) with 14 M (molarity) concentration sodium hydroxide (NaOH) at (1: 2.5) weight ratio. The used natural coarse aggregate was (5–14) mm gravel, with specific gravity of 2.62. Natural sand (<5 mm) was used as fine aggregate. All aggregates matched the Iraqi standard No.45–1986 [13].

The production of PET aggregate included these steps: (i) collected waste of plastic water bottles from local streets; (ii) labels were removed; (iii) shredded into small particles by plastic grinder machine and finally sift the shredded PET by standard sieves to comply with sand gradation. The obtained PET aggregate had a flaky to fiber-form shape with some pellet particles as shown in Fig. 1. The detailed properties of fine aggregate and shredded PET particles are presented in Table 1.

To improve the workability of fresh mixture, extra tap water and superplasticizer (Conplast SP-2000) were added with a dosage of 5% and 2% by weight of binder, respectively.

Fig. 1 Shredded fine PET aggregate

Table 1 Properties of sand and PET aggregate

Properties	Sand	PET aggregate
Bulk density, kg/m^3	1787	447
Specific gravity	2.6	1.2
Grain size, mm	0.15–5	0.15–5
Water absorption, %	1.6	0.0

2.2 Mix Proportions, Preparation and Test Methods

The reference mixture of Mk-WBP-based GPC has been designed based on a preliminary research process, which was not part of this study. In addition to the control mix (0% PET-aggregate), three mixtures have been prepared in order to substitute natural fine aggregate with shredded PET aggregate at levels 10%, 15% and 20% by volume. For all mixtures, the binder consisted of 50% Mk: 50% WBP, and the alkali activator to binder ratio was 0.65. The identification, mix proportions and quantity of materials for all mixes are tabulated in Table 2.

An electrical pan mixer was used to prepare all mixtures by thoroughly mixing solid materials (binder, gravel, sand and PET aggregate) with the liquids (alkaline solution, extra water and SP) until a homogenous fresh mixture was obtained. The fresh mixture was poured as three layers in the molds, and compacted via a vibrating table, then left to set under ambient conditions. All specimens were demoulded after 24 h and placed inside a drying oven for five hours at 80 °C, followed by exposure to sunlight until the time of test. All mixtures of concrete were tested by performing voids content, pulse velocity (UPV) and elastic dynamic modulus. The details about test standard references, specimens and test age are outlined in Table 3. The dynamic modulus (Ed) was conducted by means of measuring the highest resonance frequency of specimens that vibrated in torsional mode, as explained in previous study [3]. For this study, Ed calculated using the following formula:

$$\mathrm{Ed} = \mathrm{K} \times \mathrm{W} \times \mathrm{N}^2 \times 10^{-9}. \quad (1)$$

Table 2 Mix design of all MK-WBP geopolymer concrete mixtures

Designation	Proportion of PET (%)	Content of components in kg/m^3						
		Mk	WBP	Gravel	Sand	PET	Alkali solution	SP
Ref	0	207.5	207.5	1240	475	0	270	8.3
M10PET	10	207.5	207.5	1240	427.47	11.9	270	8.3
M15PET	15	207.5	207.5	1240	403.7	17.8	270	8.3
M20PET	20	207.5	207.5	1240	380	23.8	270	8.3

Table 3 Details of the tests

Type of test	Standard	Specimen detail	Test age (day)
Voids content	ASTM C642 [14]	Fractures pieces from prisms	7 and 28
Ultrasonic pulse velocity	ASTM C597 [15]	Cube 100 × 100 × 100 mm, (3 per age)	7 and 28
ED	ASTM C215 [16]	Prism 100 × 100 × 400 mm,	28

where

W: dry weight of specimen (kg); N: fundamental frequency (Hz) and K: correction factor is taken as 3.25 for this study.

3 Results and Discussion

3.1 Voids Contents

The volume of permeable voids within the concrete matrix directly influenced the durability. Figure 2 shows the voids content of all geopolymer concrete mixes. It was clear that the voids content for all mixtures at 28 days was higher than the 7-day values. This behavior is related to water's evaporation from pores and loss of moisture with time, consequently more voids and pores will engender within concrete matrix. It was observed that the voids content of MK-WBP-based GPC steadily reduced with increasing of PET-aggregate content. The reduction in voids at the age of 28 days was 1.0%, 5.0% and 10.0%, respectively, for M10PET, M15PET and M20PET mixtures, as compared with reference mix. This trend may be attributed to the possibility of affected PET aggregate by chemical interaction with alkaline solution, which improved the cohesion forces and minimized pores size between geopolymer paste and interface of PET particles.

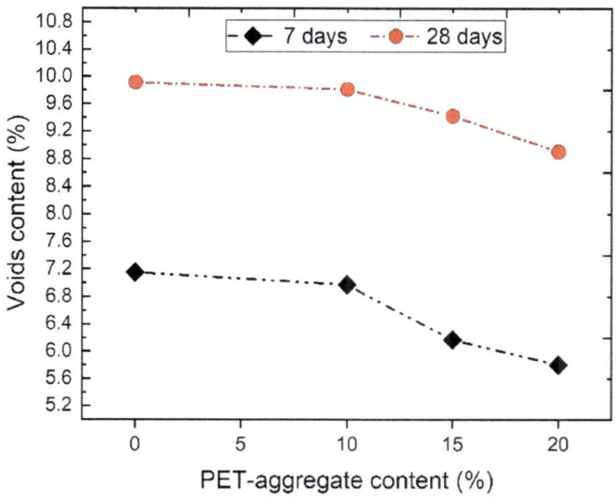

Fig. 2 Voids content with different PET percentages

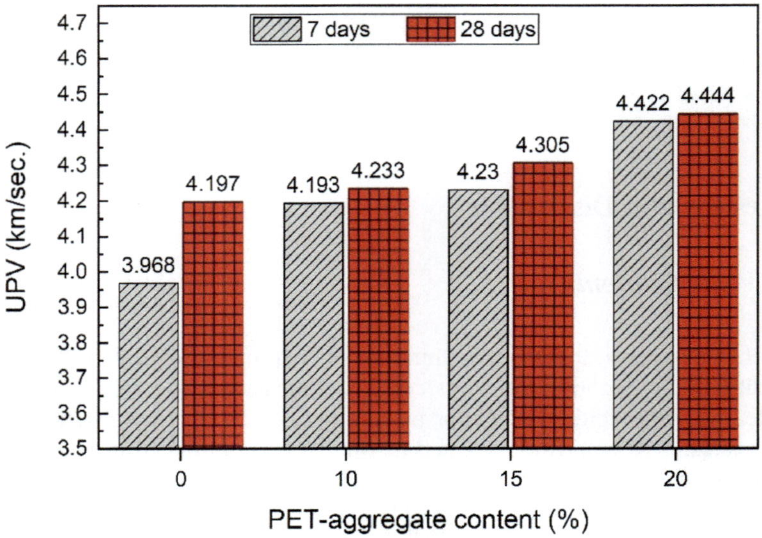

Fig. 3 UPV of geopolymer concrete with different PET percentages

3.2 UPV Test

This test is implemented for evaluating the strength, durability and quality of developed concrete without any destruction. The UPV values of all geopolymer concrete specimens at 7 and 28 days were given in Fig. 3. The results showed that the UPV for all MK-BP-GPC mixes increased with aging due to the continuation of geopolymerization process and produced denser geopolymer paste with less defects. As shown in Fig. 3, the pulse velocity of control mix (4.197 km/s) increased with increasing of PET aggregate content. The rate of increment reached 0.85%, 2.57% and 5.88% for replacement level 10%, 15% and 20%, respectively, at the age of 28-day. This accrued due to the positive effect of PET aggregate, which provided more compatibility with geopolymer paste. Consequently, dense geopolymer matrix with less voids and pores will be induced, as mentioned previously, and this will accelerate the transmission of the pulse waves. These results indicated that the matrix of all Mk-WBP-based GPC mixes was homogenous, and uniform. The mixes included PET particles can be classified as a good concrete quality grading according to IS 13311 [17].

3.3 Dynamic Modulus of Elasticity (Ed)

The test results of Ed of all mixtures at age 28-day were plotted in Fig. 4. It can be seen that the inclusion of waste PET particles enhanced the dynamic modulus of MK-WBP-based GPC at varying percent. This may be because of the effect of

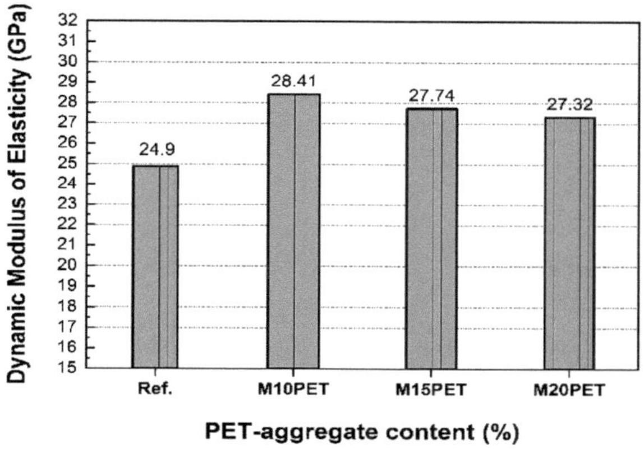

Fig. 4 Effect of PET aggregate on dynamic modulus of elasticity at 28-day

PET particles to reduce the pores and voids in geopolymer matrix as discussed previously. The elastic dynamic modulus of control mix reached 24.9 GPa, and this value increased with decreasing rate to reach 14.09%, 11.40% and 9.71% for PET content 10%, 15% and 20%, respectively. Since the density of PET aggregate was lower than natural sand, thus the mass (weight) of specimens, which was a parameter in Eq. 1, will be lighter. Consequently, the elastic dynamic modulus of mixture declined as the dosage PET replacement increased, but it was still higher than the reference mix (0% PET aggregate).

4 Conclusions

The following conclusions are drawn based on the experimental tests results:

- The voids content in blended Mk-WBP-based GPC decrease when using PET particles as fine aggregate. The voids content of control mixture (9.91%) reduced up to 8.90% when using 20% PET aggregate.
- The incorporation of shredded PET waste as fine aggregate showed an increment of the transmitted pulse velocity within the matrix of blended Mk-WBP geopolymer concrete. The mix of 20% PET aggregate has the highest UPV (4.444 km/s) and presents good concrete quality grading.
- The Mk-WBP-based GPC produced with 10–20% PET aggregate as a direct substitute for natural fine aggregate shows improvement in dynamic modulus of elasticity higher than the control mix.
- The optimum replacement level with the higher dynamic modulus is 10%.
- The achieved results detect that shredded PET waste can be used as fine aggregate to produce sustainable concrete geopolymer based on blended metakaolin and

brick powder. This will offer an ecologic solution to minimize the menace of CO_2 emission and pollution with waste brick and waste plastic bottles.

- This type of concrete needs further investigations on its durability and time-dependent properties.

References

1. Wang, B., Yan, L., Fu, Q., Kasal, B.: A comprehensive review on recycled aggregate and recycled aggregate concrete. Resour. Conserv. Recycl. **171**, 105565 (2021)
2. Ahmed, M.F.: Feasibility tests of high volume blended Metakaolin-brick powder concrete incorporating wastes of crushed brick and plastic as aggregate. Civil Eng. Architect. **9**(6), 1837–1845 (2020).https://doi.org/10.13189/cea.2021.090616
3. Ahmed, M.F., Khalil, W.I., Frayyeh, Q.J.: Effect of waste clay brick on the modulus of elasticity, drying shrinkage and microstructure of Metakaolin-based geopolymer concrete. Arabian J. Sci. Eng., 1–13 (2022). https://doi.org/10.1007/s13369-022-06611-02022
4. Akkouri, N., Bourzik, O., Baba, K., Tayeh, B.: Mechanical and thermal performances of eco-friendly mortar containing recycled pet as partial sand replacement (2021). 19 October 2021, PREPRINT (Version 1) available at Research Square https://doi.org/10.21203/rs.3.rs-932046/v1
5. Sun, K., Peng, X., Chu, S.H., Wang, S., Zeng, L., Ji, G.: Utilization of BOF steel slag aggregate in Metakaolin-based geopolymer. Constr. Build. Mater. **300**, 124024 (2021)
6. Tuyan, M., Andiç-Çakir, Ö., Ramyar, K.: Effect of alkali activator concentration and curing condition on strength and microstructure of waste clay brick powder-based geopolymer. Compos. B Eng. **135**, 242–252 (2018)
7. Mahmoodi, O., Siad, H., Lachemi, M., Dadsetan, S., Sahmaran, M.: Optimization of brick waste-based geopolymer binders at ambient temperature and pre-targeted chemical parameters. J. Clean. Prod. **268**, 122285 (2020)
8. Bamigboye, G.O., Tarverdi, K., Umoren, A., Bassey, D.E., Okorie, U., Adediran, J.: Evaluation of eco-friendly concrete having waste pet as fine aggregates. Cleaner Mater. **2**, 100026 (2021)
9. Kangavar, M.E., Lokuge, W., Manalo, A., Karunasena, W., Frigione, M.: Investigation on the properties of concrete with recycled polyethylene terephthalate (PET) granules as fine aggregate replacement. Case Stud. Constr. Mater. **16** (2022)
10. Al-Hadithi, A.I., Fawzi Al-Ani, M.: Effects of adding waste plastics on some properties of high performance concrete. 2018 11th International Conference on Developments in eSystems Engineering (DeSE) (2018)
11. Shaikh, F.U.: Tensile and flexural behaviour of recycled polyethylene terephthalate (PET) fibre reinforced geopolymer composites. Constr. Build. Mater. **245**, 118438 (2020)
12. ASTM C-618:2017, "Standard Test Method for Coal Fly Ash and Raw or Calcined Natural Pozzolan for Use in Concrete", American Society for Testing and Materials (2017)
13. Iraqi Standard Specification No. 45, "The aggregates of the natural resources used in concrete construction", Iraq (1984)
14. ASTM C642:2013,"Density, Absorption, and Voids in Hardened Concrete", American Society for Testing and Material (2013)
15. ASTM C597:2015, "Standard Test Method for Pulse Velocity through Concrete", American Society for Testing and Materials (2015)
16. ASTM C215:2014, "Fundamental Transverse, Longitudinal, and Torsional Resonant Frequencies of Concrete Specimens1", American Society for Testing and Materials (2014)
17. Indian Standard IS. 13311 (Part 1):1992. Non-destructive Testing of Concrete-Methods of Test, Part 1: Ultrasonic Pulse Velocity, Reaffirmed (2004)

Achievement of Roller Compacted Concrete Incorporating GGBS by Using Soil Compaction Approach

Sorabh Saluja, Kulwinder Kaur, and Shweta Goyal

Abstract Roller compacted concrete (RCC) gained momentum as a paving material because of economy, strength and fast placement. The basic constituents of RCC are similar to the conventional concrete. However, it requires a larger proportion of sand, lesser water and cement content in comparison with the conventional concrete. The mix design of the RCC is based on either consolidation (concrete) technique or compaction (soil, geotechnical) technique. In this paper, the mix design steps of RCC for pavement (RCCP) by using the soil approach are presented. Further, the effect of using ground granulated blast furnace slag (GGBS) and type of aggregates on the resultant properties of RCC were investigated. Different RCC mixtures were prepared with two types of aggregates and by replacing cement with GGBS. The optimum water content of each mix was finalized after conducting a compaction test for each replacement level. Experimental results show that the soil compaction approach leads to minimum voids in the finalized mix with maximum density. Limestone aggregate and cement replacement with GGBS increased the water requirement of the RCC mix. The experimental results show that GGBS inclusion had a marginal effect on 28 days compressive strength in the case of limestone aggregate, while for crushed gravel aggregate large decrease in strength was observed at 60% replacement level. In terms of 28 days of flexural strength behaviour, the better performance was shown by the 40% GGBS inclusion with limestone aggregate.

Keywords RCCP · Compaction · Limestone · GGBS · Compressive strength

S. Saluja (✉) · K. Kaur
Department of Civil Engineering, Punjabi University, Patiala 147002, India
e-mail: sorabh_ucoe@pbi.ac.in

S. Goyal
Department of Civil Engineering, Thapar Institute of Engineering and Technology, Patiala 147004, India

E. Strauss (ed.), *Proceedings of the 7th International Conference on Civil Engineering*, Lecture Notes in Civil Engineering 371, https://doi.org/10.1007/978-981-99-4045-5_2

1 Introduction

Roller compacted concrete (RCC) is a dry concrete that exhibits zero slump and gained popularity because of economy, strength and fast placement [1, 2]. The constituents of RCC are similar to those of conventional concrete while the construction and placement procedure is similar to the asphalt pavement [3]. Previously, the adaptability of RCC was limited to the dam construction but now it has gained acceptance as a pavement material also. As a pavement material, it is used for a wide variety of heavy vehicle applications such as lumber storage, log sorting yard, military vehicle roads, parking areas, bulk storage areas, forestry haul roads, airfields, secondary highways, truck and automobile parking, etc. [1, 4]. Construction with RCC is economical as compared with the conventional concrete because formworks, reinforcing and dowel bars are not required for RCC pavements. The transverse cracking in RCC pavements is prevented through the denser particle structure leading to a good interlock between aggregate and the subbase interface [5]. Additionally, low water-cement ratio used in RCC leads to lesser cement paste in the concrete matrix, thus leading to lower shrinkage which reduces the possibility of cracks in RCC pavements.

The basic ingredients of the RCC are similar with the conventional concrete, with the variation in proportioning of various ingredients. RCC requires a larger proportion of sand, lesser water and cement content on comparison with the conventional concrete. To attain strength comparable with the conventional concrete, RCC requires 20–28% less cementitious material [6, 7]. Utilization of various mineral admixtures such as micro silica fume, fly ash, ground granulated blast furnace slag (GGBS) helps in reducing the cost of RCC further and improves the resultant properties of the mix [8–10].

The design of the RCC mix can be done by using two philosophies [11]; one is based upon concrete approach while the other is based upon soil (geotechnical) approach. The basic principle involved in concrete approach is consolidation, i.e. lowering of w/c ratio; while soil approach is based on compaction, i.e. the moisture–density relationship. The resultant mixes achieved by both of these approaches should have zero slump (RCC). The RCC designed by concrete philosophy has more fluid consistency and is more workable than the mixes designed by soil approach [12, 13]. The mix design procedure for concrete approach is based upon the theory of water–cement relationship. On the other hand, in soil approach, RCC is considered as a cement-enriched processed soil with a mix design based on moisture–density relationship [11, 13]. Under this approach, optimum moisture content is determined for an individual cementitious material and aggregate type. It is said that if other factors are kept constant, RCC mixes obtained by using soil approach have higher compressive strength and better abrasion resistance as compared to mixes designed using concrete approach [11]. Also, soil approach is comparable to the field application of RCC in pavements since the compaction applied by the surcharge in laboratory can correspond to the effort applied by the rollers in the field [13].

The present study is aimed at achieving the zero slump concrete for pavement (RCCP) by using soil compaction approach. Also, the influence of aggregate type (crushed gravel, limestone) and the use of GGBS as cementitious material on the resultant RCCP mix were investigated.

2 Materials

2.1 Cementitious Material

In this present study, Ordinary Portland cement (OPC) satisfying Indian standard BIS-8112 [14] and GGBS conforming to BIS-12089 [15] were used as cementitious materials for preparing RCCP mixes. The chemical composition and physical properties of cement and GGBS are given in Table 1.

2.2 Aggregates

The selection of aggregate plays an important role in the composition of RCCP mixes. It constitutes approximately 75–85% of total volume of the RCCP and hence has a significant role in fresh and hardened concrete properties. The maximum size of coarse aggregates used in RCCP varies depending upon its application in dams or in pavements, with the smaller size aggregates used in later applications. According to ACI 325.10R-95 [1] the nominal maximum size of aggregates should be limited to 3/4 inches (19 mm) in order to avoid segregation in RCC pavement. In the present study, two types of coarse aggregates were used; viz crushed gravel (CG) and limestone (LS) aggregate, in two particle size variations, i.e. nominal maximum size of 20 mm and 10 mm. The fine aggregate used was river sand conforming to Zone-II as per BIS-383 [16]. The physical properties of both fine and coarse aggregates are presented in Table 2.

Table 1 Chemical and physical properties of cement and GGBS

Material	Chemical properties (%)								
	SiO_2	Al_2O_3	Fe_2O_3	CaO	MgO	SO_3	Na_2O	K_20	LOI
Cement	20.68	4.87	3.35	62.13	1.73	2.43	0.21	0.69	1.94
GGBS	31.61	21.70	2.50	33.21	8.00	0.18	0.85	–	0.98
	Physical properties								
	Specific gravity			Fineness (m^2/kg)			Slag activity index		
Cement	3.12			305			–		
GGBS	2.83			385			1.73		

Table 2 Physical properties of fine and coarse aggregates

Physical properties	Fine aggregates	Coarse aggregates			
		Crushed gravel (CG)		Limestone (LS)	
		20 mm	10 mm	20 mm	10 mm
Fineness modulus	2.85	6.80	6.04	6.69	5.97
Specific gravity	2.70	2.63	2.65	2.51	2.58
Water absorption (%)	1.8	1.38	1.40	1.62	1.69
Los angles abrasion (%)	–	2400*		2820*	
Impact value (%)	–	1431*		1828*	
Crushing value (%)	–	1619*		1416*	

* Values were measured after combining 20 mm and 10 mm aggregates in equal proportions

3 Experimental Programme and Results

The experimental programme consists of two parts; firstly the proportions of basic ingredients of RCC control mix were finalized. Secondly, the effect of GGBS as cementitious material and the use of either crushed gravel or limestone aggregates on the resultant strength properties of the RCCP mix were investigated.

3.1 Mix Proportioning of RCCP

Selection of combined aggregates gradation. Under the soil compaction approach, the first step involves the finalization of aggregate proportion to create a dense graded combined aggregate curve as per ACI 211.3R [17]. To achieve this target, various blends of fine aggregates, 10 mm and 20 mm coarse aggregates which fall under the ACI specified limits were tried. The finalized combined aggregate grading curve for both aggregate types viz; limestone aggregates and crushed gravel are presented in Fig. 1. Finalized gradation consists of a combination of 55% of fine aggregate, 30% of 10 mm coarse aggregate and 15% of 20 mm coarse aggregate. It can be seen that the grading curves for both aggregate types lie within the ACI recommended limits.

Selection of optimum water content. After finalizing the aggregate proportions, water content was determined by following combined guidelines laid down in ASTM D 1557 [18] and ACI 211.3R [17]. According to ACI 211.3R [17], the cementitious material content for RCC pavement range from 10 to 17% of the total dry mass of coarse and fine aggregate, depending upon strength and durability requirements. Therefore, for obtaining RCC mix, cementitious material content was varied from 11 to 17% in various trail mixes and the optimum moisture content of each mix was determined by using ASTM D 1557 [18].

For moisture–density test, 7 numbers of trail mixes were prepared with cement content varying from 11 to 17% of total dry mass of aggregates, with an increment

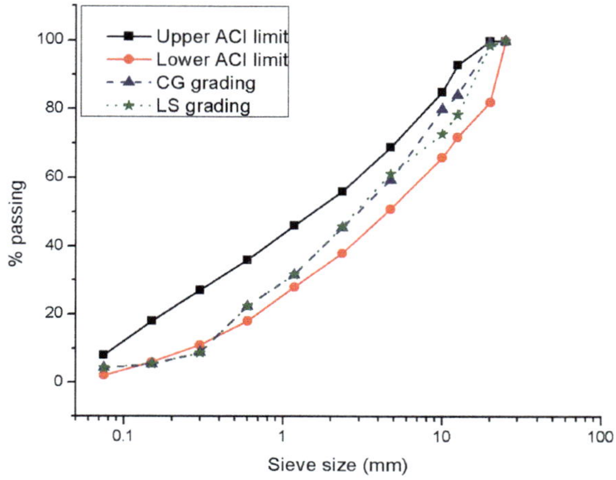

Fig. 1 Gradation curve for both aggregate types with upper and lower ACI limits

of 1% in each successive mix. Further, the water content of each mix was varied in five ranges, viz. 5%, 5.5%, 6%, 6.5% and 7% of total dry mass of RCC mixture. The optimum moisture content corresponding to each trail mix was then obtained by using the procedure laid down in ASTM D 1557 [18] and adopted by Sani et al. [19]. For obtaining optimum moisture content, all constituent materials in the required proportions were weighed and were dry mixed for 1 min. Then the required amount of water was added to the mix and was mixed thoroughly. The prepared mix was filled in 100 mm diameter mould having a volume of 920.89cm^3 and was then compacted in five layers with 25 blows per layer using an automatic compaction machine consisting of 4.54 kg circular rammer.

Thereafter, wet mass of the compacted sample was noted, which was further used to obtain wet density (γ_{wet}) of the mix. From the prepared compacted sample, a reference sample was obtained by collecting small fractions of the mix from three different locations, i.e. top, centre and bottom. This reference sample was used to obtain optimum water content. For this, the wet mass of reference sample was noted. The sample was then kept in an oven maintained at 100°C for 24 h to obtain the dry mass of the sample. Optimum water content (w) and dry density (γ_{dry}) were then measured from the obtained data, according to Eqs. (1) and (2), respectively [10];

$$w(\%) = \frac{mwet - mdry}{mdry} \times 100 \quad (1)$$

$$\gamma dry = \frac{\gamma wet}{1 + w} \quad (2)$$

where, m_{wet} is the wet mass of the reference sample, m_{dry} is the oven dry mass of the reference sample, w is the water content (%), γ_{dry} is the dry density of the sample, γ_{wet} is the wet density of sample which was calculated by dividing the wet mass of the compacted concrete by the volume of the concrete.

The variation in optimum dry density and in water content with the increment in cement content are shown in Fig. 2. Figure shows that the optimum value of dry density increases with the increase in cement content, with the subsequent decrease in optimum water content. The increase in dry density value can be attributed to the increase in particle packing density, due to higher cement content. Previous study by Choobbasti et al. [20] shows a decrease in moisture content and a marginal increase in dry density value with the increase in cement content. Shooshpasha and Shirvani [21] also reported a similar behaviour under the cement–soil stabilized system.

Finalization of cement content based on 28-day compressive strength. After getting optimum water content for all mixes having cement content in the range of 11–17%, one mix was to be finalized as a control mix having optimum cement content and maximum compressive strength. For this, cylindrical specimens of size 150 × 300 mm were prepared for seven concrete mixes with mix proportions listed in Table 3. The RCCP mix so obtained was filled into the moulds in three layers and compacted by putting a surcharge on the top of each layer. According to ASTM C 1176 [22] for cylindrical samples 4.9 kPa (9 kg) surcharge was used to facilitate consolidation. Compaction was done by using a vibrating table having a vibration frequency of 60 Hz, as specified in ASTM C 1170 [23]. All the specimens were consolidated till the formation of ring between the surcharge and the inside face of the mould. The formation of this ring shows that the RCC has reached its maximum

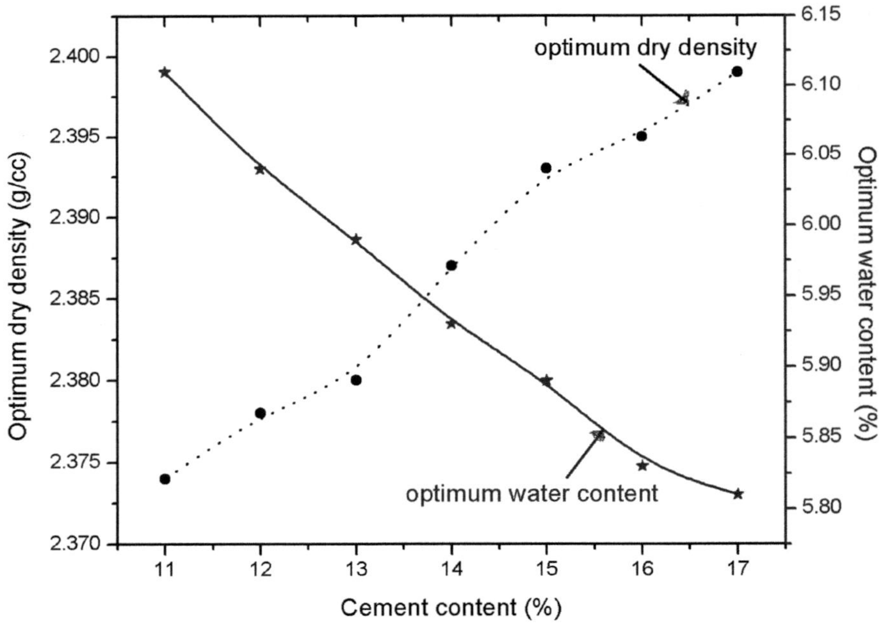

Fig. 2 Effect of cement content variation on optimum values of dry density and water content

Table 3 RCC mix proportion with cement content variation

Cement percentage (%)	Cement (kg/m^3)	Water (kg/m^3)	Sand (kg/m^3)	Aggregate (kg/m^3)	
				10 mm	20 mm
11	232.3	128.8	1161.2	633.4	316.7
12	251.6	126.5	1153.3	629.1	314.5
13	271.0	124.8	1146.6	625.4	312.7
14	290.2	122.9	1140.2	621.9	310.9
15	309.2	121.4	1133.6	618.3	309.2
16	327.8	119.3	1127.0	614.7	307.3
17	346.3	118.4	1120.6	611.2	305.6

density. The specimens were removed from the moulds after 24 h of casting and were kept in a curing tank for 28 days for compressive strength determination.

The 28-day compressive strength was measured as per BIS-516 [24] and the results are presented in Fig. 3. From the figure, it was observed that compressive strength goes on increasing with the increase in cement content up to 16% and afterwards it attains stabilized value. Therefore, based upon the 28 days strength results the mix containing 16% of cement content as total dry mass of coarse and fine aggregates was taken as the control mix for crushed gravel aggregate. In order to maintain the uniformity in all mixes the same cement content (16%) was used for the limestone aggregates.

Influence of GGBS and aggregate type on RCCP mix proportions. For evaluating the influence of GGBS and aggregate type, subsequent mixes were prepared by using GGBS as a partial replacement of cement and by taking two types of aggregates, viz. crushed gravel and limestone aggregates. For all mixes, the cementitious content was fixed at 16%, while the GGBS content was varied from 20 to 60%. For each mix, the optimum moisture content was finalized based on the procedure explained in selection of optimum water content section. The mix designation adopted for all mixes along with the obtained optimum water content and maximum dry density value is presented in Table 4. From the table it was observed that maximum dry density decreases with an increase in optimum water content of the RCC mix, irrespective of the type of aggregate. The similar observation regarding the maximum dry density behaviour with optimum water content was made by Yazici et al. [4]; Hesami et al. [6]; Aghabaglou and Ramyar [10]. Mix proportions of RCC mixes were then finalized based upon the ACI 211.3R-02 [17] guidelines, by assuming 1% air voids and are presented in Table 5.

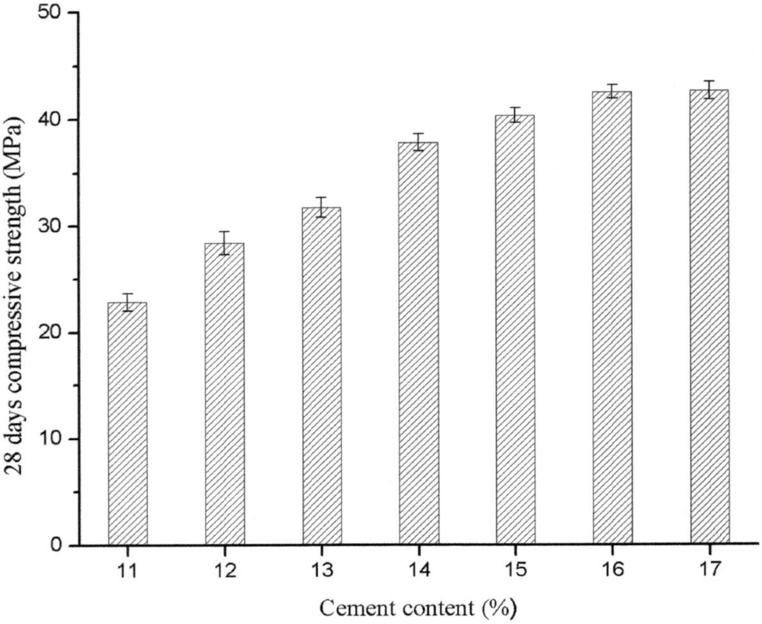

Fig. 3 Effect of variation in cement content on 28 days of compressive strength

Table 4 Optimum water content and maximum dry density of the RCCP achieved by compaction test

Mixes	CG + 0% GGBS	CG + 20% GGBS	CG + 40% GGBS	CG + 60% GGBS	LS + 0% GGBS	LS + 20% GGBS	LS + 40% GGBS	LS + 60% GGBS
Mix designation	CG-0	CG-20	CG-40	CG-60	LS-0	LS-20	LS-40	LS-60
Optimum water content (%)	5.83	6.31	6.39	6.64	6.15	6.38	6.71	6.87
Maximum dry density (g/cc)	2.395	2.373	2.368	2.347	2.368	2.345	2.330	2.318

3.2 Resultant Strength Properties of RCCP

The effect of variation in coarse aggregate type and incorporation of GGBS on strength properties of RCC like compressive strength and flexural strength is discussed in the succeeding sections. Both strength properties were investigated till 90 days of curing period.

Table 5 RCCP mix proportion for GGBS content and aggregate variation

Mix designation	Cementitious materials (kg/m³)		Water (kg/m³)	Sand (kg/m³)	Aggregate (kg/m³)	
	Cement	GGBS			10 mm	20 mm
CG-0	327.87	0	119.30	1127.06	614.76	307.38
CG-20	259.38	64.85	127.86	1114.51	607.92	303.96
CG-40	193.65	129.10	129.10	1109.46	605.16	302.58
CG-60	128.14	192.22	133.04	1101.23	600.67	300.34
LS -0	321.77	0	123.52	1105.68	603.10	301.77
LS-20	255.48	63.87	127.34	1097.76	598.78	299.61
LS-40	189.87	126.58	132.74	1087.81	593.35	296.90
LS-60	125.96	188.95	135.15	1082.52	590.46	295.45

Compressive strength. Compressive strength results obtained at 7, 28, 60 and 90 days of curing are presented in Fig. 4 for both aggregate types. From the figure, it can be observed that the mixes with limestone aggregates exhibited better compressive strength improvement with the replacement of cement with GGBS. Irrespective of type of aggregate, compressive strength behaviour changes with the addition of GGBS into RCCP mix. At early curing ages, compressive strength for RCC containing GGBS was lower than the corresponding strength of control mix. The decrease in compressive strength with the addition of GGBS into the mix can be attributed to the slower reaction between the glossy components of GGBS and water. Therefore, the secondary pozzolanic reaction between GGBS and OH^- produced during hydration of cement is also occurring at a slower rate.

However, the 28 days strength of all mixes satisfied the limit of ACI 325.10R [1], that requires a minimum value of compressive strength of 27.6 N/mm² at 28 days for RCC to be used as a surface course. Beyond the curing period of 28 days, RCCP mixes containing 20% and 40% GGBS exhibited even higher strength than the control

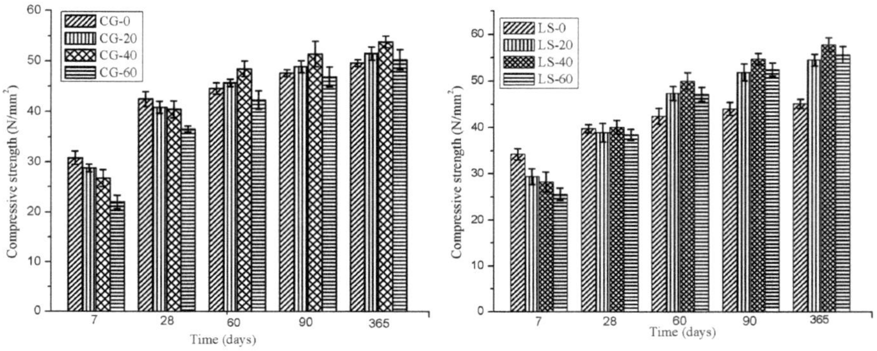

Fig. 4 Influence of GGBS content and aggregate variation on compressive strength development

RCCP, whereas the LS-60 mix shows strength comparable to the control mix. At 365 days of curing age, maximum strength was observed for CG-40 and LS-40 mix followed by 20%, 60% and 0% GGBS incorporation levels. At large replacement levels of GGBS, the whole of GGBS does not get consumed in the pozzolanic reaction due to lesser availability of $Ca(OH)_2$ from the primary hydration reaction. Previous researchers, Babu and Kumar [25]; Chidic and Panesar [26]; Patra and Mukharjee [27] also observed that the addition of GGBS into conventional concrete decreases the early age strength of concrete, but improves the later age strength and durability of conventional concrete mixes.

Flexural strength. The present study is oriented to the application of RCC as a pavement material. Since concrete pavements fail because of bending stresses, therefore flexural strength becomes important for the design of concrete pavements. According to IRC-58 [28] the 28 days flexural strength of concrete pavement should lie between 3.8 and 4.2 MPa. Therefore, the effect of coarse aggregate types and GGBS replacement levels was investigated on 7, 28 and 90 days flexural strength of RCCP. Flexural strength was measured on prism specimens of size 150 × 150 × 700 mm, as per BIS:516–1959 [24]. The 28 days flexural strength results with coarse aggregate variation and all replacement levels of cement with GGBS are shown in Fig. 5.

It was observed from the figure that RCC made with limestone aggregate presents marginally higher flexural strength as compared to crushed gravel aggregate. Ozturan and Cecen, [29] reported that normal strength concrete containing limestone yields higher flexural strength than the concrete containing gravel because of better bonding of limestone aggregate particle to the matrix. Similar with the compressive strength, inclusion of GGBS into the mix decreases the flexural strength at each replacement level for both types of aggregates till 28 days. A large decrease in flexural strength was observed at 60% replacement level of cement with GGBS followed by 20% and 40% replacement levels. For crushed gravel aggregates, the percentage decrease in strength as compared to control concrete was 13.5%, 8.7% and 23.9% at 20%, 40% and 60% replacement levels, respectively. Similarly, for limestone aggregate

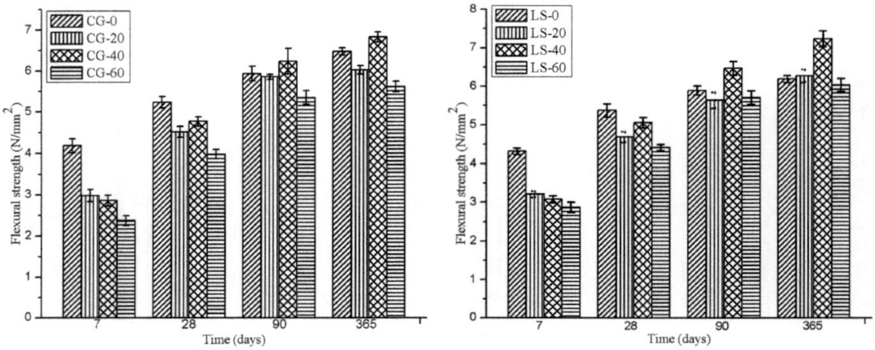

Fig. 5 Influence of GGBS content and aggregate variation on flexural strength variation

this decrease was limited to 12.6%, 5.7% and 17.6%. Previous study also shows that the pozzolanic reactions of GGBS are slow during the initial age of curing (28 days) after this period, the GGBS starts reacting with calcium hydroxide and forms CSH gel. Calcium hydroxide was formed after the hydration of cement so that it takes a longer time to develop its strength particularly at high GGBS replacement levels [30].

At 90 days, a reversal in flexural strength was registered, with a 40% replacement level of highest strength was observed. RCCP consisting of limestone aggregates and GGBS performs better than the crushed gravel aggregates RCCP mixes. Improvement in flexural strength observed in the case of RCC containing limestone aggregates with GGBS incorporation can be attributed to the fineness of GGBS particles as well as physical properties of the aggregates. The porous irregular structure of limestone aggregates can make better bonding to the mortar which was further strengthen by GGBS incorporation. It was reported that the flexural strength depends upon the surface characteristics of coarse aggregates, the strength of the mortar and the quality of the interfacial zone, which can be improved with the incorporation of finer supplementary materials [29]. However, all values lie between the IRC-58 [28] limit, i.e. higher than 3.8 MPa indicating that RCCP can be used as an effective pavement material.

4 Conclusions

1. RCC achieved by soil approach has less fluid consistency as compared to the concrete approach because in soil approach water content is used at the optimum level. RCC designed by soil compaction approach leads to minimum voids in mix with maximum density which ultimately reduces the cement and water requirement.
2. The finalized aggregate proportions, i.e. 55% of fine aggregate, 30% of 10 mm coarse aggregate and 15% of 20 mm coarse aggregate fitted well with R^2 as 0.982 in the maximum density gradation curve. Obtained higher correlation coefficient represents that the aggregate particles fit together in their densest possible arrangement along the gradation line.
3. Mix containing 16% of cement content as a total dry mass of coarse and fine aggregates was finalized for both aggregate types based upon the highest value of 28 days compressive strength.
4. Compaction results show that limestone aggregate increased the water requirement of the RCC mix as compared to the mixes containing crushed gravel aggregates because of higher water absorption. Further addition of GGBS into both mixture series increased the optimum water content because of higher fineness modulus.
5. Incorporation of GGBS into the RCC mixes led to lower compressive and flexural strength at early ages, followed by strength higher than the control mix at later ages. In the case of mixes incorporating crushed gravel aggregates, the addition

of 20% and 40% GGBS gives the strength higher than the control mix after 28 days of curing, while 60% GGBS incorporation registers strength comparable with the control mix. On the other hand, in RCC mixes incorporating limestone aggregates, strength higher than the control mix was registered at all replacement levels of cement with GGBS after 28 days of curing period.

References

1. ACI 325.10R -95. Report on roller-compacted concrete pavements. American Concrete Institute, U.S.A. (2001)
2. ACI 207.5R. Roller compacted mass concrete. ACI material journal committee report. Tilte no. 85-M44 (1988)
3. Rao, S.K., Sravana, P., Rao, T.C.: Abrasion resistance and mechanical properties of roller compacted concrete with GGBS. Constr. Build. Mater. **114**, 925–933 (2016)
4. Yazici, S., Tuyan, M., Mardani-Aghabaglou, A., Ute, A.A.: Mechanical properties and impact resistance of roller compacted concrete containing polypropylene fibre. Mag. Concr. Res. **67**(16), 867–875 (2015)
5. Zollinger, D.: Roller compacted concrete pavement. FHWA-HIF-16–003, Department of Transportation, U.S.A. (2016)
6. Hesami, S., Modarres, A., Soltaninejad, M., Madani, H.: Mechanical properties of roller compacted concrete pavement containing coal waste and lime stone powder as partial replacement of cement. Constr. Build. Mater. **111**, 625–636 (2016)
7. Vahedifard, F., Nili, M., Meehan, C.L.: Assessing the effect of supplementary cementitious materials on the performance of low-cement roller compacted concrete pavement. Constr. Build. Mater. **24**, 2528–2535 (2010)
8. Cao, C., Sun, W., Qin, H.: The analysis on strength and fly ash effect of roller-compacted concrete with high volume fly ash. Cem. Concr. Res. **30**, 71–75 (2000)
9. Atis, C.D., Sevim, U.K., Ozcan, F., Bilim, C., Karahan, O., Tanrikulu, A.H., Eksi, A.: Strength properties of roller compacted concrete containing a non-standard high calcium fly ash. Mater. Lett. **58**, 1446–1450 (2004)
10. Aghabaglou, A.M., Ramyar, K.: Mechanical properties of high volume fly ash roller compacted concrete designed by maximum density method. Constr. Build. Mater. **38**, 356–364 (2013)
11. Hansen, K.D., Reinhardt, W.G.: Roller Compacted Concrete Dams. McGraw–Hill, New York (1991)
12. Aslan, O.: Predicting long term strength of roller compacted concrete containing natural Pozzolan by steam curing. M.Sc Thesis, Graduate School of Natural and Applied Sciences; Middle East Technical University (2006)
13. U.S. Reclamation. Design and construction considerations for hydraulic structures Roller Compacted Concrete. Second edition, U.S. Department of the interior Bureau of reclamation, Denver, Calorado (2017)
14. BIS- 8112. Specification for 43 Grade Ordinary Portland Cement. Bureau of Indian Standards, New Delhi, India (1989)
15. BIS-12089. Specification for Granulated Slag for the manufacture of Portland Slag cement. Bureau of Indian Standards, New Delhi, India (1987)
16. BIS-383. Specification for Coarse and Fine Aggregates from Natural sources for concrete. Bureau of Indian Standards, New Delhi, India (2002)
17. ACI 211.3R -02. Guide for selecting proportions for No- slump concrete. American Concrete Institute, U.S.A. (2002)

18. ASTM D1557. Standard test methods for Laboratory compaction characteristics of soil using modified efford (2.700 kN.m/m^3). American Society of Testing and Materials, West Conshohocken, U.S.A. (2009)
19. Sani, J.E., Yohanna, P., Chukwujama, I.A.: Effect of rice husk ash admixed with treated sisal fibre on properties of lateritic soil as a road construction material. J. King Saud Univ. Eng. Sci. **32**(1), 11–18 (2020)
20. Choobbasti, A.J., Vafaei, A., Kutanaei, S.S.: Mechanical properties of sandy soil improved with cement and nanosilica. DE GRUYTER **5**, 111–116 (2015)
21. Shooshpasha, I., Shirvani, R.A.: Effect of cement stabilization on geotechnical properties of sandy soils. Geomech. Eng. **8**(1), 17–31 (2015)
22. ASTM C 1176/C1176 M. Standard practice for making roller compacted concrete in cylinder molds using a vibrating table. American Society of Testing and Materials, West Conshohocken, U.S.A. (2008)
23. ASTM C 1170/C1170M. Standard test method for determining consistency and density of roller compacted concrete using a vibrating table. American Society of Testing and Materials, West Conshohocken, U.S.A. (2008)
24. BIS-516:1959. Indian Standard Methods of Tests for Strength of Concrete. Bureau of Indian Standards, New Delhi, India (Reconfirmed 2004)
25. Babu, G.K., Kumar, V.S.R.: Efficiency of GGBS in concrete. Cem. Concr. Res. **30**, 1031–1036 (2000)
26. Chidiac, S.E., Panesar, D.K.: Evolution of mechanical properties of concrete containing ground granulated blast furnace slag and effects on the scaling resistance test at 28days. Cement Concr. Compos. **30**(1), 63–71 (2008)
27. Patra, R.K., Mukharjee, B.B.: Fresh and hardened properties of concrete incorporating ground granulated blast furnace slag-a review. Adv. Concrete Constr. **4**(4), 283–303 (2016)
28. IRC-58. Guidelines for the design of plain and jointed rigid pavements for highways. Second edition, The Indian Road Congress, New Delhi, India (2002)
29. Ozturan, T., Çeçen, C.: Effect of coarse aggregate type on mechanical properties of concretes with different strengths. Cem. Concr. Res. **27**(2), 165–170 (1997)
30. Zhou, X.M., Slater, J.R., Wavell, S.E., Oladiran, O.: Effects of PFA and GGBS on early ages engineering properties of Portland cement systems. J. Adv. Concr. Technol. **10**, 74–85 (2012)

Effect of Paraformaldehyde Fibers on Mechanical and Shrinkage Properties of High-Strength Concrete

Ying Hu, Sheng Zhang, Junlong Jin, Bo Chen, Shenlin Hu, Yang Li, and Yixin Hong

Abstract The influence of polyformaldehyde fibers on the mechanical and shrinkage performance of concrete was studied. The research was conducted to investigate the early performance of paraformaldehyde fiber concrete using the concrete mechanical properties test method and early shrinkage test apparatus under different test conditions of fiber length, fiber admixture, and curing conditions. The results demonstrated that the improvement of compressive strength of high-strength concrete by paraformaldehyde fibers was not significant, and the compressive strength was still slightly reduced compared to that of the baseline concrete. The compressive strength of concrete was affected by different curing conditions. The improvement of shrinkage properties of high-strength concrete by paraformaldehyde fiber is more significant. Curing conditions have a greater effect on the shrinkage properties of paraformaldehyde fiber concrete, and the better the curing conditions, the more adequate the hydration reaction of concrete. Meanwhile, this study proposes shrinkage and anti-cracking control measures for high-strength concrete such as preferential selection of raw materials, optimization of matching ratio, reasonable configuration of reinforcement, and strengthening of maintenance.

Keywords High-strength concrete · Polyformaldehyde fiber · Mechanical properties · Shrinkage properties

---

Y. Hu · J. Jin
School of Civil Engineering, Chongqing University, Chongqing 400045, China

S. Zhang · B. Chen · S. Hu · Y. Li
Chengdu First Construction Engineering of CDCEG, No. 123 Yusha Street, Chengdu 610017, Sichuan, China

Y. Hong (✉)
Department of Civil Engineering, Xi'an Jiaotong-Liverpool University, Suzhou 215123, China
e-mail: goodchan99@126.com

E. Strauss (ed.), *Proceedings of the 7th International Conference on Civil Engineering*, Lecture Notes in Civil Engineering 371, https://doi.org/10.1007/978-981-99-4045-5_3

1 Introduction

Polyoxymethylene (POM) fiber is a comprehensive high-performance organic synthetic fiber with excellent performance, high strength and high elastic modulus and excellent abrasion resistance, alkali resistance, high elongation response, scratch fracture resistance, and long-term stability, can be widely used in toughening and crack resistance of cement-based materials, with better dispersion performance, toughening and crack resistance and durability in concrete, more effective in solving the early plastic cracking of the concrete, good control of the drying shrinkage deformation of concrete [1–3].

In recent years, many scholars have paid attention to the research and development of polyacetal fibers [4]. Some researchers found that the main mechanical indexes of polyvinyl alcohol fiber concrete such as compressive strength, splitting tensile strength, and modulus of elasticity all increased to different degrees with the increase of loading rate, and the test results showed that the incorporation of a certain amount of polyvinyl alcohol fiber in concrete could enhance its brittle and residual strength, making it have better mechanical properties and toughness [5–7]. In their investigation of the impact of polypropylene fibers on concrete under uniaxial cyclic compression, Xu et al. [8] found that the plastic strain of concrete was not significantly affected by the addition of polypropylene fibers, but the deterioration of the elastic stiffness and strength of concrete was reduced, and at the same time, the ultimate strain and toughness of concrete were also improved. Padmanabha et al. [9] studied the effect of basalt fiber on the performance of concrete, and the results showed that the compressive strength and modulus of elasticity at break of basalt fiber concrete were significantly increased under short-cut fiber conditions, while the concrete in the experimental group with high fiber dosing showed the phenomenon of lumping and delamination, and the quality of concrete was seriously affected.

The above studies illustrate that paraformaldehyde fibers can enhance the mechanical properties of concrete, but the amount and legality of the impact of paraformaldehyde fibers on the mechanical and shrinkage properties of concrete are unknown, and there are less investigations on the mechanical and shrinkage properties of paraformaldehyde fiber concrete. According to the existing literature, the performance of concrete is greatly improved when filled with paraformaldehyde fiber in ordinary concrete, which makes up for the shortcomings and deficiencies of concrete [10]. Therefore, to reduce the risk of cracking high-strength concrete due to self-shrinkage and drying shrinkage, the effect of the application of paraformaldehyde fibers on the early mechanical and shrinkage properties of high-strength concrete should be further investigated.

In this study, the mechanical properties of concrete were tested under three different fiber lengths of 6, 9, and 12 mm, three different fiber admixtures of 0.6, 1.2, and 1.8 kg/m^3, and three different curing methods of dry curing, sealed curing, and wet curing to investigate the impact of paraformaldehyde fiber on the mechanical and shrinkage properties of high-strength concrete.

2 Experimental Program

2.1 Material Mixing Ratio

C60 concrete was used for the test, and its detailed mix is shown in Table 1. The cement used is ordinary silicate cement; the coarse aggregate is solid force crushed stone and tork crushed stone with the ratio of 3:7; the fine aggregate (mechanism sand) has a fineness modulus of 2.9; and the high-performance water-reducing agent poly-carboxylate water-reducing agent to improve the compatibility. The basic properties of the polycarboxylate fibers are listed in Table 2.

2.2 Design of Mechanical Properties Test for Paraformaldehyde Fiber Concrete

The test refers to the Standard for Physical and Mechanical Properties of Concrete Test Method (GB/T50081-2019) for the cubic compressive strength test of concrete specimens.

The test piece is a standard cube of 150 mm, each group is set up with 3 test pieces, and the test time is 28d. The result of each group of test pieces is calculated as the arithmetic average of the measured values of three test pieces. If the maximum or minimum value of the three measured values differs from the median value by more than 15% of the middle value, the median value is used as the measured value. If the maximum or minimum value differs from the median value by more than 15% of the median value, the group test results were invalid. After calculation, all measured values of this test are within the valid range.

Table 1 C60 concrete mix ratio

Water-to-glue ratio	Sand rate (%)	Matching ratio (g/m^3)					
		Water	Cement	Sand	Small stones	Boulders	Water reducing agent
0.28	0.36	146	530	605	322.5	752.5	12.4

Table 2 Basic properties of polyformaldehyde fiber

Fiber	Diameter (μm)	Density (g/cm^3)	Elastic modulus (GPa)	Elongation at break (%)
Polyformaldehyde	30	1.41	7–9	20–30

2.3 Experimental Design of Shrinkage Performance of Paraformaldehyde Fiber Concrete

The shrinkage test mold used in this test is a plexiglass mold, and the size of the test block chosen is 100 mm × 100 mm × 300 mm; the holes for pre-buried nuts are specially opened in the middle of the side plate of the mold for connecting the screws after demolding and as measurement points for subsequent shrinkage measurements, and the design of the test mold is shown in Fig. 1.

This test used three fiber lengths of 6, 9, and 12 mm, three different fiber doses of 0.6, 1.2, and 1.8 kg/m^3, and three curing conditions of dry, sealed, and wet curing conditions. The concrete specimens in the dry curing group can be removed from the side molds after the initial setting; while the specimens in the sealed curing group are demolded after the final setting and sealed with aluminum foil to avoid moisture exchange between the specimens and the external environment; the specimens in the wet curing group are demolded after the initial setting of the concrete and covered with a wet sponge to keep the moisture inside the concrete continuously replenished, and the sponge is checked every 12 h for the degree of wetting and timely replenishment. The degree of wetting and timely replenishment of water.

According to the initial setting time of the concrete in the ratio given by the commercial concrete mixing plant company, the concrete test block is demolded after the initial setting. After demolding, the concrete test blocks are treated in accordance with the corresponding maintenance method. After the concrete specimen is processed, the plastic screw is installed in the pre-buried nut as the measurement point, and then all the ready concrete specimens are placed lightly on the horizontal steel testing platform, and the micrometer base is adjusted so that the micrometer probe is in phase with the measurement point, and the data acquisition can be started

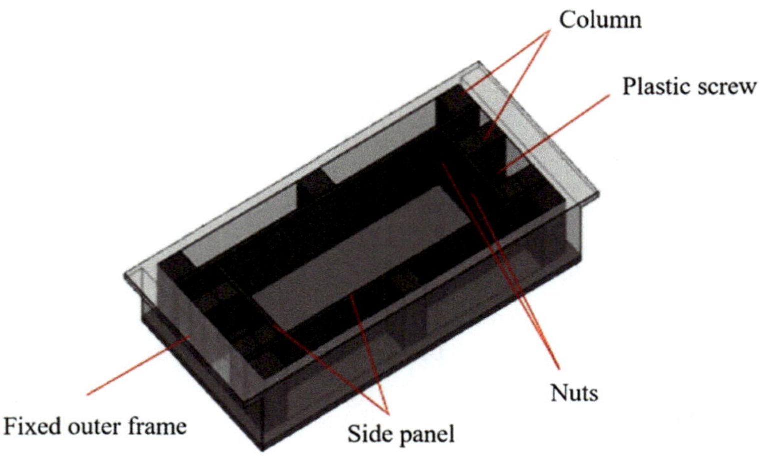

Fig. 1 Schematic diagram of shrinkage test mold

after zeroing the micrometer. The frequency of shrinkage data collection is once every half hour.

As this test site is in the Chongqing area, the humidity in this area is high, in order to maintain stable maintenance conditions (ambient temperature is 20 ± 2 °C, relative humidity is 60 ± 5%), a high power air conditioner is selected for temperature regulation and a dehumidifier is selected for humidity regulation to ensure that all specimens can be maintained in the same conditions in the closed room.

3 Results and Discussion

3.1 Analysis of Mechanical Performance Test Results

3.1.1 Effect of Fiber Length on the Compressive Properties of Concrete

Figure 2 shows the effect of different lengths of paraformaldehyde fibers on the compressive strength of concrete during 28 d age. The concrete compressive strength increases with the length of the formaldehyde fiber, but when the length of the formaldehyde fiber increases to 9 mm, the compressive strength of the concrete-enhanced formaldehyde fiber reduces with the increase of polydehyde fibers.

Therefore, the compressive strength of concrete will vary when a particular amount of paraformaldehyde fiber is added. However, the concrete compressive strength of the concrete that incorporates the formaldehyde fiber is slightly reduced compared to the benchmarking concrete resistance, but considering that the greater impact on the compressive strength of concrete is the modulus of the reinforcing material, it is possible that the higher modulus of paraformaldehyde fiber-reinforced concrete compressive strength change trend is another trend, and the later research is yet to be further in-depth.

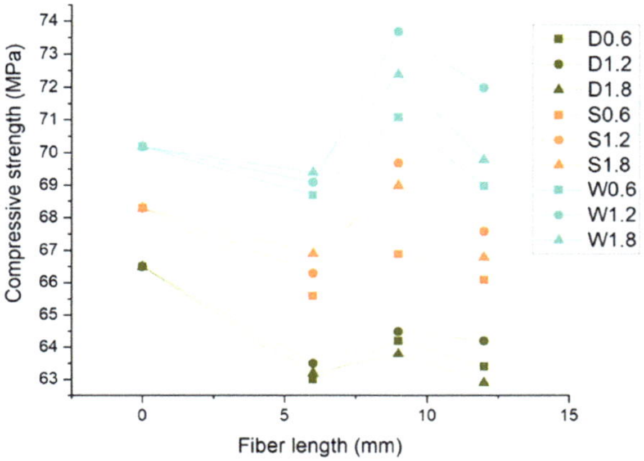

Fig. 2 Effect of different POM fiber length on compressive strength of concrete at 28 days

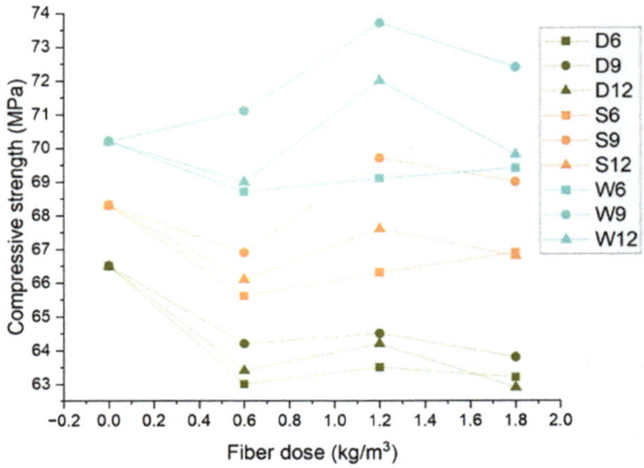

Fig. 3 Effect of different POM fiber amount on compressive strength of concrete at 28 days

3.1.2 The Impact of Fiber Admixture on the Compressive Properties of Concrete

Figure 3 shows the effect of the compressive strength of concrete with different lengths of paraformaldehyde fiber admixture. From the figure, the compressive strength of paraformaldehyde fiber-reinforced concrete at the age of 28 d increases with the increase of paraformaldehyde fiber admixture from 0.6 to 1.2 kg/m^3, but when the paraformaldehyde fiber admixture reaches 1.2 kg/m^3, the compressive strength of paraformaldehyde fiber-reinforced concrete falls as the amount of paraformaldehyde fiber increases. In addition, the longer the average length of fibers, the less the number of fibers, and the weaker the reinforcing effect of fibers on concrete at the same admixture level.

3.1.3 The Impact of Maintenance Conditions on the Compressive Properties of Concrete

Figure 4 shows the impact of various curing conditions on polycarbonate fiber concrete's compressive strength. The compressive strength of concrete is more affected by the various curing conditions, with wet curing having the greatest impact and dry curing having the least.

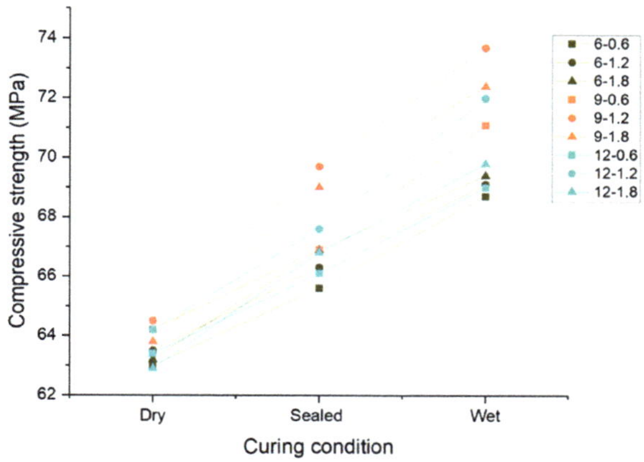

Fig. 4 Effect of various curing conditions on the 28-day compressive strength of POM fiber concrete

3.2 Shrinkage Performance Test Results and Analysis

3.2.1 Effect of Fiber Length on Concrete Shrinkage

The results from Fig. 5 show that the overall shrinkage strain of the prepared paraformaldehyde fiber concrete is between 230 and 410 μz, and the incorporation of paraformaldehyde fiber in concrete can reduce the degree of shrinkage of concrete at 28 d. When the length of the incorporated paraformaldehyde fiber gradually grows, the shrinkage of the concrete further decreases, but when the length of the incorporated paraformaldehyde fiber reaches 9 mm, the shrinkage formation slows down and there is a tendency for the shrinkage to become larger. This is because the concrete system is mixed with paraformaldehyde fiber, the fiber in the concrete can form a uniform, chaotic three-dimensional support system, effectively inhibiting the sinking of the aggregate, reducing its inherent defects, improving the uniformity and microstructure of concrete, while blocking the channel of water overflow inside the concrete, or make the curvature of the water seepage channel increased, reducing or delaying the dissipation of water, and then effectively reduce the concrete dry shrinkage strain.

3.2.2 The Impact of Fiber Admixture on the Shrinkage of Concrete

The results from Fig. 6 show that the overall shrinkage strain of the concrete with different admixtures of paraformaldehyde fibers is between 230 and 410 μz, and the admixture of different admixtures of paraformaldehyde fibers in concrete can obviously reduces the degree of shrinkage of concrete at 28 d. When the admixture of paraformaldehyde fiber is gradually increased, the shrinkage of concrete will be further reduced, in addition to the fact that paraformaldehyde fiber can significantly improve the homogeneity and microstructure of concrete, reduce or delay the water

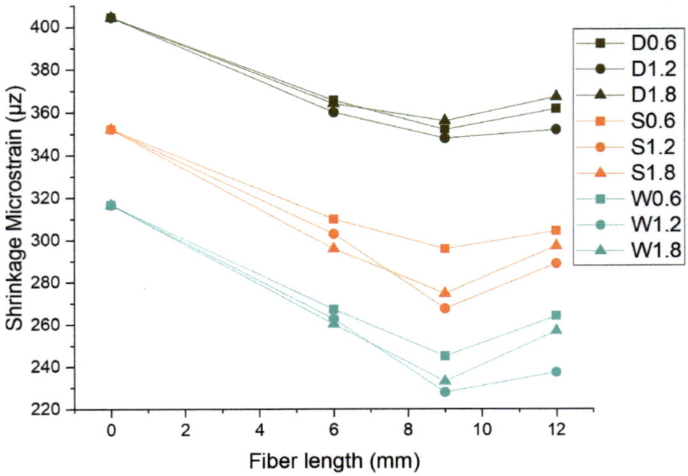

Fig. 5 Effect of different POM fiber length on shrinkage strain of concrete at 28 days

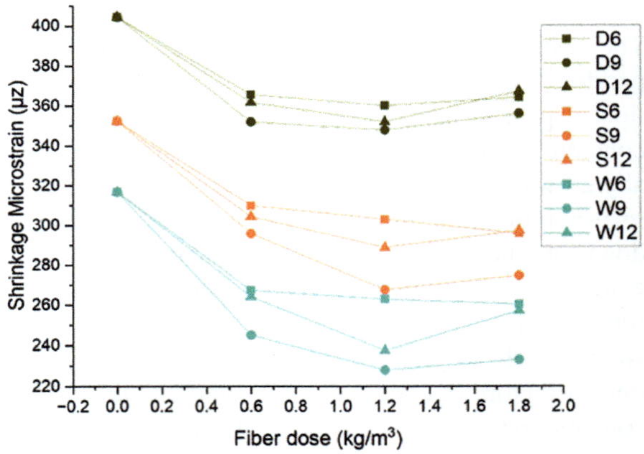

Fig. 6 Effect of different POM fiber amount on shrinkage strain of concrete at 28 days

dissipation, and the inherent defects in paraformaldehyde fiber concrete are less, which is more conducive to reducing the shrinkage strain of concrete.

3.2.3 Effect of Maintenance Conditions on Concrete Shrinkage

The results in Fig. 7 show that sealed curing, dry curing, and wet curing after 28 d curing all inhibit the shrinkage of the polyformaldehyde fiber concrete, and the concrete shrinkage microstrain is between 230 and 370 μz. The results indicate that the better the curing conditions, the more adequate the hydration reaction of the

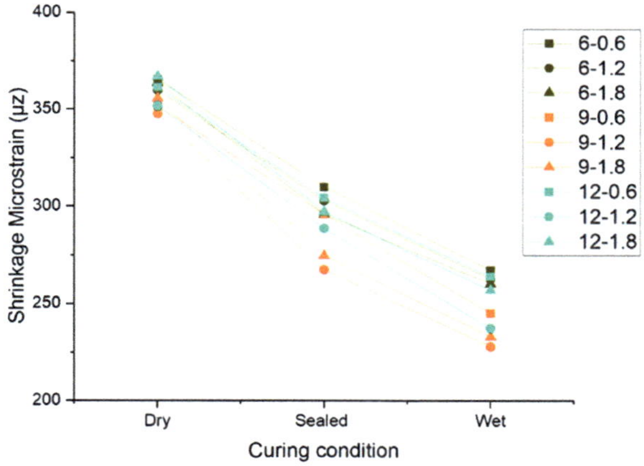

Fig. 7 Effect of different curing methods on shrinkage strain of POM fiber concrete for 28 days

concrete, and the development of microscopic pores of concrete forms a more fine-grained pore structure, which is more conducive to the reduction of shrinkage degree of concrete.

3.3 Shrinkage and Crack Resistance Control of High-Strength Concrete

Shrinkage cracking of concrete is the main reason for the reduction of its structural durability and service life. The appearance of cracks not only reduces the impermeability of buildings and affects their functions, but also causes corrosion of reinforcement, carbonation of concrete, and affects the strength of concrete. Thus, the shrinkage anti-cracking measures of high-strength concrete are summarized and proposed from four aspects: preferential selection of raw materials, optimization of matching ratio, reasonable configuration of reinforcement, and strengthening of maintenance according to the causes of concrete cracks.

(1) Preferential selection of raw materials

To achieve excellent anti-cracking properties of concrete, the raw materials used to formulate concrete should have the following characteristics: on the one hand, the raw materials themselves have a large tensile capacity, ultimate tensile deformation capacity, and other mechanical properties; at the same time, they have a small adiabatic temperature rise, linear expansion coefficient, thermal strength, very small self-expansion, and other temperature characteristics.

(2) Matching ratio optimization

Compared with the concrete used in concrete projects 50 years ago, the current concrete project structural elements are more prone to early cracking, which is mainly due to the increasing fineness of cement, dosage, and tricalcium silicate content, coupled with the addition of a large number of fly ash, silica fume, and other ultrafine particles of admixtures.

The design of an anti-cracking ratio is very necessary for the components that need to consider the early cracking of concrete. In the process of ratio optimization, the designer should fully consider the following three points: on the basis of ensuring the design strength of the concrete structure and the working performance of the mix, trying to control the amount of cementitious materials to the minimum, and increasing the volume of coarse aggregates relatively; use coarse and fine aggregates with reasonable continuous gradation, so that the internal voids of the aggregates are filled densely, control the aggregate bulk density in the maximum, also reduce the amount of cementitious materials; choose the appropriate water–cement ratio to meet the requirements of strength and durability, the higher or lower water–cement ratio will cause the concrete shrinkage is strong, too much crack development.

(3) Reasonable configuration of reinforcement

For reinforced concrete, the greater the reinforcement arrangement, and as the reinforcing ratio rises, the likelihood of concrete cracking increases, due to the fact that the configuration of the reinforcement will result in a higher risk of cracking the concrete element at the location of the reinforcement. The shrinkage cracking behavior of reinforced concrete is influenced by the larger role of the way the reinforcement is configured. To better reduce the risk of shrinkage cracking of the concrete, during the planning and building phases of concrete buildings, the reinforcement configuration should not only ensure the necessary structural load-bearing capacity but also make its distribution uniform and improve the crack resistance of reinforced concrete as much as possible.

(4) Strengthen conservation

After the concrete is poured and formed, the concrete in the internal continuous hydration stage has not yet reached the expected strength index and resistance to deformation, and the improper maintenance conditions, temperature, and humidity cannot adapt to the concrete hydration reaction, which is affected by cold shrinkage and dry shrinkage, and cracks will appear on the surface of the concrete that affects the quality. By strengthening the temperature and humidity control in the process of concrete curing, the continuous hydration reaction of cement is maintained, which in turn improves the ultimate tensile strength of concrete, and then through temperature control, it helps to slow down the duration and speed of concrete's heating and cooling to avoid the emergence of uncontrollable temperature stresses that lead to cracking.

4 Conclusions

This research looked into how paraformaldehyde fibers affected the compressive strength and early shrinkage characteristics of high-strength concrete. The compressive strength and shrinkage strain of concrete specimens with different fiber lengths and fiber admixtures under dry, sealed, and wet conditions were investigated. The study's findings support the following opinions.

(1) The paraformaldehyde fiber has a negligible improvement effect on the compressive strength of high-strength concrete, and the compressive strength is still somewhat lower than that of the base concrete. After the paraformaldehyde fiber was added, the shrinkage strain of high-strength concrete was reduced to some amount, but the improvement effect of the paraformaldehyde fiber on the shrinkage performance of high-strength concrete is more notable.

(2) When paraformaldehyde fiber length is increased from 6 to 9 mm, the compressive strength of the paraformaldehyde fiber concrete increases. However, when paraformaldehyde fiber length is increased to more than 9 mm, the compressive strength of the paraformaldehyde fiber concrete declines; when mixed with paraformaldehyde fiber length gradually increases, the shrinkage of concrete further decreases, but when mixed with paraformaldehyde fiber length reaches more than 9 mm, the trend of shrinkage decreases gradually slows down, and turns to the trend of shrinkage becomes larger.

(3) When the amount of paraformaldehyde fiber admixture is from 0.6 to 1.2 kg/m^3, the compressive strength of high-strength concrete increases with the addition of paraformaldehyde fiber, but when the amount of paraformaldehyde fiber admixture is greater than 1.2 kg/m^3, the compressive strength of paraformaldehyde fiber concrete reduces with the addition of paraformaldehyde fiber; when the amount of paraformaldehyde fiber admixture is from 0.6 to 1.2 kg/m^3, the shrinkage strain of high-strength concrete reduces with an increase in polyformaldehyde fiber additive, but when the admixture of polyformaldehyde fiber increases above 1.2 kg/m^3, the shrinkage strain of polyformaldehyde fiber concrete increases with an increase in polyformaldehyde fiber admixture, but it is still smaller than the shrinkage strain of base concrete.

(4) Different curing conditions have a larger effect on the compressive strength of concrete, with wet curing having the greatest impact and sealed curing coming in second; the shrinkage performance of polyformaldehyde fiber concrete is more affected by curing conditions, the better the curing condition, the more fully the concrete hydrates, which is more conducive to reducing the degree of shrinkage of concrete, of which wet curing has the best improvement on the shrinkage performance of concrete, and sealed curing is the second.

References

1. Rui, Y., et al.: Comparative study on the effect of steel and polyoxymethylene fibers on the characteristics of ultra-high performance concrete (UHPC). Cement Concr. Compos. **127**, 104418 (2022)
2. Andrzejewski, J., et al.: The influence of the hybridization process on the mechanical and thermal properties of Polyoxymethylene (POM) composites with the use of a novel sustainable reinforcing system based on Biocarbon and basalt Fiber (BC/BF). **13**(16), 3496 (2020)
3. Zhou, S., et al.: Tribological properties of PTFE fiber filled polyoxymethylene composites: the influence of fiber orientation. Compos. Commun. **28**, 100918 (2021)
4. Zhang, W., et al.: Experimental and numerical analysis of interfacial bonding strength of polyoxymethylene reinforced cement composites. Constr. Build. Mater. **207**, 1–9 (2019)
5. Xu, H., et al.: Experimental study on mechanical properties of fiber reinforced concrete: effect of cellulose fiber, polyvinyl alcohol fiber and polyolefin fiber. Constr. Build. Mater. **261**, 120610 (2020)
6. Shen, D., et al.: Effect of polyvinyl alcohol fiber on the cracking risk of high strength concrete under uniaxial restrained condition at early age. Constr. Build. Mater. **300**, 124206 (2021)
7. He, J., et al.: Mechanical properties of high strength POM-FRCC and its performance under elevated temperatures. Constr. Build. Mater. **290**, 123177 (2021)
8. Xu, L., et al.: Experimental investigation on damage behavior of polypropylene fiber reinforced concrete under compression. Int. J. Concrete Struct. Mater. **12**(1), 1–20 (2018)
9. Iyer, P., Kenno, S.Y., Das, S.: Mechanical properties of fiber-reinforced concrete made with basalt filament fibers. J. Mater. Civ. Eng. **27**(11), 04015015 (2015)
10. Benabdallah, H., Olender, D.: Finite element simulation of the wear of polyoxymethylene in pin-on-disc configuration. Wear **261**(11–12), 1213–1224 (2006)

Comparing Between Crushed and Fine Aggregate Recycled in Concrete

Ebraheem Alrashidi and Abdulaziz Almutairi

Abstract This paper is based on investigating the effect of using fine and coarse crushed bricks in the compression strength of concrete. Detailed research has been done from the current literature that is based on previous laboratory-based tests of the use of bricks in concrete. In order to investigate this particular area 10 mixes of concrete were prepared. Mixes included 0%, 25%, 50%, 75% and 100% use of fine and coarse crushed aggregates. Bricks were crushed to achieve fine and coarse crushed bricks. 3 specimens of each mix were casted and cured in the water tank. Compression strength test was conducted after 28 days of curing for each specimen. Results were analysed and compared between the mixes with coarse crushed aggregates and fine crushed aggregates. Finally, it was concluded that the mean target strength of concrete is achievable by replacing up to 50% of gravel and sand with coarse and crushed bricks. However, physical properties of bricks like absorption reduce the amount of water in concrete by leaving less water to react with cement that results in higher compression strength due to a reduction in water–cement ratio.

Keywords Concrete · Aggregate · Mix design · Crushed aggregate · Fine aggregate

1 Introduction

Debieb and Kenai [1] have stated that crushed bricks as fine and coarse aggregates are becoming a serious interest to civil engineers, as their use significantly decreases the issue of waste storage and protects natural aggregate resources. Hence, it is clear that using recycled construction materials as replacement material in concrete

---

E. Alrashidi
School of Architecture, Technology and Engineering, University of Brighton, Cockcroft Building Lewes Road, Brighton BN2 4GJ, UK

A. Almutairi (✉)
College of Engineering, Design and Physical Sciences, Civil and Environmental Engineering Department, Brunel University London, Kingston Lane, Uxbridge UB8 3PH, UK
e-mail: Abdulaziz.almutairi@Brunel.ac.uk

E. Strauss (ed.), *Proceedings of the 7th International Conference on Civil Engineering*, Lecture Notes in Civil Engineering 371, https://doi.org/10.1007/978-981-99-4045-5_4

reduces the need for natural aggregates. As natural aggregates become more costly, the demand of using recycled aggregate is increasing and these materials are available from demolished concrete [2] and bricks. Nowadays, many countries such as USA, Japan, Hong Kong, Scotland and India have started using recycled aggregates for construction projects as one of the components materials in concrete [3].

Sonawane and Pimplikar [3] investigated the physical properties of crushed bricks based on experimental laboratory tests and concluded that "the use of recycled aggregate generally increases the drying shrinkage creep and porosity to water and decreases the compression strength of concrete compared to that of natural aggregate in concrete". The authors have stated that recycling decreases life cycle cost (LCC) approximately from 34 to 41% and reduces Co_2 emission ($LCCO_2$) approximately from 23 to 28% for putting at public and private removal services.

1.1 Crushed Bricks as Aggregate

Concrete is the most commonly used material in the construction industry because it is durable, strong and can be used as a structural material [4]. However, concrete, or rather the component materials in concrete, has an adverse impact on carbon emissions and climate change [5] as well as reducing natural resources such as aggregate and sand. There are also indirect impacts from the use of these natural materials in concrete, including negative impacts on local ecosystems during the extraction process and land consumption for quarrying the materials [2]. Ajayi et al. [6] add that the industry is wasteful, with a high portion of materials sent to landfill. These impacts are the driving force behind the need to improve the sustainability of concrete and the key components of the material including sand, coarse aggregate and cement [6]. One effective way of decreasing the environmental impact of concrete and potentially decreasing the cost of the material is to replace the component materials with sustainable recycled alternatives. For example, there is research on replacing aggregate with rubber to create rubberised concrete [2] or using fly ash [7].

According to Debieb and Kenai [1] and Adamson et al. [8], crushed bricks are a good source of replacement material for more sustainable concrete as these materials reduce waste associated with the construction process and reduce the volume of materials consumed and the need to store waste.

1.2 Mechanical Properties of Concrete Using Crushed Brick

Debieb and Kenai [1] conducted research on the mechanical properties of concrete testing samples with coarsely crushed brick and finely crushed brick. The tests included comparing the performance of concrete with natural coarse (NG) and fine (NS) calcareous aggregates to concrete with coarse (RG) and fine (RS) crushed bricks aggregates.

It is noted that the different coarse aggregates were of similar grading while RS samples were coarser than the NS. In addition, the bulk density of the recycled brick aggregates was lower than the natural materials with higher water absorption. The samples were subjected to a range of testing and the results of the compression tests are summarised in Figs. 1 and 2.

These tests show that compressive strength is lower for crushed brick concrete, whereby the higher the proportion of brick, the lower the compressive strength of the concrete, with a 10–35% decrease in compressive strength at 28 days. This decrease was attributed to the high-water absorption of recycled fine aggregates compared to conventional fine sand. The research concluded that recycled brick aggregates should be used with caution.

These findings are confirmed in other research, where for example [9] studied the mechanical properties of concrete with crushed waste clay, testing five samples of coarse aggregates which contained 0, 25%, 50%, 75% and 100% crushed bricks. This material was mixed with sand and treated with 5% cement. The research found that the cement-treated crushed clay brick aggregate had a lower unconfined compressive strength compared to conventional mixes [10] also found that using crushed recycled brick as aggregate results in concrete of less density and lower strength than conventional mixes, which limit the use of this material in the construction industry. In another study, [11] studied the use of recycled clay brick powder as a replacement for Portland cement and fly ash in self-consolidating concrete mix. The results point out that partial replacement of cementitious material with crushed brick reduced the workability of the concrete and it was also found that the crushed brick concrete had higher 28-day and 56-day strength than the conventional reference concrete. It appears, therefore, that crushed brick is useful but as a replacement for cement rather than aggregate [12] obtained similar results and concluded that the impact on compressive strength is dependent on the proportion of crushed brick in the concrete.

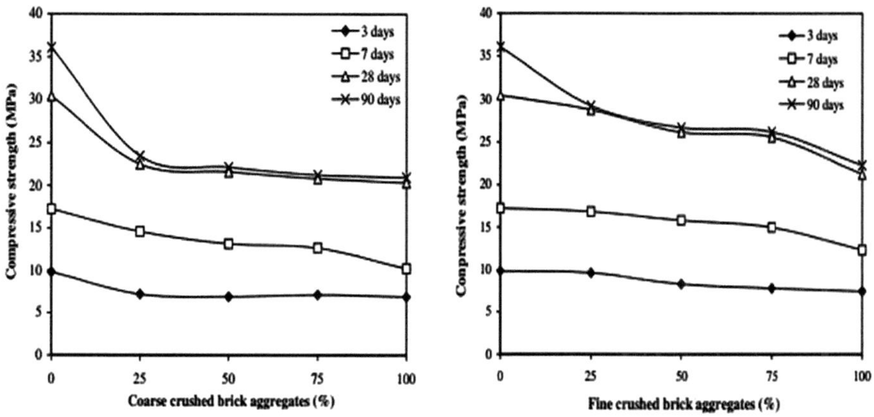

Fig. 1 Compare concrete compressive strength with coarse crushed brick and fine crushed brick

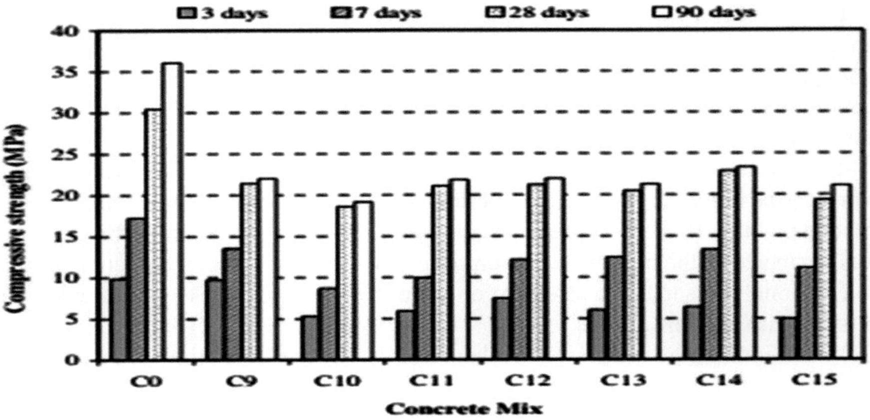

Fig. 2 Concrete compressive strength with concrete mix

2 Methodology

In this experimental project, 10 different mix designs will be used. The number of cubes are 50, cube size is 100*100 mm. In order to provide recycled aggregates 10 bricks will be crushed. These will be used as recycled aggregates in the mixes at different percentages. The coarse aggregate will replace gravel at the following percentages; 0%, 25%, 50%, 75%, 100%. Moreover, the fine aggregate will replace sand at the same proportions. To begin with, bricks were brought to the laboratory and crushed, producing recycled aggregate. Bricks are crushed by hammer to be used as recycled aggregate. After crushing bricks two sieves are placed on top of one another with dimensions of 10 mm and the other one is 5 mm. A plane sieve is placed at the bottom to control dust and take the fine aggregate in order to sieve them later. After this procedure is completed recycled aggregates (coarse aggregate) are saturated in water inside a bucket at certain time then drain the bucket from water and take out recycled aggregate and put in a container to surface dry them. While the fine aggregates were not saturated because they are less than 2 mm, which means the amount of fine would soak in water and possibly lose a lot of fine aggregates. Therefore, fine aggregates have been kept in storage at dry environment.

2.1 Mix Design

The first mix has no recycled aggregate 0%, just sand, gravel, cement and water were used. Before mixing, the moisture content for the sand and gravel was found in order to know how much adjusted water mass needed for the first mix.

The moisture content for gravel and sand was found using the following Eq. (1):

$$MC\% = (W - D)/D \times 100 \quad (1)$$

where,

MC = Moisture content expressed as percentage (%).
W = Wet weight.
D = Weight after drying.

Find how much water is inside sand and gravel using Eq. (2), then find adjusted water mass.

$$\text{Water inside S, G} = \text{Weight } S, G \times \text{Moisture Content }\% \quad (2)$$

Then, use the following equation to find adjusted water mass:

$$\text{Adjusted water mass} = \text{Water weight} - \text{Water inside S, G} \quad (3)$$

2.2 *Coarse Crushed Aggregate*

To find the density of gravel and crushed bricks Eq. (4) was used:

$$\text{Density} = \text{Weight}/\text{Volume} \quad (4)$$

2.3 *Fine Crushed Aggregate*

The same principle was followed, but this time sand was replaced by fine aggregate at the same percentages as above. And then cubes were left for curing for 34 days to be ready for compression strength test.

3 Result and Discussion

In this paper the results and analysis of compression strength test will be presented, followed by a discussion of results. After curing for 28 days, the cubes are ready for undergoing compression test.

3.1 *Cube Crasher Machine (Compressive Test)*

In this project, a cube crasher machine as shown in Fig. 3, will be used for compression strength test to find max load and compression strength stresses.

3.2 Brick Compression Strength Test

Two small cubes of brick were placed and ready to be tested under cube crasher machine as shown in Fig. 3. Each cube has been tested separately, also before testing make sure to find the cross-sectional area of each cube as they are not symmetric. Once, the max load of the cubes is reordered, find the compressive stress for each cube.

Then, find the average compressive stress for each cube using Eq. (5) compressive strength was found according to BS EN 12390-3:2009

$$\text{Compressive Strength} = \text{Max load applied (N)} - \text{Area of Cube (mm}^2) \qquad (5)$$

Table 1 shows measurements taken for cubes (1) & (2) to find the cross-sectional area for each cube. Lengths and widths of the cubes were not symmetric, so by taking a different measurement for the cubes, the average lengths and widths were found using equation (6). After taking all necessary measurements, apply the load at the cubes until failure as shown in Fig. 3. The failure of the cubes was satisfactory according to BS 1881-116:1983 as shown in Fig. 4.

$$\text{L1} * \text{W2} * \text{L2}/3 = \text{Average mm} \qquad (6)$$

where,

L1 = length of the cube.

W2 = width of the cube.

L3 = length of the cube.

Area was found by using Eq. (7).

$$\text{Area} = \text{Average (mm)} * \text{Average (mm)} = \text{mm}^2 \qquad (7)$$

1. Max load cube (1) = 55.40 KN.
2. Max load cube (2) = 45.40 KN

Compression test result of brick is listed in Table 2. Cubic bricks were failed under compressive strength of 26.86 and 22.07 MPa.

Fig. 3 Cube crasher machine, cube during test and cube failure (after test)

Table 1 Cross-sectional area calculations for both cubes

Cube number	L1 (mm)	W2 (mm)	L3 (mm)	Average (mm)	Area (mm^2)
Cube (1)	47.69	40.05	48.44	45.50	2062.50
	47.80	39.96	48.23	45.33	
Cube (2)	48.66	39.92	48.57	45.70	2056.50
	48.45	37.74	48.74	45.00	

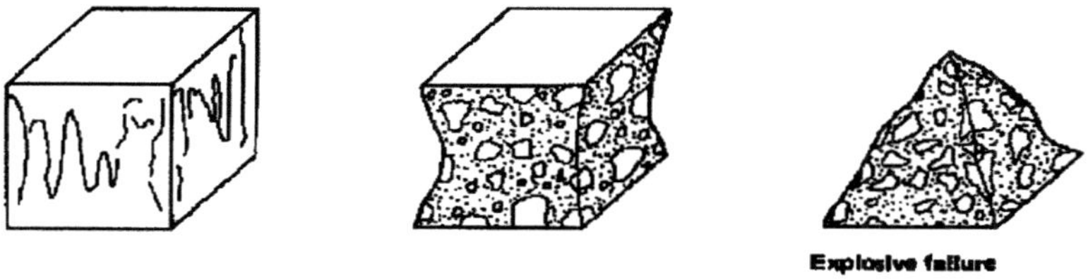

NOTE All four exposed faces are cracked approximately equally, generally with little damage to faces in contact with the platens.

Fig. 4 Different failure mode of concrete cubes, BS 1881–116:1983

Table 2 Compressive strength stress for brick

Cube number	Compressive strength stress (Mpa)
Cube (1)	26.86
Cube (2)	22.07
Average compressive stress for cubes (1) & (2)	24.47 brick strength

From Table 3 it can be seen that the mix designed required more water content with increasing the content of coarse aggregate in samples. This could be due to more void space in concrete.

Results of compression strength are shown in Table 4. From the data obtained it can be said that 25% content of crushed bricks has increased the compression strength by 10% approximately. However, the increase of crushed brick content from 50 to 100% has decreased the compression strength of the samples from 49.7 MPa to 26 MPa.

Table 3 Water content of coarse aggregate mix

Mix percentage (%)	0	25	50	75	100
Water content (g)	411.75	476	524.50	497.14	516.50

Table 4 Compressive strength of concrete cubes with coarse aggregates

Coarse aggregate					
Mix (%)	Compressive strength (Mpa)	Compressive strength (Mpa)	Compressive strength (Mpa)	Average compressive (Mpa)	Medium value (Mpa)
0	42	40	40	40.7	40
25	49	50	50	49.7	50
50	40	41	39	40.0	40
75	33	33	33	33.0	33
100	26	26	26	26.0	26

3.3 *Graphical Presentation of Compression Test Results for Samples with Coarse Aggregates*

From Fig. 5 it can be observed that the sample with 0% content of coarse crushed aggregate has failed on its mean target strength which is 40 MPa. However, 25% addition of coarse crushed aggregate has increased its compression strength by 10 MPa. This could be due to 64.25 g increase of water content in the mix design. Moreover, three more increments of 25% coarse crushed aggregate have dramatically decreased the compression strength of the sample from 50 to 26 MPa. The results also show a small difference in mean values and medium values, the maximum difference that has been noticed is 0.7 MPa.

From Table 5 it can be seen that the mix designed required more water content with increasing the content of fine aggregate in samples. This could be due to more void space in concrete.

Results from compression tests of samples with fine aggregates are shown in Table 6. It can be observed that the compression strength of samples with fine aggregates has shown the same trend as the samples with coarse aggregates. As the compressive strength has been increased by adding 25% content of fine crushed bricks. The

Fig. 5 Compressive strength versus coarse crushed aggregates content

Table 5 Water content of fine aggregate mix

Mix percentage (%)	0	25	25	50	75	100
Water content (g)	411.75	573.68	467.60	483.68	625.10	569.90

Table 6 Compressive strength of concrete cubes with fine aggregates

Fine aggregate					
Mix percentage (%)	Compressive strength (Mpa)	Compressive strength (Mpa)	Compressive strength (Mpa)	Average compressive (Mpa)	Medium value
0	42	40	40	40.7	42
25	60	59	56	57.5	59
50	53	53	54	53.3	54
75	16	17	17	16.7	17
100	36	38	35	36.3	36

compression strength has been decreased from 57.5 MPa to 36.3 MPa. However, the mix sample with 75% content has not followed the trend by showing a fluctuation in the compressive strength of sample that could be because of the maximum water content as shown in Table 5.

3.4 Graphical Presentation of Compression Test Results for Samples with Fine Aggregates

Figure 6 shows the compressive strength of concrete samples with different content of fine crushed aggregates. Results from the tests show that the maximum difference between the mean and medium value is 1.5 MPa which is in sample with 25% content of fine crushed aggregates. In addition, it can be seen that sample with 0% content of fine crushed aggregate has achieved target mean strength with 0.7 MPa difference according to mean value. While, sample with 25% fine crushed aggregate has increased compression strength to 57.5 MPa. Furthermore, samples with 50% and 100% have followed the trend of decreasing compression strength by increasing fine crushed aggregate gradually. However, sample with 75% fine crushed aggregate has fluctuated down the trend to 16.7 MPa. This could be because of higher water content in mix design as shown in Table 5.

3.5 Compression Strength of Concrete Cubes with Different Content of Coarse Crushed Aggregates

Compression strength of concrete cubes with 0%, 25%, 50%, 75% and 100% content of coarse crushed aggregates are shown in Fig. 5. It is observed that the compression strength has been increased from 40.7 MPa to 50 MPa by adding 25% coarse crushed aggregates. This is because of the physical property of crushed brick, as they absorb more water than the gravels that leaves less water in water–cement ratio.

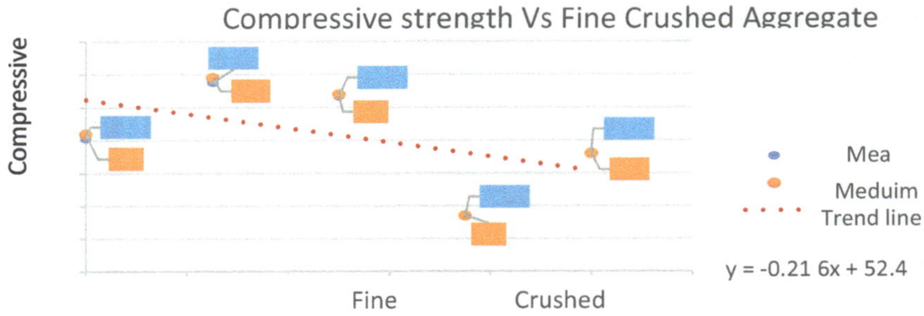

Fig. 6 Compressive strength versus fine crushed aggregates content

However, from the results it can be seen that after 25% of coarse crushed aggregate, the compression strength of the concrete cubes has dropped down. Samples with 50%, 75% and 100% content of coarse crushed aggregates have failed at compression strength of 40 MPa, 33 MPa and 26 MPa, respectively, the same trend was found by [1, 10] while investigating the mechanical properties of concrete cubes with use of coarse crushed bricks instead of natural gravels. In addition, the results show that there is a small difference between the mean values and medium values of compression strength for all samples. This implies that there was consistency in casting samples. The maximum difference that has been noted is 0.7 MPa in sample with 0% and 25% content of coarse crushed aggregates. Figure 5 also illustrates that the sample with 0% and 50% content of coarse crushed aggregate has failed with compression strength of 40 MPa, which is the target mean strength of the mix. This shows that in practical the conventional sample has the same compression strength as the sample with 50% coarse crushed aggregates. While, in theory the equation of the trend line of the graph which is $y = -0.18x + 46.8$ shows that the highest value of Y can be only achieved if the X is considered 0. Meaning that conventional sample will have the maximum compression strength of 46.8 MPa. On the other hand, practically the sample with 25% of coarse crushed aggregate will absorb water more due to crushed brick properties and thus leaving less water for cement hydration. This results as a reduction of water to react with cement thus less water has been reacted with cement than preferred amount in mix design that increases the compression strength of the sample an investigation on impact of crushed bricks in concrete cubes by [12] also concluded that the compression strength of concrete cubes are dependent on the physical properties of crushed bricks used. While, conventional sample has normal aggregates that do not absorb water and leave all the water to react in cement hydration process.

Finally, it can be argued that the increase in compression strength of sample with 25% coarse crushed aggregate is due to the reduction of water by crushed bricks that have more moisture than the ordinary aggregates. While, decrease in compression strength of samples with coarse crushed aggregate with the volume of 50%, 75% and 100% portrays that although there will be less water present for water–cement ratio due to a higher amount of coarse crushed aggregates that should increase compression strength. But, practically such higher amount of replacement of natural aggregates with crushed bricks will definitely impact the compression strength.

3.6 Compression Strength of Concrete Cubes with Different Content of Fine Crushed Aggregates

Fine crushed aggregates were sieved by 2 mm sieve size. Therefore, the fine particles can be considered as sandy silt according to BS 5930. The results that obtained from compression test are shown in Fig. 6. The graph demonstrates that the conventional sample has failed at compression strength of 40.7 MPa. There is an increase of 16.8 MPa with addition of 25% as a replacement of sand. And then the graph has dropped down gradually by increasing the volume of fine crushed aggregates. A fluctuation of compression strength has been noticed by the sample with 75% fine crushed aggregates with compression strength of 16.7 MPa. Samples with 0%, 25%, 50%, 75% and 100% content of coarse crushed aggregates are shown in Fig. 6. That has failed with compression strength of 40.7 MPa, 57.5 MPa, 53.3 MPa, 16.7 MPa and 36.3 MPa, respectively. In addition, the graph has represented the same trend as the graph of coarse crushed aggregates in Fig. 5. The maximum difference between the mean values and medium values in compression strength is 1.5 MPa in a sample with 25% content of fine crushed aggregates. The sample with 25% and 50% has shown higher compression strength than the conventional sample. That could be due to the physical properties of fine crushed bricks in terms of moisture content or porosity [1, 10] also concluded that the use of crushed brick in concrete will result in less strength. This has resulted in less water in the mix to react with cement therefore, the compression strength has been increased [12] obtained similar results and concluded that the impact on compressive strength is dependent on the proportion of crushed brick in the concrete. However, the sample with 75% content of fine crushed aggregate has shown the minimum strength even than the one with 100% content of fine crushed aggregate. This is due to the highest requirement of water in mix design compare to other samples as shown in Table 5.

Summarising the impact of different content of fine crushed aggregate, it can be up to 50% replacement of sand by fine crushed aggregate will increase the compression strength of concrete. Increasing its content more than 50% will not only reduce the compression strength of concrete but it will also result an unpredictable ultimate failure strength.

3.7 Comparison of Using Fine and Coarse Aggregate in Concrete

Compression test results of coarse crushed aggregates as a replacement of gravels and fine crushed aggregates as a replacement of sand in concrete are shown in Table 7.

From the above table, it can be said that the use of fine or coarse crushed aggregate in concrete up to 50% will not decrease the compression strength of concrete from the mean target strength. However, increasing the content more than 50% will decrease the compression strength greatly.

Water–cement ratio is inversely proportional to the compression strength of concrete [13]. From Table 7 it can be seen that replacing natural aggregate and sand with coarse and fine crushed aggregates has increased the water–cement ratio of the mix. This implies that the component of cement has been graded poorly after replacing aggregates and sand thus more surface area is exposed to be covered by concrete paste that had demanded more water in mix. However, the compression strength of concrete cubes has increased by adding 25% coarse crushed brick. This is because natural aggregates have an absorption characteristic up to 3%, while London bricks can absorb water from 25 to 35% [14]. Therefore, the amount of water needed for targeted mean strength has been divided into coarse brick absorbtion and reaction with cement that has resulted more compression strength due to less water available to react with cement that makes the water–cement ratio less than the conventional mix.

Furthermore, from Table 7 it can be said that the use of fine crushed aggregate as a replacement of sand increases the compression strength more than the use of coarse crushed aggregate as a replacement of gravels in concrete. Samples with 25% and 50% content fine crushed aggregate have shown an increase of 16.8 MPa and 12.6 MPa, respectively. This could be because natural aggregates that are the main component of concrete for compression strength were not replaced but only the sand was replaced. On the other hand, 25% content of coarse crushed brick has increased the compression strength of concrete cube by 9 MPa, which could be due to physical properties of crushed bricks that absorb more water and result less water in water–cement hydration process.

Table 7 Compression strength of fine and coarse crushed aggregates

Mix (%)	Coarse crushed aggregate mean compression strength (MPa)	W/C ratio	Fine crushed aggregate mean compression strength (MPa)	W/C ratio
0	40.7	0.28	40.7	0.28
25	49.7	0.32	57.5	0.32
50	40.0	0.36	53.3	0.33
75	33.0	0.34	16.7	0.42
100	26.0	0.35	36.3	0.39

In summary, it can be said that the replacement of sand by fine crushed aggregates increases compression strength more than replacing gravels with coarse crushed aggregates up to 50% content. From this result, it can also be argued that gravels are the main component that plays a vital role in compression strength of concrete.

4 Conclusion

The main conclusions from this study are presented below.

1. For crushed brick aggregate it is possible to replace both the fine and coarse fraction since much fine material is produced in the crushing process.
2. The test study was in general agreement with other published studies from which the following findings were observed.
   - By replacing natural aggregate and sand up to 50% by coarse crushed aggregates and fine crushed aggregates, the target means strength of concrete can be achieved. However, it will decrease the durability.
   - The fine crushed aggregates that are achieved in breaking bricks can be used as a replacement of sand in concrete mix up to 50%.
   - Using fine crushed aggregate does not increase compression strength, however it helps to decrease voids in the sample that makes it denser and thus increases compression strength of concrete.
3. Concrete with 100% content of coarse and fine brick in concrete has more compressive strength than the brick itself. From this paper, it can be said that coarse crushed aggregate and fine crushed aggregate can replace natural aggregate and sand completely in concrete applications that required 26 MPa and 36 MPa compressive strength like sidewalks and residential slabs.

References

1. Debieb, F., Kenai, S.: The use of coarse and fine crushed bricks as aggregate in concrete. Constr. Build. Mater. **22**, 886–893 (2008)
2. Hall, M.R., Najim, K.B.: Structural behaviour and durability of steelreinforced structural plain/self-compacting rubberised concrete (PRC/SCRC). Constr. Build. Mater. **73**, 490–497 (2014)
3. Sonawane, T.R., Pimplikar, S.S.: Use of recycled aggregate concrete. IOSR J. Mech. Civil Eng. **52**(59) (2013)
4. Yang, J., Du, Q., Bao, Y.: Concrete with recycled concrete aggregate and crushed clay bricks. Constr. Build. Mater. **25**(4), 1935–1945 (2011)
5. Vatopoulos, K., Tzimas, E.: Assessment of CO2 capture technologies in cement manufacturing process. J. Clean. Prod. **32**, 251–261 (2012)
6. Ajayi, S.O., Oyedele, L.O., Bilal, M., Akinade, O.O., Alaka, H.A., Owolabi, H.A., Kadiri, K.O.: Waste effectiveness of the construction industry: Understanding the impediments and requisites for improvements. Resour. Conserv. Recycl. **102**, 101–112 (2015)

7. Husain, A.H., Abas, N.F.: The mechanical performance of waste concrete ash as cement replacement materials in concrete. Appl. Mech. Mater. **747**, 234–237 (2015)
8. Adamson, M., Razmjoo, A., Poursaee, A.: Durability of concrete incorporating crushed brick as coarse aggregate. Constr. Build. Mater. **94**, 426–432 (2015)
9. Hu, L., Hao, J., Wang, L.: Laboratory evaluation of cement treated aggregate containing crushed clay brick. J. Traffic Transp. Eng. **1**(5), 371–382 (2014)
10. Jankovic, K., Nikolic, D., Bojovic, D.: Concrete paving blocks and flags made with crushed brick as aggregate. Constr. Build. Mater. **28**(1), 659663 (2012)
11. Sun, R., Huang, D., Ge, Z., Hu, Y., Guan, Y.: Properties of self-consolidating concrete with recycled clay-brick-powder replacing cementitious material. J. Sustain. Cement-Based Mater. **3**(3–4), 211–219 (2014)
12. Ge, Z., Gao, Z., Sun, R., Zheng, L.: Mix design of concrete with recycled clay-brick-powder using the orthogonal design method. Constr. Build. Mater. **31**, 289–293 (2012)
13. Felekoglu, B., Turkel, S., Baradan, B.: Build and environment effect of water/cement ration on the fresh and hardened properties of self-compacting concrete **42**(1), 1795–1802 (2006)
14. MPA: Brick and block production (2013). Available: http://www.mortar.org.uk/documents/LT05-Bricks-and-Blocks.pdf. Accessed 10 April 2016

An Overview on Utilization of Steel Slag as Road Construction Materials

Ashkar Rahman Aquib, Zarrin Tasnim Probha, and Md. Arifin Haque

Abstract The transportation and highway industries are starting to use sustainable, recycled materials in order to reduce their reliance on natural sources. One of these materials is steel-furnace slag, which is a by-product of steel manufacturing that is usually disposed of in the environment, causing pollution. Due to its high disposal cost and the positive features of steel slag, many countries have recognized it as a useful construction material rather than an industrial waste. Reusing steel slag can save natural resources and reduce the need for landfills for its disposal. It has been found to be a good material for road construction, as it outperforms natural aggregates in terms of skid resistance, rutting, moisture damage resistance, and fatigue. This paper examines the advantages of using steel slag in road construction, along with any potential detrimental effects. It also provides a comparison between different types of slag in terms of their suitability for road construction.

Keywords Steel slag · Road construction materials · Pavement performance

1 Introduction

Steel slag is a solid waste derived from the steelmaking process, which accounts for approximately 15% of the total crude steel output [1, 2]. The annual global production of iron and steel slag exceeds 400 million tons. Nowadays, the focus is on preventing trash generation, recycling, and reusing waste, as well as reducing the harmful effects of disposal on the environment. Among all the solid/liquid wastes, slag generated at iron making and steelmaking units are created in the largest quantities [3]. Heavy metals, particularly lead (Pb), cadmium (Cd), mercury (Hg), chromium (Cr), and arsenic (As), can leak from steel slags when they are piled up or landfilled, using valuable land resources [4]. Various initiatives have been made to successfully utilize steel slag while lowering the danger involved [5]. Steel slag can be directly used as sinter flux and iron smelting flux for the production of steel [6]. The pretreated

A. R. Aquib (✉) · Z. T. Probha · Md. A. Haque
Bangladesh University of Engineering and Technology, Dhaka, Bangladesh
e-mail: araquib21@gmail.com

E. Strauss (ed.), *Proceedings of the 7th International Conference on Civil Engineering*, Lecture Notes in Civil Engineering 371, https://doi.org/10.1007/978-981-99-4045-5_5

steel slag can be utilized directly for road construction (around 49.7% in the US), as aggregate in the manufacture of concrete or asphalt (the main recycling route), or both [6]. Steel slag can also be recycled by being ground into fine fillers that are used as raw materials for cement manufacturing or as a concrete admixture. [7]. The main use for steel slag and air-cooled iron slag are as feed for cement kilns and as aggregates for asphaltic pavement, fill, and road bases. Slag that has been air-cooled can also be utilized as a concrete aggregate [5, 8]. Specific characteristics of steel slags could affect the characteristics of materials used in road building. This research examines multiple papers on the physical, chemical, radioactive, and mineralogical properties of steel slags. In addition, how steel slag improves the attributes of the pavement, such as skid resistance, rutting, moisture damage, fatigue failure, creep, permeability, durability, bearing capacity, affinity of steel slag with binder, and California bearing ratio is also discussed in this study. Following that, the use of steel slags in asphalt mixtures and the influence on roadway qualities are investigated. This review also discusses a few negative effects of steel slag utilized as road pavement materials.

2 Characteristics of Steel Slag

2.1 Physical Properties

Steel slags can have numerous physical properties shown in Table 1. The properties which have a certain amount of interest are Los Angeles abrasion value (LAAV), water absorption, specific gravity, crushing value, soundness, polished stone value, and porosity.

2.2 Chemical Properties

Steel slag can be obtained from different furnaces. The ones of interest are Basic Oxygen Furnace (BOF), Electric Arc Furnace (EAF), Ladle Furnace (LF). The chemical composition of steel slag may vary from process to process or furnace to furnace. The chemical composition obtained from different literature is shown in Table 2. It is seen that CaO has the major share of percentage in steel slags. This ranges from 16.90% to 58.4%. Another notable thing is that Ladle Furnaces have significant amount of SiO_2 greater than that of the other furnaces.

Table 1 Physical properties of steel slag

References	LA abrasion (%)	Water absorption (%)	Specific gravity (g/cc)	Crushing value	Soundness	Polished stone value	Porosity
[9]	18.60	2.5	3.25	–	–	–	–
[10]	13.1	1.18	3.29	12%	–	57	5.76%
[11]	24	5.46	–	23%	2.07%	54	–
[12]	9.80	1.20	–	26.05%	–	56.60	–
[13]	14	2.4	–	–	–	56	–
[14]	20.1	1.68	3.310	–	2.7%	–	–
[15]	13	>1	–	–	–	70	–
[16]	40.9	2.9	3.45	–	0.1%	–	–
[17]	11	1.8	3.41	–	4.2%	11	–
[18]	29	3.8	–	–	–	–	–
[19]	5.10	3.896	2.816	–	0.71%	55.3	–
[20]	20	0.85	3.34	–	–	–	2.5%

Table 2 Chemical properties of steel slag

References	Furnace	SiO_2	MgO	MnO	P_2O_5	Al_2O_3	TiO_2	Fe_2O3	CaO
[21]	BOF	9–11	1–2	4–5	2.3–3.2	0.7–1.4	–	10–13	50–57
[22]		15.30	1.10	0.39	3.10	1.30	–	–	53.30
[23]		17.80	6.32	6.52	1.33	2.04	–	6.58	40.10
[24]		12.50	4.30	6.10	1.10	2.40	0.80	31.20	41.30
[25]		10.07	4.75	4.40	1.38	2.33	0.92	–	45.77
[26]		18.45	4.67	3.03	1.69	4.59	2.44	15.57	45.17
[27]		11.50	9.30	3.70	1.30	2.30	0.37	27.30	37.40
[28]		10.24	7.16	7.16	1.77	1.70	0.54	24.73	40.15
[29]		11.10	9.60	3.10	–	1.90	–	10.90	45.00
[30]		11.80	6.30	1.90	2.70	2.00	0.50	22.60	47–50
[31]	EAF	10–40	5–10	–	–	5–15	–	5–30	20–50
[32]		26.40	1.86	2.66	–	4.84	–	43.40	16.90
[33]		13.10	5.03	4.18	–	5.51	0.60	36.80	33.0
[34]	LF	30.31	7.41	–	–	1.31	1.09	–	58.4
[35]		26.80	3.2	1.0	–	5.2	0.3	1.59	57.0

2.3 *Mineralogical Properties*

Steel slags comprise of different phases shown in Table 3. These phases are important elements of steel slag.

Table 3 Mineralogical properties of steel slag

Phases	References
Alite (Ca_3SiO_5)	[29]
Periclase (MgO)	[29]
Larnite (Ca_2SiO_4)	[29]
Fluorite (CaF_2)	[36]
Dicalcium aluminoferrite ($Ca_2(Fe, Al, Ti)_2O_5$	[36]
Lime phase (Ca, Fe) O	[36]
Hematite (Fe_2O_3)	[37]
Magnetite (Fe_3O_4)	[38]
Quartz (SiO_2)	[39]
Dolomite ($CaMg(CO_3)_2$)	[40]
Calcium ferrite ($CaFe_2O_4$)	[41]
Wustite (FeO)	[42]
Calcium Hydroxide ($Ca(OH)_2$)	[37]
Magnesium Iron Oxide ($Mg_{0.239}\ Fe_{0.761}O$)	[43]

2.4 Radioactive Properties

${}^{40}_{19}K$, ${}^{232}_{90}Th({}^{226}_{86}Ra)$, ${}^{226}_{86}Ra$ and ${}^{238}_{92}U$ isotopes are generally present in steel slag shown in Table 4. These are natural isotopes and are regarded to be usual. These may come from raw materials and accessory materials related to steelmaking [44].

Measured and calculated values of radioactivity of the slag aggregate are very much below than the values recommended by the Nuclear Safety Authority (STUK) Guide ST 12.2 [45].

Table 4 Radioactive properties of steel slag [45]

Slag aggregate	Activity concentration ($Bqkg^{-1}$)		
	${}^{232}_{90}Th$	${}^{226}_{86}Ra$	${}^{40}_{19}K$
0–4 mm	14.4 ± 0.9	24.0 ± 0.8	22.0 ± 2.8
4–8 mm	10.4 ± 1.2	26.8 ± 3.0	19.3 ± 1.7
8–16 mm	9.7 ± 2.1	16.9 ± 2.2	14.2 ± 6.2
16–32 mm	0.2 ± 2.1	14.8 ± 2.0	14.1 ± 6.8
0–32 mm	8.6 ± 0.9	22.0 ± 1.2	23.3 ± 1.4
Allowed value (Road construction)	500	700	8000

3 Effect of Steel Slag on Pavement Properties

3.1 Skid Resistance

The amount of friction between a tire and the road surface is a major factor in determining the skid resistance of the surface courses, and is commonly associated with accident rates [46]. The friction between the tires and the asphalt surface helps to reduce the chances of skidding, and steel slag is said to have a friction angle of 40–50° [47]. The texture of steel slag is rougher and more angular than natural aggregates, giving it better interlocking properties and higher skid resistance. Bituminous mixes containing EAF aggregates and limestone sand have been found to be successful in terms of skid resistance [48]. Incorporating steel slag into hot mix asphalt as a surface/ wearing course increases skid resistance on both dry and wet conditions [49]. The polishing effect of traffic on pavement skid resistance is frequently assessed using the British pendulum tester (BPT) and the dynamic friction tester (DFT) [50]. When EAF (Electric Arc Furnace) slag is used as coarse aggregate and LFS (Ladle Furnace Slag) is used as filler and fine aggregate in a mixture, it shows better skid resistance performance since it has a higher British Pendulum Number than other mixtures [51].

3.2 Rutting

Rutting resistance of steel slag aggregate (SSA) mixtures has increased with the addition of SSA (Steel Slag Aggregate) because of aggregate structure's increased interlocking. The effectiveness of the improvement declines as the SSA content rises since more asphalt cement is required, which serves as a lubricant and lowers the shear resistance of the AC (Asphalt Concrete) mixes [52]. In comparison to the reference mixtures under examination, every parameter of the rutting resistance was improved in the mixtures incorporating steel slag. The mixture containing EAF (Electric Arc Furnace) slag as the coarse aggregate and LFS (Ladle Furnace Slag) as filler and fine aggregate shows the best rutting resistance performance since it has a lower rut depth (2.4 mm) than other mixtures [51]. The repeated axial load test was conducted to assess the performance of limestone asphalt mix with different proportions(0%, 30%, 60%, and 90%) of EAF steel slag. It was observed that as the ratio of steel slag in the mix increased, the permanent deformation caused by the axial load decreased, signifying a better resistance to rutting [53]. Cyclic axial loading and Hamburg Wheel Tracking tests were conducted to compare the rutting performance of a limestone asphalt mix to a steel slag-limestone mix. It was discovered that the mix created by substituting the coarse fraction of limestone with BOF steel slag exhibited improved rutting resistance in comparison to the limestone mix [54]. The steel-slag mix was able to resist 6.1 mm of rutting after 5000 cycles, whereas the conventional mixture only lasted for 2019 cycles before rutting had reached 15 mm.

Furthermore, in the wheel tracking test, the steel-slag mixture had higher resistance to permanent deformation than the conventional mixture [11].

3.3 Fatigue Failure

It was seen that the fatigue resistance increased as the amount of steel slag in the mix increased. This could be attributed to the angular shape of steel slag and the improved adherence of asphalt to steel slag aggregate [55]. To guarantee a wide variety of findings, tests were run at various initial tensile strain levels. Results showed that the 25% steel slag aggregate (SSA) mix had the greatest fatigue life, as the angularity of the SSA improved the interlocking of the aggregate structure. Although increasing the SSA quantity past 25% decreased the fatigue life of the mixes, they were still higher than that of the 0% SSA mixes [52]. In a Four-point bending test the researchers observed an increase in fatigue life with an increase in the percent replacement of limestone coarse fraction by EAF steel slag. It was observed that BOF steel-slag mix was superior to the basalt mix at all levels of stress in the fatigue test. Moisture infiltrating both mixtures decreased their fatigue resistance, but the steel-slag mix still performed better than the basalt mix [56]. Recycled asphalt mixtures (RAM) have a preferred level of fatigue resistance. The best fatigue resistance performance is found in steel slag virgin asphalt mixture. By adding steel slag to recycled asphalt mixtures made of basalt, fatigue resistance can be increased. This is due to the strong angularity and microstructure of steel slag, which appears better for interfacial adhesion [57].

3.4 Moisture Damage Resistance

Limestone asphalt mixes with different proportions(0%, 30%, 60%, and 90%) of EAF steel slag were tested. The mixture with 90% of EAF steel slag had the highest TSR value. In both wet and dry conditions, the mixture with 90% of EAF steel slag exhibited the highest indirect tensile strength (ITS) value. As the proportion of EAF slag increases in the mixture, the resistance to moisture damage increases [53]. The combination of BOF slag aggregates and the Sasobit additive produced a level of short-term moisture damage resistance that was sufficient. The boiling test demonstrated a low stripping percentage of 4.4% for a 2.5% concentration of Sasobit, and BOF slag warm mix asphalt (WMA) had both high tensile strength ratios (TSR) and residual Marshall stability (RMS) ratios of above 80%. Despite the increase in moisture sensitivity due to the additive, BOF slag WMA mixtures showed strong resistance to moisture damage [54]. The Tensile Strength Ratios of mixtures with Basic Oxygen Furnace (BOF) are higher than those made with basalt, suggesting that they are better able to resist moisture damage [56]. The Marshall stability test and the freeze–thaw split test showed that substituting cement (CE), slaked lime (SL), and steel slag powder (SSP) for limestone filler in asphalt mixtures yielded greater moisture damage resistance. The asphalt mixture's susceptibility to moisture gradually declines as the amount of SSP replacing LF (limestone filler) increases. In this test, 25% of the total volume of fillers was the ideal amount of SSP to substitute [58]. The addition of steel slag increases the residual Marshall stability (RMS), tensile strength ratio (TSR) and reduces the Cantabro spatter loss of recycled asphalt mixtures (RAM), leading to greater moisture resistance than basalt RAM. Steel slag RAM with 50% reclaimed asphalt pavement (RAP) has an RMS of 90.5%, a TSR of 89.3%, and a spatter loss of 5.5%, all of which are higher than basalt RAM with 50% RAP. Incorporating steel slag into pavements helps improve moisture resistance, an essential factor for successful pavement construction [57].

3.5 Permeability

In comparison to the normal combinations, steelmaking slag mixtures were more porous and permeable [56]. In this work, a permeable steel slag-asphalt mix was produced using a combination of road petroleum asphalt and steel slag material. This mixture was evaluated in accordance with the Technical Regulations (CJJT 190–2012) for Permeable Asphalt Pavement and compared to a permeable limestone-asphalt mixture. The permeable steel slag–asphalt mix had a permeability coefficient of 60.61 mL/s, a 0.5 h stability of 9.12 kN, a 48 h stability of 8.27 kN, and a Marshall stability of 9.12 kN, demonstrating that the mix was of good quality in terms of permeability, water stability, and Marshall stability [59]. In this study, the specified mixing ratio was initially followed to create a permeable steel-slag-bitumen mixture (PSSBM). The properties of the interaction between bitumen and steel slag were then investigated. The study showed that the mixture had excellent permeability, water stability, and Marshall stability, with a permeability coefficient, flow value, and residual stability of 55.56 mL/s, 9.41 kN, 2.56 mm, and 91.18%, respectively. In addition, it had high-temperature (HT) stability and a low volume-expansion rate [60].

3.6 Durability

With the addition of slags, the chosen durability indexes were improved, making these pavements less prone to aging and thermal cracking [56]. The mixes with EAF slag have a low water damage rate, indicating strong durability [53]. Tests and measurements were conducted to assess the long-term quality of steel slag aggregate. The results showed that the slag aggregate had significantly higher durability after three years of traffic use. Skid resistance was also measured in 1987 and 1990 on national roads at four points 500 m apart [48].

3.7 Bearing Capacity

Steel slag can be a successful way to enhance the bearing capacity of soil. It is a viable option to construct road pavement in areas with weak bearing capacity such as black cotton soil and swampy lands. As the percentage of steel slag in the soil increases, the bearing capacity also rises. However, when the steel slag is present in more than 30–33%, economical and technical issues arise. The project's cost increases significantly in comparison to other soil stabilization methods. In addition, excessive steel slag can lead to corrosion and thermal expansion [61].

3.8 Affinity of Steel Slag with Binder

Steel slags have a strong attraction to bitumen binders, which are usually acidic and have a pH value of less than 7. The alkalinity of steel slags, with a pH value of around 12, helps to create a strong bond between the two and resists against stripping. This can be tested by placing a sample in boiling water to assess the level of stripping [62].

3.9 California Bearing Ratio (CBR)

The results of the tests indicate that as the percentage of steel slag in the soil is increased, the CBR value also increases. Sample 1, which was composed of 0% steel slag and 100% soil, had a CBR of 1.75% for 2.5 mm penetration and 1.69% for 5 mm penetration. As the percentage of steel slag was increased to 10%, the CBR value for Sample 2 increased to 2.07% for 2.5 mm penetration and 2.01% for 5 mm penetration. The CBR value for Sample 3 (20% steel slag and 80% soil) was 2.30% for 2.5 mm penetration and 2.28% for 5 mm penetration, and the CBR value for Sample 4 (30% steel slag and 70% soil) was 3.5% for 2.5 mm penetration and 3.39% for 5 mm penetration. Finally, Sample 5 (40% steel slag and 60% soil) had a CBR of 4.69% for 2.5 mm penetration and 4.56% for 5 mm penetration. The CBR value of Sample 6, which had a 50/50 mix of steel slag and soil, was 5.88% when tested at a 2.5 mm penetration, and 5.83% when tested at a 5 mm penetration [61]. It was discovered that the Steel Slag Aggregate mixtures created a base course with incredibly high CBR values. Furthermore, the CBR of the steel slag aggregates when combined with different kinds of natural aggregates varied, with a range of roughly 400 [63].

3.10 Creep

The experiment showed that the steel-slag mixture can withstand permanent deformation and strain better than the usual mixture, with permanent deformation of 0.044 mm and strain of 0.377% as opposed to 0.605 mm and 0.946%, respectively. It is most likely because the steel-slag mixture has better interlocking and adhesive properties that it can take more deformation and last longer [11]. The Steel Slag Powder (SSP) 0–Limestone Powder (LSP) 100 sample showed the largest accumulated strain during 10 cycles, indicating that the strain of the four asphalt mastics increases when the period is increased. In contrast to the other samples, the Steel slag powder (SSP) 100-Limestone powder (LSP) 0 asphalt mastic has a lower strain [64]. The creep resistance of the Asphalt Concrete mixes was improved when up to 75% of the coarse limestone aggregate was replaced by Steel Slag Aggregate (SSA). However, the creep resistance diminished when the SSA was increased to 100% due to the necessity of larger amounts of asphalt cement. The most effective creep resistance was found in the 25% SSA mix, as the amount of asphalt cement was kept low while the interlocking of the aggregate structure was improved [52].

3.11 Improve Slurry Seal Performance

This study was aimed at improving the performance of slurry seal in the laboratory by utilizing a better combination of aggregates. It was found that utilizing only limestone or slag aggregate separately did not yield optimal results due to certain issues that came with their use. The mixture of slag and limestone that passed through a no. 30 sieve size and had a 15% bitumen emulsion content provided the optimum performance [65].

4 The Use of Steel Slag as Different Road Construction Materials and Its Effect (Table 5)

See Table 5.

5 Comparison Among Different Types of Slags (Table 6)

See Table 6.

Table 5 The use of steel slag as different road construction materials and its effect

References	Example	Composition of aggregate	Effect
[66]	Weathered steel slag fine aggregate and modified weathered steel slag fine aggregate	Simultaneously using gneiss coarse aggregate and 47% steel slag fine aggregate (size below 4.75 mm) with asphalt mixture	The mechanical properties and deformation resistance of the bituminous mixture were greatly enhanced due to the modified steel slag. This type of slag has an 8% higher fracture energy than asphalt mix prepared with limestone, and its cohesive strength is higher too. To be used in asphalt, the steel slag must be aged for a long period of time, even though it had already been aged for 12 months, as it still showed signs of expansion
[67]	Warm mix asphalt incorporated with coarse steel aggregate and RAP (reclaimed asphalt pavement)	40% coarse steel slag along with 40% fine RAP materials	The substitution of steel slag caused the need for bitumen to rise by 13.7% due to the porous nature of the slag. Greater impact on the resilient modulus at lower temperatures than the interlocking of the aggregates. Even though the resilient modulus of the slag mixture was lower at low temperatures, it was higher at medium and high temperatures
[68]	Stone matrix asphalt (SMA) mixtures asphalt was 60/70 penetration grade	One type of limestone aggregate (as coarse, fine, and filler fraction) and two types of steel slag (as a substitute for the coarse or fine fraction of limestone aggregate)	The resilient module of the asphalt mix increased by 31% when steel slag was used as the coarse aggregate instead of limestone Also, the Marshall stability value of the mix with slag was 11.08 kN, which is significantly higher than the 8.4 kN obtained with limestone. The Marshall Quotient value of the asphalts containing slag was also much higher, reaching up to 45%. This indicates that the slag provides superior deformation resistance and strength against creep deformation

(continued)

Table 5 (continued)

References	Example	Composition of aggregate	Effect
[17]	Stone mastic asphalt with BOF (SMA-BOF)	Natural aggregate (19-mm maximum aggregate size gradation)	According to the study, among all other combinations, SMA asphalt mixed with steel slag had the lowest rut depth and the maximum dynamic stability(3553 cycles/mm). The driving comfort and friction qualities of the roads constructed with slag were on par with or better than those of roads constructed with other aggregates, in spite of the significant stresses that were exerted on the asphalt by cars braking, accelerating, and turning. The slag-built roads lasted three years without cracking or experiencing any other problems. The usage of steel slag was successful on highways with lots of braking and turning
[69]	Bituminous paving mixture	30% as Natural aggregate	Slag-based mixtures have the following characteristics: 22–23% fragmentation, 6.5–9.5% fragmentation, and abrasion resistance With various natural aggregates, values of 1.5–2.2% water absorption and polishing coefficient were approximately obtained Slag-based samples' Marshall stability ratings produced results that were close to those of natural aggregate and had a favorable impact on mechanical qualities
[70]	SMA-13 (normal maximum aggregate size 13 mm)	9.5–16 mm in grain size:48% 4.75–9.5 mm: 24% as Naturel Agg approximately 80% of the whole blended aggregates were substituted by steel slag of various sizes	The Marshall test results of basalt and steel slag showed that the two materials boast similar density, air gaps, and stability Steel slag, however, has a porous structure which allows it to absorb more bitumen from the asphalt mixture than basalt. This should be taken into account when preparing asphalt mixtures. When the samples were submerged in water, a slight expansion was noticed, but the value remained below 1% even after 7 days. With adequate aging, the slag can reach a satisfactory level of stability. Furthermore, the use of slag can improve the deformation resistance of asphalt under high temperature, wear resistance, and skid resistance

(continued)

Table 5 (continued)

References	Example	Composition of aggregate	Effect
[10]	Asphalt mixture	>4.75 mm aggregate	Because of its angular structure, the steel slag is tightly attached to the asphalt matrix. Slag-prepared samples withstand higher traffic loads and temperatures than other samples. When compared to other aggregates, steel slag's critical compressive strain energy density value was the greatest. As a result, it demonstrates that a large amount of energy is required for the slag to reach a material failure state
[71]	Asphalt mortar	>0.75 mm size, as filler	Although the addition of steel slag as a filler in the asphalt mixture boosted deformation resistance, it resulted in a modest loss in low-temperature cracking resistance. The use of slag as a filler has enhanced the complicated shear modulus of asphalt. In experiments, utilizing slag and limestone combined yielded better results than using other components. When combined with limestone, the ideal slag ratio value should be less than 75%
[19]	Asphalt mixture	80% Steel Slag, 20% Copper Mine Tailings	With the application of slag in the provided ratio, the Marshall stability rose from 13.5 kN to 15.75 kN and the Marshall Quotient value increased from 4.63 kN/mm to 7.29 kN/mm. The interlocking with the asphalt matrix increased due to the slag's strong internal friction, angular form, and high specific gravity, indicating that this material is suited for asphalt pavements. The resilience modulus was greater in samples containing copper and slag than in ones containing granite
[19]	Asphalt mixture	40% Granite, 40% EAF steel slag, and 20% CMT	The Marshall stability test properties demonstrate that binder PG 76 performed better than 80–100 bitumen. The ideal bitumen concentration for 80–100 binder and PG 76 is 5.13% and 5.16%, respectively

(continued)

Table 5 (continued)

References	Example	Composition of aggregate	Effect
[72]	Asphalt mixture	100% as Coarse Aggregate (in the mixture coarse aggregate 57%)	The inclusion of steel slag as coarse aggregate and recycled concrete waste as fine aggregate in asphalt mixture boosted Marshall stability by 34.4%. These particles also reduced the flow. The angular structure of these two aggregates provides abrasion resistance, friction resistance, and persistent adhesion. The amount of distortion has risen. When compared to the reference sample, mixtures of the same aggregate type raised the indirect tensile modulus of elasticity by 2.35 times

Table 6 Comparison among different types of slags in terms of suitability for road construction

References	Criteria of comparison	EAF	BOF and/or others
[73, 74]	Specific gravity	2.8–3.9	3.08–3.19
[75]	Cost	More expensive	Less expensive
[51]	Skid resistance	When used as coarse aggregate in a mixture it shows better skid resistance performance	When LFS (Ladle Furnace Slag) is used as fines and filler in a mixture, it shows comparatively worse skid resistance performance
[52, 53]	Fatigue failure	An increase in fatigue life with an increase in the percent replacement of limestone coarse fraction by steel slag	Steel slag collected from united iron and steel manufacturing company Jordan showed that the 25% SSA mix had the greatest fatigue life. Increasing the Steel Slag Aggregate quantity past 25% decreased the fatigue life of the mixes
[53, 58]	Moisture damage resistance	Moisture damage resistance increases with increasing EAF slag	Moisture damage resistance increases as the amount of Steel Slag Powder replacing LF (limestone filler) increases up to 25% of the total volume of fillers. But after that moisture damage resistance decreases

6 Conclusion and Recommendation

The reviewed literature led to the conclusion that steel slag has numerous advantages for use as a road-building material. Its physical and chemical properties have been determined to be beneficial in this regard, with tests such as L.A. abrasion, crushing value, and soundness showing its superiority to natural aggregates. Steel slag has been found to possess superior skid resistance when compared to natural aggregate. Additionally, it has been demonstrated to have greater resistance to moisture damage and rutting. Furthermore, fatigue failure studies have indicated that steel slag performs better than natural aggregate. Literature suggests that steel slag is a favorable choice for use as a coarse aggregate in road construction, and it may also serve as a good filler and fine aggregate. The production of steel slag leads to a greater emission of CO_2, and the use of steel slag aggregate can be expensive to transport. These factors can create undesirable environmental outcomes. In this study, EAF slag was found to be the best choice for road construction due to its improved skid resistance, moisture damage resistance, fatigue failure properties, and specific gravity. Further research should be conducted to evaluate the suitability of different types of slag for road construction. There are already some studies being conducted on the topic, but additional research is necessary.

References

1. Furlani, E., Tonello, G., Maschio, S.: Recycling of steel slag and glass cullet from energy saving lamps by fast firing production of Ceramics. Waste Manage. **30**(8–9), 1714–1719 (2010). https://doi.org/10.1016/j.wasman.2010.03.030
2. Pang, B., Zhou, Z., Hou, P., Du, P., Zhang, L., Xu, H.: Autogenous and engineered healing mechanisms of carbonated steel slag aggregate in concrete. Constr. Build. Mater. **107**, 191–202 (2016). https://doi.org/10.1016/j.conbuildmat.2015.12.191
3. Kumar, S.: Recent trends in slag management & utilization in the steel industry. Minerals Metals Rev., 94–102 (2019)
4. Zhou, X., Xiang, X., Li, C.: Ecological security analysis and advances in innocent treatment of steel slag. Multipurpose Utilization Mineral Resour. **3**, 8–11 (2014)
5. Song, Q., Guo, M., Wang, L., Ling, T.: Use of steel slag as sustainable construction materials: a review of accelerated carbonation treatment. Resour. Conserv. Recycl. **173**, 105740 (2021). https://doi.org/10.1016/j.resconrec.2021.105740
6. Guo, J., Bao, Y., Wang, M.: Steel slag in China: treatment, recycling, and Management. Waste Manage. **78**, 318–330 (2018). https://doi.org/10.1016/j.wasman.2018.04.045
7. Gao, T., Dai, T., Shen, L., Jiang, L.: Benefits of using steel slag in cement clinker production for environmental conservation and economic revenue generation. J. Clean. Prod. **282**, 124538 (2021). https://doi.org/10.1016/j.jclepro.2020.124538
8. Oss, H.: U.S. Geological Survey, Mineral Commodity Summaries (2007)
9. Wen, H., Wu, S., Bhusal, S.: Performance evaluation of asphalt mixes containing steel slag aggregate as a measure to resist studded tire wear. J. Mater. Civil Eng. **28**(5) (2016). https://doi.org/10.1061/(asce)mt.1943-5533.0001475
10. Xue, Y., Wu, S., Hou, H., Zha, J.: Experimental investigation of basic oxygen furnace slag used as aggregate in asphalt mixture. J. Hazard. Mater. **138**(2), 261–268 (2006). https://doi.org/10.1016/j.jhazmat.2006.02.073
11. Hainin, M.R., Yusoff, N.I., Mohammad Sabri, M.F., Abdul Aziz, M.A., Sahul Hameed, M.A., Farooq Reshi, W.: Steel slag as an aggregate replacement in Malaysian hot mix asphalt. ISRN Civil Eng. **2012**, 1–5 (2012). https://doi.org/10.5402/2012/459016
12. Yi, T.: Performance evaluation of steel slag as natural aggregates replacement in asphaltic concrete. Ph.D. diss., M.Sc. Thesis, Universiti Sains Malaysia (2008)
13. Díaz-Piloneta, M., Terrados-Cristos, M., Álvarez-Cabal, J.V., Vergara-González, E.: Comprehensive analysis of steel slag as aggregate for road construction: experimental testing and environmental impact assessment. Materials **14**(13), 3587 (2021). https://doi.org/10.3390/ma14133587
14. Kim, K., Haeng Jo, S., Kim, N., Kim, H.: Characteristics of hot mix asphalt containing steel slag aggregate according to temperature and void percentage. Constr. Build. Mater. **188**, 1128–1136 (2018). https://doi.org/10.1016/j.conbuildmat.2018.08.172
15. Sofilic, T., Mladenovic, A., Sofilic, U.: Characterization of the EAF steel slag as aggregate for use in road construction. Chem. Eng. Trans. **19**, 117–122 (2010)
16. Rohde, L., Peres Núñez, W., Augusto Pereira Ceratti, J.: Electric arc furnace steel slag: base material for low-volume roads. Transp. Res. Record **1819**(1), 201–207 (2003)
17. Chen, J.S., Wei, S.H.: Engineering properties and performance of asphalt mixtures incorporating steel slag. Constr. Build. Mater. **128**, 148–153 (2016)
18. Maghool, F., Arulrajah, A., Du, Y.J., Horpibulsuk, S., Chinkulkijniwat, A.: Environmental impacts of utilizing waste steel slag aggregates as recycled road construction materials. Clean Technol. Environ. Policy **19**(4), 949–958 (2017)
19. Oluwasola, E.A., Hainin, M.R., Aziz, M.M.A.: Evaluation of asphalt mixtures incorporating electric arc furnace steel slag and copper mine tailings for road construction. Transp. Geotechn. **2**, 47–55 (2015)
20. Behiry, A.E.A.E.M.: Evaluation of steel slag and crushed limestone mixtures as subbase material in flexible pavement. Ain Shams Eng. J. **4**(1), 43–53 (2013)

21. Ameri, M., Shahabishahmiri, H., Kazemzadehazad, S.: Evaluation of the use of steel slag in concrete. In: ARRB Conference, 25th, 2012, Perth, Western Australia, Australia (2012, September)
22. Reddy, A.S., Pradhan, R.K., Chandra, S.: Utilization of basic oxygen furnace (BOF) slag in the production of a hydraulic cement binder. Int. J. Miner. Process. **79**(2), 98–105 (2006)
23. Nicolae, M., Vîlciu, I., Zaman, F.: X-ray diffraction analysis of steel slag and blast furnace slag viewing their use for road construction. UPB Sci. Bull **69**(2), 99–108 (2007)
24. Chaurand, P., Rose, J., Briois, V., Olivi, L., Hazemann, J., Proux, O., … Bottero, J.: Environmental impacts of steel slag reused in road construction: A Crystallographic and molecular (XANES) approach. J. Hazardous Mater. **139**(3), 537–542 (2007). https://doi.org/10.1016/j.jhazmat.2006.02.060
25. Abdi, M.R.: Effects of basic Oxygen steel slag (BOS) on strength and durability of kaolinite (2011)
26. Ziaee, S.A., Kavussi, A., Jalili Qazizadeh, M., Mohammadzadeh Moghadam, A.: Evaluation of long term ageing of asphalt mixtures containing EAF and BOF steel slags. Int. J. Transp. Eng. **2**(3), 245–265 (2015)
27. Lizarazo-Marriaga, J., Claisse, P., Ganjian, E.: Effect of steel slag and portland cement in the rate of hydration and strength of blast furnace slag pastes. J. Mater. Civ. Eng. **23**(2), 153–160 (2011)
28. Gumieri, A.G., Dal Molin, D.C., Vilela, A.C.: Characteristics of steel slag granulated in a steel plant—valuation of the microstructure through electron probe micro analysis. In: International RILEM Conference on the Use of Recycled Materials in Building and Structures (Vol. 2, pp. 1067–1075) (2004)
29. Tossavainen, M., Engstrom, F., Yang, Q., Menad, N., Larsson, M.L., Bjorkman, B.: Characteristics of steel slag under different cooling conditions. Waste Manage. **27**(10), 1335–1344 (2007)
30. Mahieux, P.Y., Aubert, J.E., Escadeillas, G.: Utilization of weathered basic oxygen furnace slag in the production of hydraulic road binders. Constr. Build. Mater. **23**(2), 742–747 (2009)
31. Mombelli, D., Mapelli, C., Barella, S., Di Cecca, C., Le Saout, G., Garcia-Diaz, E.: The effect of chemical composition on the leaching behaviour of electric arc furnace (EAF) carbon steel slag during a standard leaching test. J. Environ. Chem. Eng. **4**(1), 1050–1060 (2016)
32. Roslan, N.H., Ismail, M., Abdul-Majid, Z., Ghoreishiamiri, S., Muhammad, B.: Performance of steel slag and steel sludge in concrete. Constr. Build. Mater. **104**, 16–24 (2016)
33. Hekal, E.E., Abo-El-Enein, S.A., El-Korashy, S.A., Megahed, G.M., El-Sayed, T.M.: Hydration characteristics of Portland cement–electric arc furnace slag blends. HBRC J. **9**(2), 118–124 (2013)
34. Salman, M., Cizer, Ö., Pontikes, Y., Snellings, R., Vandewalle, L., Blanpain, B., Van Balen, K.: Cementitious binders from activated stainless steel refining slag and the effect of alkali solutions. J. Hazard. Mater. **286**, 211–219
35. Shi, C., Hu, S.: Cementitious properties of ladle slag fines under autoclave curing conditions. Cement Concrete Res. **33**(11), 1851–1856 (2015)
36. Das, B., Prakash, S., Reddy, P.S.R., Misra, V.N.: An overview of utilization of slag and sludge from steel industries. Resour. Conserv. Recycl. **50**(1), 40–57 (2007)
37. Miraoui, M., Zentar, R., Abriak, N.E.: Road material basis in dredged sediment and basic oxygen furnace steel slag. Constr. Build. Mater. **30**, 309–319 (2012)
38. Björkman, B., Eriksson, J., Nedar, L., Samuelsson, C.: Waste reduction through process optimization and development. JOM **48**(3), 45–49 (1996)
39. Belhadj, E., Diliberto, C., Lecomte, A.: Characterization and activation of basic oxygen furnace slag. Cement Concr. Compos. **34**(1), 34–40 (2012)
40. Yildirim, I. Z., Prezzi, M.: Use of steel slag in subgrade applications (2009)
41. Engström, F., Adolfsson, D., Yang, Q., Samuelsson, C., Björkman, B.: Crystallization behaviour of some steelmaking slags. Steel Res. Int. **81**(5), 362–371 (2010)
42. Vlcek, J., Tomková, V., Ovcacikova, H., Ovcacik, F., Topinkova, M., Matejka, V.: Slags from steel production: properties and their utilization. Metalurgija **52**(3), 329–333 (2013)

43. Ko, M.S., Chen, Y.L., Jiang, J.H.: Accelerated carbonation of basic oxygen furnace slag and the effects on its mechanical properties. Constr. Build. Mater. **98**, 286–293 (2015)
44. Sofilić, T., Barišić, D., Rastovčan Mioč, A., Sofilić, U.: Radionuclides in steel slag intended for road construction. J. Radioanal. Nucl. Chem. **284**(1), 73–77 (2010)
45. Sofilić, T., Barišić, D., Sofilić, U.: Natural radioactivity in steel slag aggregate. Arch. Metall. Mater. **56**, 627–634 (2011)
46. Asi, I.M.: Evaluating skid resistance of different asphalt concrete mixes. Build. Environ. **42**(1), 325–329 (2007)
47. Noureldin, A.S., McDaniel, R.S.: Evaluation of surface mixtures of steel slag and asphalt. Transp. Res. Record (1269) (1990)
48. Kehagia, F.: Skid resistance performance of asphalt wearing courses with electric arc furnace slag aggregates. Waste Manage. Res. **27**(3), 288–294 (2009)
49. Kumar, H., Varma, S.: A review on utilization of steel slag in hot mix asphalt. Int. J. Pavement Res. Technol. **14**(2), 232–242 (2021)
50. Chu, L., Cui, X., Zhang, K., Fwa, T.F., Han, S.: Directional skid resistance characteristics of road pavement: Implications for friction measurements by British pendulum tester and dynamic friction tester. Transp. Res. Rec. **2673**(10), 793–803 (2019)
51. Skaf, M., Pasquini, E., Revilla-Cuesta, V., Ortega-López, V.: Performance and durability of porous asphalt mixtures manufactured exclusively with electric steel slags. Materials **12**(20), 3306 (2019)
52. Asi, I.M., Qasrawi, H.Y., Shalabi, F.I.: Use of steel slag aggregate in asphalt concrete mixes. Can. J. Civ. Eng. **34**(8), 902–911 (2007)
53. Pasetto, M., Baldo, N.: Mix design and performance analysis of asphalt concretes with electric arc furnace slag. Constr. Build. Mater. **25**(8), 3458–3468 (2011)
54. Amelian, S., Manian, M., Abtahi, S.M., Goli, A.: Moisture sensitivity and mechanical performance assessment of warm mix asphalt containing by-product steel slag. J. Clean. Prod. **176**, 329–337 (2018)
55. Kavussi, A., Qazizadeh, M.J.: Fatigue characterization of asphalt mixes containing electric arc furnace (EAF) steel slag subjected to long term aging. Constr. Build. Mater. **72**, 158–166 (2014)
56. Xie, J., Chen, J., Wu, S., Lin, J., Wei, W.: Performance characteristics of asphalt mixture with basic oxygen furnace slag. Constr. Build. Mater. **38**, 796–803 (2013)
57. Wang, Z., Wu, S., Yang, C., Xie, J., Xiao, Y., Zhao, Z., ... & Zhang, L.: Quantitative assessment of road performance of recycled asphalt mixtures incorporated with steel slag. Materials **15**(14), 5005 (2022)
58. Xiao, Z., Chen, M., Wu, S., Xie, J., Kong, D., Qiao, Z., Niu, C.: Moisture susceptibility evaluation of asphalt mixtures containing steel slag powder as filler. Materials **12**(19), 3211 (2019)
59. Liu, W., Li, H., Zhu, H., Xu, P.: Properties of a steel slag–permeable asphalt mixture and the reaction of the steel slag–asphalt interface. Materials **12**(21), 3603 (2019)
60. Liu, W., Li, H., Zhu, H., Xu, P.: Effects of steel-slag components on interfacial-reaction characteristics of permeable steel-slag–bitumen mixture. Materials **13**(17), 3885 (2020)
61. Patil, R.N., Phalak, K.T., Bidkar, K.: Use of steel slag to increase bearing capacity of soil in road construction. Int. J. Emerg. Trends Technol. **2**(1) (2019)
62. Aziz, M.M.A., Hainin, M.R., Yaacob, H., Ali, Z., Chang, F.L., Adnan, A.M.: Characterisation and utilisation of steel slag for the construction of roads and highways. Mater. Res. Innov. **18**(sup6), S6-255 (2014)
63. Joshi, V.K., Arenicz, R.M.: Use of Slag: A Direct Benefit to Our Environment. National Slag Association, MF, 203–11 (2011)
64. Zhang, Q., Luo, J., Yang, Z., Wang, J., Zhao, Y., Zhang, Y.: Creep and fatigue properties of asphalt mastic with steel slag powder filler. Case Studies Constr. Mater. **18**, e01743 (2023)
65. Khan, M.I., Wahhab, H.A.A.: Improving slurry seal performance in Eastern Saudi Arabia using steel slag. Constr. Build. Mater. **12**(4), 195–201 (1998)

66. Chen, Z., Wu, S., Wen, J., Zhao, M., Yi, M., Wan, J.: Utilization of gneiss coarse aggregate and steel slag fine aggregate in asphalt mixture. Constr. Build. Mater. **93**, 911–918 (2015)
67. Fakhri, M., Ahmadi, A.: Recycling of RAP and steel slag aggregates into the warm mix asphalt: a performance evaluation. Constr. Build. Mater. **147**, 630–638 (2017)
68. Behnood, A., Ameri, M.: Experimental investigation of stone matrix asphalt mixtures containing steel slag. Scientia Iranica **19**(5), 1214–1219 (2012)
69. Sorlini, S., Sanzeni, A., Rondi, L.: Reuse of steel slag in bituminous paving mixtures. J. Hazard. Mater. **209**, 84–91 (2012)
70. Wu, S., Xue, Y., Ye, Q., Chen, Y.: Utilization of steel slag as aggregates for stone mastic asphalt (SMA) mixtures. Build. Environ. **42**(7), 2580–2585 (2007)
71. Tao, G., Xiao, Y., Yang, L., Cui, P., Kong, D., Xue, Y.: Characteristics of steel slag filler and its influence on rheological properties of asphalt mortar. Constr. Build. Mater. **201**, 439–446 (2019)
72. Arabani, M., Azarhoosh, A.R.: The effect of recycled concrete aggregate and steel slag on the dynamic properties of asphalt mixtures. Constr. Build. Mater. **35**, 1–7 (2012)
73. Electric arc furnace slag. (n.d.). Retrieved January 24, 2023, from https://encyclopedia.pub/entry/2519 (2023)
74. Cikmit, A.A., Tsuchida, T., Kang, G., Hashimoto, R., Honda, H.: Particle-size effect of basic oxygen furnace steel slag in stabilization of dredged Marine Clay. Soils Found. **59**(5), 1385–1398 (2019). https://doi.org/10.1016/j.sandf.2019.06.013
75. Bof and EAF Steels: What are the differences?—economics 243 fall 2018. (n.d.). Retrieved January 24, 2023, from https://econ243.academic.wlu.edu/2016/03/07/bof-and-eaf-steels-what-are-the-differences/ (2023)

Study on the Performance of Ultra-Fine Cement Slurry Reinforced Coral Aggregates and Coral Concrete

Shuang Li, Zhuolin Xie, Jianmin Hua, Lepeng Huang, Jian Kang, and Xuran Liu

Abstract Natural coral has irregular shape, dense internal holes and low cylinder compressive strength, which restrict the mechanical properties of coral concrete. The coral coarse aggregate was innovatively reinforced by grouting with ultra-fine cement slurry based on low concentration acetic acid pretreatment in this study. In order to study the influence of ultra-fine cement slurry treatment methods on coral coarse aggregates, 6 different types of coral aggregates and a total of 90 corresponding coral aggregate concrete test blocks were designed, including the physical and mechanical performance test of aggregate, working performance test and mechanical performance test of concrete. The test results showed that the ultra-fine cement slurry could significantly enhance the cylinder compressive strength of the coral coarse aggregate, which increases by 140% after treatment than before treatment. Concrete prepared with reinforced coral coarse aggregates has increased slump and expansion, resulting in improved workability. The cubic compressive strength, axial compressive strength and splitting tensile strength of reinforced coral aggregate concrete increase as the water to slurry ratio of ultra-fine cement slurry decreases and the age of aggregate curing increases. The elastic modulus tends to increase under the influence of water-slurry ratio, but has no significant effect on Poisson's ratio.

Keywords Coral aggregate concrete · Ultra-fine cement · Reinforced coral coarse aggregate · Mechanical performance

---

S. Li · Z. Xie · J. Hua (✉) · L. Huang · J. Kang
School of Civil Engineering, Chongqing University, Chongqing 400045, China
e-mail: cqu2022@126.com

J. Hua · L. Huang
Key Laboratory of New Technology for Construction of Cities in Mountain Area, Ministry of Education, Chongqing University, Chongqing 400045, China

X. Liu
China Construction Third Engineering Bureau Group Co. Ltd., No. 799, East Lake High-Tech Development Zone, Wuhan 430073, Hubei, China

E. Strauss (ed.), *Proceedings of the 7th International Conference on Civil Engineering*, Lecture Notes in Civil Engineering 371, https://doi.org/10.1007/978-981-99-4045-5_6

1 Introduction

The ocean is the future direction of development, and a large amount of coral debris will be produced in the process of dredging waterways and harbor construction [1]. The production of concrete using these coral debris as coarse aggregate can not only solve the shortage of concrete materials, but also shorten the construction cycle, reduce economic costs [2–4]. The research on the properties related to coral aggregate concrete (CAC) has gradually become the focus of many researchers for the last few years [5, 6]. In fact, the use of CAC can be dated back to as early as the 1940s. After decades of development, some achievements have been made in the research of CAC properties [7]. Although the research results show that the mechanical properties and ductility of concrete are weakened when coral aggregates are used to replace ordinary aggregates in the same mix proportion, overall the solution of using coral aggregates to replace ordinary aggregates for concrete preparation still has the potential to be used in practical engineering applications. If the relevant properties of CAC can be further optimized, then CAC will have a broader application prospect.

Today, it has been shown that the mechanical properties of the CAC can be affected by the physical and mechanical properties of the coral aggregate, but there have been rare researches on the enhancement of coral aggregate itself, and most of them use acid solution or organic solution to modify coral aggregate [8]. Natural coral aggregate is light and porous, with a porosity close to 50% [9]. The cylinder compressive strength is generally about 2.0–3.0 MPa, much lower than that of ordinary crushed stone [10]. In general, the pore structure of concrete has an important impact on its mechanical properties. The lower the porosity of concrete, the higher its internal compactness, and the corresponding mechanical properties will be better [11]. If these pores of coral aggregates can be filled by suitable substances, the performance of coral aggregates as well as CAC will potentially be enhanced. Ultra-fine cement is a cement that has been rapidly developed in recent years, and it is a new type of cement with an average particle size of 3.1 μm and a specific surface area of 780 m^2/kg [12]. In recent years, it has been widely used to fill and repair cracks in concrete with good results. Therefore, in this study, ultra-fine cement was chosen as a reinforcing material for coral aggregate.

In order to systematically elucidate the specific effects of ultra-fine cement slurry on coral aggregates and CAC related performance indexes, this study designed six types of coral aggregate for comparative test, and explored the changes of water-slurry ratio and curing age of aggregate on the physical and mechanical properties of coral aggregate and the working and mechanical properties of CAC. In Sect. 2, the reinforcement method of coral aggregate, composition of raw materials, CAC mix proportions and specimen preparation are introduced. In Sect. 3, the physical and mechanical properties of coral aggregate are firstly discussed, then the working properties of reinforced CAC are discussed, and finally, the mechanical property indexes of CAC such as cube compressive strength, axial compressive strength, splitting tensile strength, elastic modulus and Poisson's ratio are discussed.

2 Materials and Methods

2.1 Coral Aggregates

The study used the common coral aggregate and the reinforced coral aggregate for testing, Fig. 1 is the common coral coarse aggregate. After an initial washing of the common coral coarse aggregate using fresh water, the coral aggregate was soaked in 3% acetic acid solution for 60 min, after which it was rinsed with tap water for use.

For reinforcing coral aggregate, cement slurry composed of ultra-fine cement was used to reinforce the coral aggregate. Three kinds of reinforced cement slurry mix were designed: 1.2, 1.4 and 1.6. In accordance with the above water-slurry ratio, the ultra-fine cement is thoroughly mixed with water, and then put into the coral aggregate, which must ensure that the coral aggregate can be completely submerged by the reinforced cement slurry within the submerged time period. Then the mixture was put into a mixing drum with a pressure of 0.2 MPa and pressurized for 30 min. After the pressurized mixing was completed, the coral aggregate was fished out of the slurry with a fine-mesh fishing net, and sprinkled with water to keep the surface moist during later aggregate maintenance.

2.2 Other Raw Materials

Fine aggregate: In this study, river sand with a fineness modulus of 2.7 was used, which is medium sand.

Fig. 1 Coral coarse aggregate

Cement: The strength grade of 42.5 ordinary silicate cement was used.

Water: Fresh water was used for the test.

Water reducing agent: In this paper, polycarboxylic acid superplasticizer of Zhongjian New Material is selected.

2.3 Mix Proportion and Specimen Production

A total of 6 types of aggregates (ordinary coral aggregate and 5 different types of reinforced coral aggregate) were used to prepare the concrete, and a total of 6 types of specimen types were included in the study, and the mix proportions are shown in Table 1.

A total of 90 test specimens were prepared for this test, as shown in Table 2.

3 Results and Discussion

3.1 Physical and Mechanical Properties of Coral Aggregate

3.1.1 Particle Size and Gradation

This test uses coral coarse aggregate, after crushing and vibrating screen machine screening, particle size 5–16 mm, of which 4.75–9.5 mm accounted for about 50%, 9.5–16 mm accounted for about 50%, in line with the "light aggregate and its test method Part 1: Light Aggregate" (GB/T 17,431.1–2010) [13]. As can be seen from Fig. 2, the coarse aggregate gradation curve selected for this experiment meets the requirements for use.

Table 1 Coral concrete mix proportions

Group	Water-to-glue ratio	Cement	Coarse aggregate	Sand	Water	Mineral powder	Fly ash
NT-CAC	0.21	420	615	1030	150	140	140
G-1.2-7d-CAC	0.21	420	615	1030	150	140	140
G-1.4-7d-CAC	0.21	420	615	1030	150	140	140
G-1.6-7d-CAC	0.21	420	615	1030	150	140	140
G-1.2-3d-CAC	0.21	420	615	1030	150	140	140
G-1.2-28d-CAC	0.21	420	615	1030	150	140	140

Note NT stands for ordinary coral coarse aggregate. G-X–Y, G stands for reinforced coral coarse aggregate, X stands for reinforced cement slurry ratios (ω) used of 1.2, 1.4 and 1.6, respectively. Y represents the curing ages of the reinforced coral coarse aggregate used in the preparation of concrete of 3d, 7d and 28d, respectively. CAC stands for coral concrete

Table 2 Specimen specifics

Type of test	Specimen size	Number of specimens					
		NT	G-1.2-7d	G-1.4-7d	G-1.6-7d	G-1.2-3d	G-1.2-28d
Cube compressive test	100 × 100 × 100 mm	3	3	3	3	3	3
Axial compressive test	150 × 150 × 300 mm	3	3	3	3	3	3
Splitting tensile test	100 × 100 × 100 mm	3	3	3	3	3	3
Modulus of elasticity test	150 × 150 × 300 mm	3	3	3	3	3	3
Poisson's ratio test	150 × 150 × 300 mm	3	3	3	3	3	3

Fig. 2 Gradation curves of coral coarse aggregate

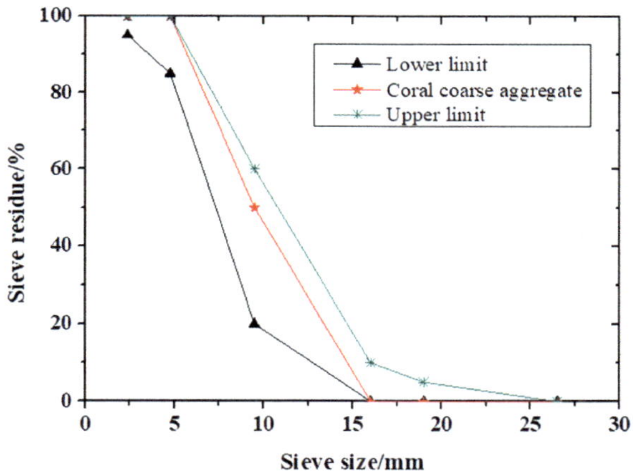

3.1.2 Water Absorption

As shown in Fig. 3, the water absorption rate of coral aggregate enhanced by ultra-fine cement slurry decreases to a greater extent after immersion. This is mainly because the hydration products generated after grouting of aggregate fill the pores, reduce the number of connected pores, thus improving the density of aggregate, resulting in different degrees of water absorption. The analysis of the reinforced coral aggregate after 28 days of curing showed that when the water-slurry ratio was 1.6, the water absorption rate of the aggregate in 1 h was reduced by about 21.1% compared with the ordinary coral aggregate. $\omega = 1.4$, the water absorption rate of 1 h decreased by 36.6% compared with that of ordinary coral aggregate. $\omega = 1.2$, the water absorption rate of 1 h was 43.1% lower than that of ordinary coral aggregate. Therefore, it can

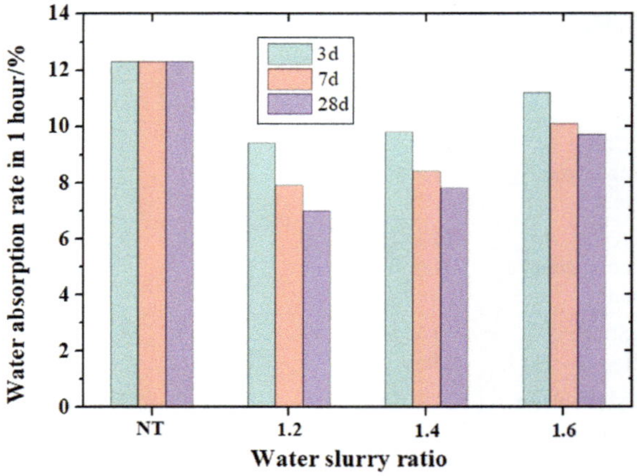

Fig. 3 Water absorption of coral aggregate before and after reinforcement

be found that with the decrease of water-slurry ratio, 1 h water absorption rate of reinforced coral aggregate also decreases gradually, and 1 h water absorption rate of aggregate is significantly correlated with the slurry water-slurry ratio.

3.1.3 Cylinder Compressive Strength

From Fig. 4, it can be seen that the cylinder compressive strength of coral aggregate enhanced by ultra-fine cement slurry is improved in different degrees after immersion. When the water-slurry ratio was 1.6, the enhanced coral aggregate after 28 days of curing increased by about 63.3% compared with the ordinary coral aggregate. When $\omega = 1.4$, the enhanced coral aggregate after 28 days of curing increased about 130.0% compared with the ordinary coral aggregate. $\omega = 1.2$, the enhanced coral aggregate after 28 days of curing increased about 140.0% compared with the ordinary coral aggregate. Therefore, it can be found that with the decrease of water-slurry ratio, the cylinder compressive strength of reinforced coral aggregate gradually increases.

3.1.4 Working Performance of CAC

As can be seen from Fig. 5, the CAC prepared with ultra-fine cement slurry reinforced coral aggregate has significantly improved working performance compared with that prepared with ordinary coral aggregate. As can be seen from Fig. 5a, when $\omega = 1.6$, slump increases from 175 to 205 mm and expansion increases from 305 to 370 mm, and the workability of CAC increases significantly. With the further decrease of water-slurry ratio, the slump and expansion of CAC are further increased. Compared with NT group, when $\omega = 1.2$, the slump and expansion of CAC increased by 80 and 175 mm, respectively, by 45.7 and 57.4%. As shown in Fig. 5b, under the same water-slurry ratio, the longer the curing age of reinforced coral aggregate, the better

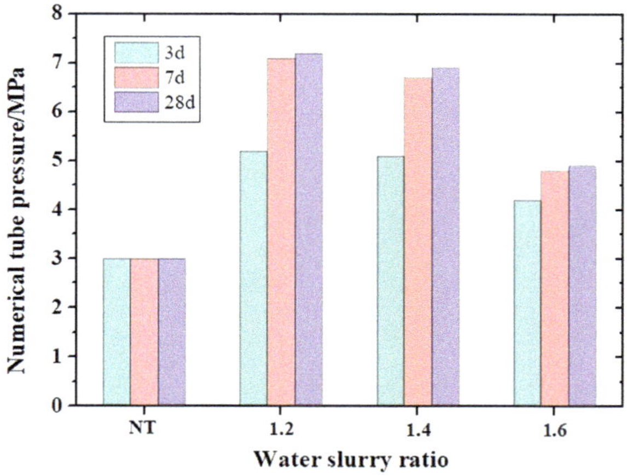

Fig. 4 Changes in cylinder compressive strength before and after coral aggregate reinforcement

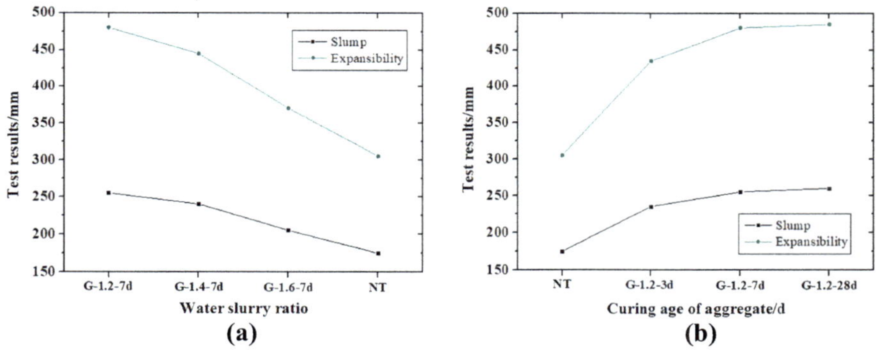

Fig. 5 Performance test results: **a** Water-slurry ratio, slump and expansion; **b** Curing age, slump and expansion degree of aggregate

the working performance of prepared CAC. Compared with 3d aggregate curing age, slump and expansion of CAC prepared at 28d aggregate curing age increased by 25 mm and 50 mm, respectively, by 10.6% and 11.5%. Therefore, the ultra-fine cement slurry filling coral aggregate pores and the denser slurry shell formed with the growth of aggregate curing age can have different degrees of improvement on the working performance of CAC.

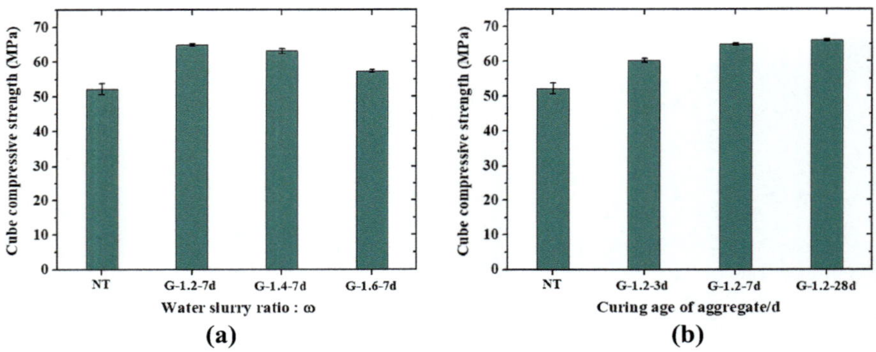

Fig. 6 f_{cu} of CAC specimens: **a** Water slurry ratio and f_{cu}; **b** Curing age of aggregate and f_{cu}

3.2 Mechanical Properties of CAC

3.2.1 Cube Compressive Property

After conducting cube compressive strength (f_{cu}) tests on all CAC test blocks, the results are shown in Fig. 6. There were 3 CAC samples in each group. The average value of 3 test blocks in each group was calculated as f_{cu} of water-slurry ratio CAC. It can be found that the f_{cu} value of the CAC of coral aggregate reinforced by ultra-fine cement slurry is higher than that of the CAC of ordinary coral aggregate. As can be seen from Fig. 6a, the smaller the water-slurry ratio, the larger the f_{cu} of CAC. When $\omega = 1.2$, the fcu of CAC increases the most, and the f_{cu} of CAC is 64.9 MPa, and the 95% confidence interval ranges from 64.4 MPa to 65.5 MPa, which is 24.1% higher than that of untreated CAC. When $\omega = 1.4$, the fcu is 63.1 MPa, and the f_{cu} of CAC is increased by 20.7%. When $\omega = 1.6$, the f_{cu} of CAC increases the least, only 9.8%, which may be related to the degree of aggregate pore filling and the thickness of slurry shell on the aggregate surface. As shown in Fig. 6b, the enhancement of the curing age of coral aggregate also improves the f_{cu} of CAC in different degrees. With the growth of the curing age of reinforced coral aggregate, the f_{cu} of CAC increases gradually. The f_{cu} of CAC at the curing age of 3d and 7d aggregate is 91.2% and 98.2% of the CAC at the curing age of 28d aggregate, respectively. Therefore, the age of aggregate curing compared to the water to slurry ratio can be seen to have no significant effect on the f_{cu} of CAC.

3.2.2 Axial Compressive Property

Figure 7 shows the axial compressive strength (f_c) test results of CAC test blocks with different water-slurry ratios and different curing ages of aggregates. Similarly, the f_c of CAC was measured by the average of the three test blocks. According to the test results, with the decrease of water-slurry ratio and the increase of aggregate curing

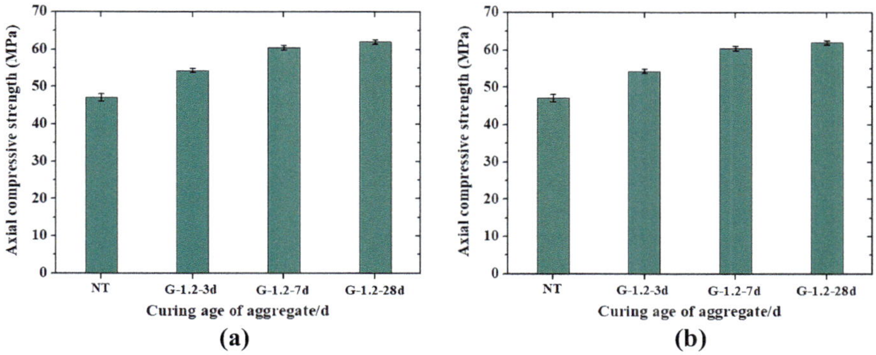

Fig. 7 f_c of CAC specimens: **a** Water slurry ratio and f_c; **b** Curing age of aggregate and f_c

age, the f_c of CAC test block gradually increases. When $\omega = 1.2$–1.4, f_c changes little, f_c is 59.1–60.5 MPa, but compared with NT-CAC group, f_c increases by 25.5–28.5%. When the curing age of aggregate is 7d–28d, the change of f_c is not much, f_c is 60.5–62.0 MPa, which is 11.4–14.2% higher than that of 3d. Compared with NT-CAC, the f_c increase of specimens with water-slurry ratio of 1.2 and aggregate curing age of 28 days is the largest, reaching 31.6%.

3.2.3 Splitting Tensile Property

Figure 8 shows the splitting tensile strength (f_t) test results of CAC test blocks with different water-slurry ratios and different curing ages of aggregate. The f_t of CAC is still the average of the three test blocks. As can be seen from the test results, f_t of CAC test block shows an increasing trend with the decrease of water-slurry ratio and the increase of aggregate curing age. When $\omega = 1.2$ and aggregate curing age is 28d, $f_{t\text{-}mean} = 5.12$ MPa is increased by 19.9%, 95% confidence interval is 4.98–5.25 MPa. Compared with NT-CAC, G-1.2-7d-CAC, G-1.4-7d-CAC and G-1.6-7d-CAC increased by 17.3%, 15.2% and 7.7%, respectively. In conclusion, coral aggregate enhanced with a water-slurry ratio of 1.2 can better improve the anti-splitting failure ability of CAC. Compared with G-1.2-7d-CAC, the f_t of G-1.2-28d-CAC increased by only 2.2%. It can be seen that the improvement of aggregate performance brought by the curing age of aggregate after 7d has no significant effect on the f_t improvement of CAC.

3.2.4 Elastic Modulus

Figure 9 shows the elastic modulus (E_c) test results of CAC test blocks with different water-slurry ratios and different curing ages of aggregates. The E_c of CAC is calculated as the average of the test results of the three test blocks. The difference between

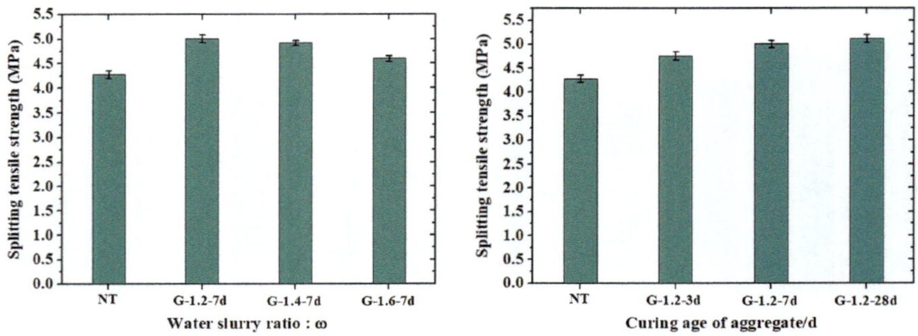

Fig. 8 f_t of CAC specimens: **a** Water slurry ratio and f_t; **b** Curing age of aggregate and f_t

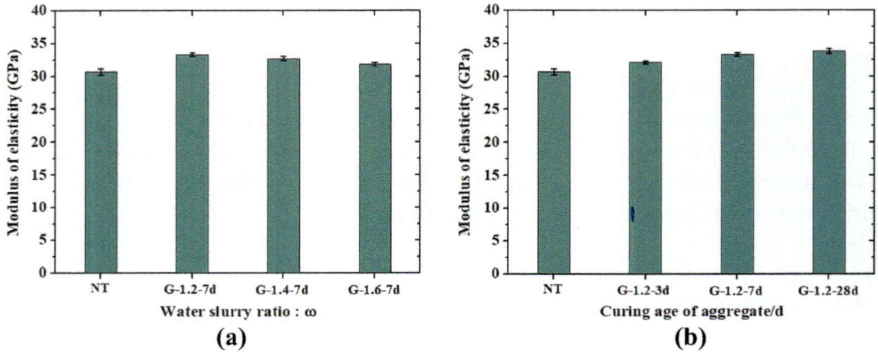

Fig. 9 E_c of CAC specimens: **a** Water slurry ratio and E_c; **b** Curing age of aggregate and E_c

the axial compressive strength of 3 test blocks in each group and the axial compressive strength used to test the control load is not more than 20% of the latter. As can be seen from the test results, the E_c of CAC test block is at a low level, but presents an increasing trend with the decrease of water-slurry ratio and the growth of aggregate curing age. The maximum value was reached when $\omega = 1.2$ and aggregate curing age was 28 days, $E_{c\text{-}mean} = 33.8$ GPa, and 95% confidence interval was 33.3–34.4 GPa. Compared with NT-CAC, G-1.2-28d-CAC, G-1.2-7d-CAC, G-1.4-7d-CAC and G-1.6-7d-CAC increased by 10.1%, 8.5%, 6.5% and 3.6%, respectively. In conclusion, both the water-slurry ratio and the curing age of aggregate have improved the E_c of CAC, and the water-slurry ratio has a more significant improvement effect than the curing age of aggregate, but the overall improvement range is smaller than that of f_{cu}, f_c and f_t.

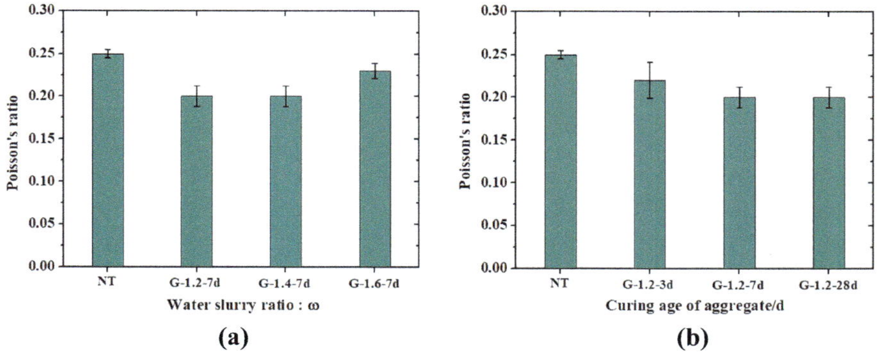

Fig. 10 μ of CAC specimens: **a** Water slurry ratio and μ; **b** Curing age of aggregate and μ

3.2.5 Poisson's Ratio

Figure 10 shows the Poisson's ratio (μ) test results of CAC test blocks with different water-slurry ratios and different aggregate curing ages. The μ of CAC was calculated as the average of the test results of three test blocks. The difference between the axial compressive strength of 3 test blocks in each group and the axial compressive strength used to test the control load is not more than 20% of the latter. It can be seen from the test results that the μ of all CAC blocks in the experimental group has little change, ranging from 0.20 to 0.25. Compared with $\mu = 0.25$ in the NT-CAC group, μ decreased to varying degrees in the remaining test groups, but the overall decrease was small, with μ of 0.2 in the G-1.2-7d-CAC, G-1.4-7d-CAC and G-1.2-28d-CAC groups with 95% confidence intervals of 0.18 to 0.22. It can be seen that the water-slurry ratio and the curing age of aggregate have a decreasing trend on the Poisson's ratio of CAC, but the influence is small.

4 Conclusions

In this study, ultra-fine cement slurry was innovatively applied to reinforce coral aggregate by grouting, and the physical and mechanical properties of coral aggregate before and after reinforcement, as well as the working and mechanical properties of reinforced coral aggregate concrete were experimentally studied. Its main conclusions are as follows:

(1) After the coral coarse aggregate is reinforced by ultra-fine cement slurry, the water absorption rate of the aggregate at 1 h is reduced, and the cylinder compressive strength of the aggregate is greatly improved.

(2) The decrease of water absorption of reinforced coral aggregate makes the working performance of CAC significantly improved, the slump at $\omega = 1.6$ was increased from 175 to 205 mm, and with the decrease of water to slurry ratio, the slump and expansion further increased.
(3) Through cube compressive, axial compressive and splitting tensile tests, it is found that the reinforced coral aggregate with different water-slurry ratios and aggregate curing ages can improve the f_{cu}, f_c and f_t of CAC. The maximum f_{cu}, f_c and f_t of G-1.2-28d-CAC are 66.1, 62.0 and 5.12 MPa, which are 26.4, 31.6 and 19.9% higher than those of NT-CAC.
(4) Through the elastic modulus and Poisson's ratio tests, it was found that the E_c of CAC increased with different water-slurry ratios and enhanced coral aggregate curing ages, but had no significant effect on μ.

References

1. Chen, L., et al.: Green construction for low-carbon cities: a review. Environ. Chem. Lett. (2023)
2. Chen, L., et al.: Strategies to achieve a carbon neutral society: a review. Environ. Chem. Lett. **20**(4), 2277–2310 (2022)
3. Liu, J., et al.: Literature review of coral concrete. Arab. J. Sci. Eng. **43**(4), 1529–1541 (2018)
4. lyu, B., et al.: Coral aggregate concrete: Numerical description of physical, chemical and morphological properties of coral aggregate. Cem. Concr. Compos. **100**, 25–34 (2019)
5. Li, F., et al.: Axial compressive behavior of GFRP-confined seawater coral aggregate concrete incorporating slag-based alkali-activated materials. Constr. Build. Mater. **347**, 128437 (2022)
6. Wang, A., et al.: The development of coral concretes and their upgrading technologies: a critical review. Constr. Build. Mater. **187**, 1004–1019 (2018)
7. Liu, B., et al.: Mechanical properties of hybrid fiber reinforced coral concrete. Case Stud. Constr. Mater. **16**, e00865 (2022)
8. Wang, A., et al.: A gentle acid-wash and pre-coating treatment of coral aggregate to manufacture high-strength geopolymer concrete. Constr. Build. Mater. **274**, 121780 (2021)
9. Zhou, L., et al.: Mechanical behavior and durability of coral aggregate concrete and bonding performance with fiber-reinforced polymer (FRP) bars: a critical review. J. Clean. Prod. **289**, 125652 (2021)
10. Ma, L., et al.: Mechanical properties of coral concrete subjected to uniaxial dynamic compression. Constr. Build. Mater. **199**, 244–255 (2019)
11. Bu, J., Tian, Z.: Relationship between pore structure and compressive strength of concrete: Experiments and statistical modeling. Sādhanā **41**(3), 337–344 (2016)
12. Chen, J.J., Kwan, A.K.H.: Superfine cement for improving packing density, rheology and strength of cement paste. Cem. Concr. Compos. **34**(1), 1–10 (2012)
13. General Administration of Quality Supervision, I.Q.o.t.P.s.R.o.C., Its Test Methods-Part 2: Test Methods for Lightweight Aggregates (2010)

Development of Pavement Condition Index for Philippine Asphalt National Roads

Jamie Alea B. Ramos, Lea B. Bronuela-Ambrocio, Hilario Sean O. Palmiano, John Paul T. Dacanay, Lestelle V. Torio-Kaimo, and Jonas Christian R. Quero

Abstract The Pavement Condition Index (PCI) is one of the key performance indices used to evaluate pavements. In the Philippines, the Department of Public Works and Highways (DPWH) is currently using a similar evaluation called the Visual Condition Index (VCI) to assess the condition of national roads. It was adapted from the Road Condition Manual of the Road and Traffic Authority of New South Wales, Australia. However, published literature on how the VCI was localized to Philippine conditions cannot be traced. This study attempts to address this concern by deriving a new condition rating based on the experience and knowledge of the field experts in the country. Asphalt pavement sections are selected from the historical data on the condition of the road networks in the country, and simulated using photographs collected in the field. With the aid of an electronic survey, field experts from different parts of the country subjectively evaluated the overall condition of selected pavement sections. Correlation and multiple regression analysis are employed to develop the PCI for Philippine asphalt national roads using the available data acquired from field experts of several DPWH District Engineering Offices (DEOs). The developed PCI model for asphalt pavements obtained a coefficient of determination (R^2) of 0.58, implying that the resulting PCI model captures 58% of the variability in the expert-based dataset. Comparisons with the VCI are also conducted to possibly aid future improvements on pavement condition assessments in the country.

Keywords Asphalt pavements · Pavement condition index · Pavement surface defects · Philippine national roads

---

J. A. B. Ramos (✉) · J. P. T. Dacanay
University of the Philippines National Center for Transportation Studies, U.P. Campus, 1101 Diliman, Quezon City, Philippines
e-mail: jbramos1@up.edu.ph

L. B. Bronuela-Ambrocio · H. S. O. Palmiano · L. V. Torio-Kaimo · J. C. R. Quero
Institute of Civil Engineering, University of the Philippines-Diliman, U.P. Campus, 1101 Diliman, Quezon City, Philippines

E. Strauss (ed.), *Proceedings of the 7th International Conference on Civil Engineering*, Lecture Notes in Civil Engineering 371, https://doi.org/10.1007/978-981-99-4045-5_7

1 Introduction

The Pavement Condition Index (PCI) is one of the key performance indices used to evaluate the condition of pavements. It is a numerical value that quantifies the surface condition of pavements based on the existing surface defects [1].

In the Philippines, the Department of Public Works and Highways (DPWH) is currently using a similar parameter called the Visual Condition Index (VCI) to assess the Philippine national roads. The index is a part of the DPWH Visual Road Condition Assessment (RoCond) Manual, which was patterned from the Road Condition Manual of the Road and Traffic Authority of New South Wales, Australia [2]. However, published literature on how the VCI was localized to Philippine conditions cannot be traced.

1.1 Objectives

This study attempts to address this concern by developing a new pavement condition rating based on the knowledge and experience of local field experts. The main objective of the study is to develop the PCI for Philippine asphalt national roads. To achieve this, the specific objectives are set as follows:

- Establish a methodology of zero to minimum cost for the development of expert-based PCI based on related literature
- Identify the particular surface defects that significantly influenced the condition of the pavement based on the field experts' ratings
- Develop PCI estimation models for asphalt pavements in the Philippines based on the knowledge and experience of local field experts

The development of the PCI for Philippine asphalt national roads consists of two main components: pavement section selection and pavement condition assessment. Pavement sections that will represent Philippine national roads are first selected from the historical data on the pavement condition of road networks in the country. These selected pavement sections are simulated using photographs of the surface defects obtained from the field. With the aid of an electronic survey, field experts from different parts of the country subjectively evaluated the overall condition of the selected pavement sections. Based on the electronic survey results, the specific surface defects that significantly influenced the field experts' ratings are determined using Pearson correlation analysis. The PCI estimation model for Philippine asphalt national roads is then developed using multiple linear regression analysis. The developed PCI model is also compared to the existing VCI used by the DPWH.

1.2 Scope and Limitations

This paper will focus on developing the PCI model for asphalt pavements. The study will be extended in the future to develop the PCI model for concrete pavements. In addition, the developed PCI model will be limited to a linear equation.

Since the study is currently a work in progress, results will be derived from the available data. Currently, 80 field experts from several DPWH District Engineering Offices (DEOs) answered the electronic survey on pavement condition assessment and were able to assess 30 asphalt pavement sections in total.

2 Review of Related Literature

2.1 Visual Condition Index (VCI)

The VCI is the numerical rating currently used by the DPWH to quantify the pavement surface defects into a single value ranging from 0 to 100. The VCI formula varies depending on the road surface type (asphalt, concrete, and grave/earth). In particular, the surface defects considered and the corresponding weights of the defects are different for each road surface type. Equation 1 shows the general formula for calculating the VCI of asphalt and concrete roads, where SDWf equates to the sum of the weighted defects [2].

$$\mathrm{VCI} = \max\left(0, \left(100 * \left(1 - \sqrt{1 - \left(\left(100 - \left(\frac{\min(300, \mathrm{SDWf})}{3}\right)\right) * 100\right)^2}\right)\right)\right) \tag{1}$$

Surface defects considered in the VCI for asphalt pavements include crocodile cracks, transverse cracks, edge breaks, patches, potholes, surface failures, rutting, and wearing surfaces. Each defect (and its severity) has a corresponding weight factor. Table 1 summarizes the weight factors for the surface defects considered in the VCI for asphalt pavements [2].

For concrete pavements, the surface defects included in the VCI are multiple cracks, transverse cracks, spalling, faulting, shattered slabs, scaling, and joint sealant deterioration. Similar to asphalt pavements, each defect (and its severity) has a corresponding weight factor. Table 2 summarizes the weight factors for the surface defects considered in the VCI for concrete pavements [2].

Table 1 Defect weight factors for asphalt VCI

Pavement Defects	Severity	Weight factors
Crocodile cracks	Narrow	3.50
	Wide	5.90
Transverse cracks	Narrow	3.30
	Wide	5.50
Edge breaks	Small	0.41
	Medium	0.82
	Large	1.25
Patching	–	1.25
Potholes	–	0.36
Surface failures	–	0.18
Rutting	–	4.00
Wearing surfaces	Minor	0.55
	Severe	1.20

Table 2 Defect weight factors for concrete VCI

Pavement defects	Severity	Weight factors
Multiple cracks	Narrow	3.60
	Wide	–
Transverse cracks	Narrow	3.50
	Wide	5.50
Spalling	–	3.00
Faulting	–	4.20
Shattered slabs	–	1.36
Scaling	Minor	0.55
	Severe	1.20
Joint sealant deterioration	–	0.13

2.2 Expert-Based Pavement Condition Indices

Pavement condition indices can be developed based on the opinions or subjective ratings of selected experts or panel members. Some of the existing indices that used this particular methodology are the Present Serviceability Index (PSI) of America, the Maintenance Control Index (MCI) of Japan, and the National Highway Pavement Condition Index (NHPCI) and Highway Pavement Condition Index (HPCI) of South Korea.

Present Serviceability Index (PSI). The PSI is a parameter developed to quantify the serviceability of pavements. It is one of the results of the American Association of State Highway Officials (AASHO) Road Test. In the road test, the rating panel was established based on the objective of representing all highway users (such as highway maintenance workers, highway materials suppliers, automotive manufacturers, etc.). The panel evaluated the serviceability of the selected pavement sections using a 5-point rating system after driving and/or walking on the pavement section. Aside from the panel's ratings, the longitudinal and transverse profiles of the selected pavement sections were also identified by measuring their defects, such as cracking, spalling, patching, rutting, etc. [3].

A total of 74 asphalt pavement sections and 49 concrete pavements were assessed during the road test. By using multiple linear regression analysis on the panel's ratings and measured defects, the PSI for asphalt pavements (Eq. 2) and PSI for concrete pavements (Eq. 3) were formulated. Parameters included in the equations are mean wheel path slope variance (SV), mean wheel path rut depth (RD), cracking (C), and patching (P) [3].

$$\mathrm{PSI}_{\mathrm{asphalt}} = 5.03 - 1.91\log(1 + \mathrm{SV}) - 1.38\mathrm{RD}^2 - 0.01\sqrt{C + P} \tag{2}$$

$$\mathrm{PSI}_{\mathrm{concrete}} = 5.41 - 1.78\log(1 + \mathrm{SV}) - 0.09\sqrt{C + P} \tag{3}$$

Maintenance Control Index (MCI). With the PSI as a reference, the Ministry of Construction of Japan developed the MCI based on the expertise and opinion of road pavement managers. The maintenance cost, the rehabilitation strategies needed, and the surface condition of 1808 asphalt pavement locations were monitored for the development of the MCI. Road pavement managers also visually inspected and evaluated these pavement sections using a 10-point rating system. The resulting index (Eq. 4) is based on road surface characteristics, such as the percentage of cracking (C), the average rut depth (D), and the standard deviation of the longitudinal profile (σ) [4, 5].

$$\mathrm{MCI} = \min\begin{pmatrix} 10 - 1.48C^{0.3} - 0.29D^{0.7} - 0.47\sigma^{0.2} \\ 10 - 1.51C^{0.3} - 0.3D^{0.7} \\ 10 - 2.23C^{0.3} \\ 10 - 0.54D^{0.7} \end{pmatrix} \tag{4}$$

National Highway Pavement Condition Index (NHPCI). The NHPCI was developed to improve the network-level Pavement Management System (PMS) of asphalt national highways in South Korea. It was developed by applying multiple regression analysis between the measured defects of the selected pavement sections and the corresponding ratings of pavement evaluation panel members. As such, two datasets are collected for the development of the index.

The first dataset is the defects (such as cracking, rutting, roughness, patching, etc.) of the 40 selected pavement sections, which were measured using the Automated Road Analyzer (ARAN). The second dataset, on the other hand, is the ratings of the pavement evaluation members. Through field investigation, the 10 pavement evaluation members from the government, industry, research, and academe qualitatively evaluated the condition of the road from 0 to 10.

The developed index is given by Eq. 5, with the crack ratio (C), the rut depth (RD), and the international roughness index (IRI) identified and set as the independent variables [5, 6].

$$\text{NHPCI} = (0.33 + 0.003C + 0.004\text{RD} + 0.0183\text{IRI})^{-2} \tag{5}$$

Highway Pavement Condition Index (HPCI). The HPCI is another condition index used in South Korea. In particular, the HPCI was developed to assess the condition of expressways in the country using a method similar to that used to develop the NHPCI. The method for the development of the HPCI focused on the application of multiple regression on the subjective panel rating and the measured defects of actual pavement sections. The developed HPCI for asphalt and concrete pavements are shown in Eqs. 6 and 7, respectively. Variables needed to solve the HPCI are the international roughness index (IRI), the rut depth (RD), and the surface distress or crack quantity (SD) [6].

$$\text{HPCI}_{\text{asphalt}} = 5 - 0.54\left(\text{IRI}^{0.8}\right) - 0.75\left(\text{RD}^{1.2}\right) - 0.9\log(1 + \text{SD}) \tag{6}$$

$$\text{HPCI}_{\text{concrete}} = 5 - 0.8\left(\text{IRI}^{0.7}\right) - 0.85\log(1 + 2.5\text{SD}) \tag{7}$$

3 Methodology for the Development of PCI for Philippine National Roads

Based on the expert-based pavement condition indices discussed in the previous section, the development of an index requires two primary data: the measured defects of the pavement section, and the experts' ratings of the pavement section. These datasets are typically acquired by conducting field surveys which incur costs. As such, the typical methodology of previously developed indices is adjusted due to the budget restrictions of this study.

In this study, available historical data on the pavement condition of road networks and existing electronic survey platforms are taken advantage of to reduce or eliminate the costs. The methodology consists of two main components: pavement section selection and pavement condition assessment.

Table 3 Defects considered in the asphalt pavement section selection

Asphalt pavement defects	Measurement	Units of measure
Crocodile cracks	Area	m^2
	Crack width	mm
Longitudinal Cracks	Area	m^2
	Crack width	mm
Transverse cracks	Area	m^2
	Crack width	mm
Edge breaks	Length	m
	Average width	mm
Patching	Area	m^2
Potholes	Count	–
Surface failures	Count	–
Rutting	Mean rut depth	mm
Wearing surface	Area	m^2
	Texture	–

3.1 Pavement Section Selection

Pavement sections are selected from the 2-year historical data on the condition of road networks in the country, acquired from the DPWH. Since the available historical data includes the measurements of the defects of the pavement sections, the need to physically measure the defects can be avoided. Asphalt pavement defects available in the historical data and the corresponding parameters presented to the field experts are summarized in Table 3.

The expected range of values for each defect considered is divided into four bins. Pavement sections are selected such that each bin of each defect is represented. However, there are cases wherein some of these bins are not observed in the historical data and are not represented in consequence.

Each pavement section is established as a two-lane 1-km pavement section. Selected pavement sections are then simulated using photographs of the defects obtained in the field. Figure 1 shows a sample of the representation of a selected asphalt pavement section.

3.2 Pavement Condition Assessment

Instead of bringing the experts to the pavement sections, pavement sections are delivered to the experts. The selected pavement sections are simulated and presented to field experts with the aid of an electronic survey. Given the representation of the pavement section (Fig. 1) and the assumptions on the pavement section length

Asphalt Pavement Defects ***(Mga Sira ng Aspaltong Kalsada)***	**Sample Picture of Defects*** ***(Halimbawang Larawan ng Mga Sira*)***
Longitudinal Cracks Area *(Lawak)* = 134 m^2 Crack Width *(Lapad ng Bitak)* ≤ 3 mm	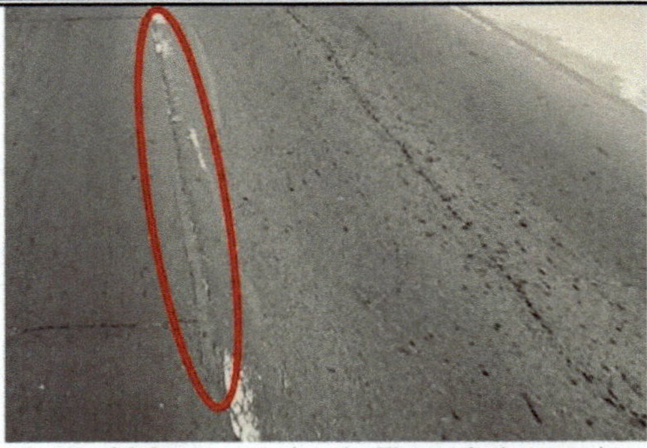 *Picture of defect is a sample only. The cracked area varies from that of the picture. *(Ang larawan ng sira ay halimbawa lamang. Ang lawak ng bitak ay maaaring iba sa nasa larawan.)*
Wearing Surface Area *(Lawak)* = 134 m^2 Texture *(Tekstura)*: rough with loss of both fine and coarse aggregates	 *Picture of defect is a sample only. The wearing surface area varies from that of the picture. *(Ang larawan ng sira ay halimbawa lamang. Ang lawak ng wearing surface ay maaaring iba sa nasa larawan.)*

Fig. 1 Sample of asphalt pavement section representation

and width, field experts are asked to assess the overall surface condition of the pavement section based on their best judgment. A ten-point system is implemented, with 0 indicating the worst possible pavement condition and 10 indicating the perfect pavement condition without any defects. Figure 2 shows a portion of the electronic survey.

The electronic survey on pavement condition assessment is distributed nationwide to field experts on pavement maintenance and management. Data obtained from the pavement condition assessment is then used to establish the relationship between the asphalt pavement defects and the pavement condition (PCI). The developed PCI model for Philippine asphalt national roads is also compared to the parameter currently used in the country (VCI of DPWH).

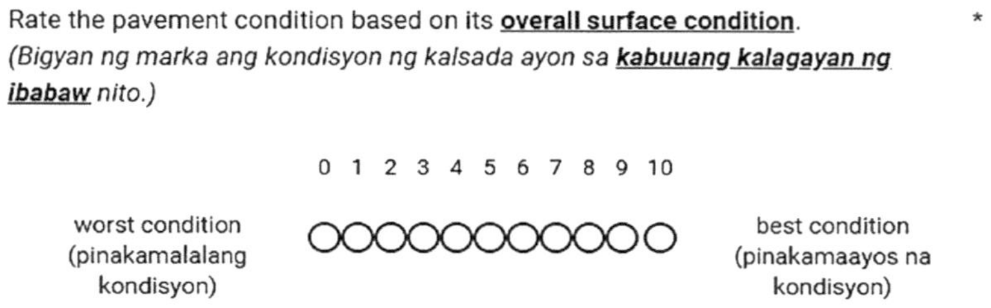

Fig. 2 Portion of electronic survey on pavement condition assessment

4 Results and Discussion

4.1 Pavement Condition Ratings of Field Experts

Currently, 80 field experts respond to the pavement condition assessment survey. They represent several DPWH DEOs from 15 out of 17 regions in the Philippines. The relevant years of experience of the respondents range from 1 to 30 years, with the majority having 6–10 years of experience. Approximately 14% of the field experts are postgraduate degree holders, while the remaining are bachelor's degree holders.

In total, 30 pavement sections are evaluated by 80 field experts. On average, eight field experts evaluated each pavement section. The final pavement condition rating that will be used for developing the PCI is set to be the average values of the field experts' ratings. Table 4 summarizes the average of the pavement condition ratings for the 30 asphalt pavement sections. Field experts' ratings range from 0 to 10, with 0 as the worst and 10 as the best surface condition.

Table 4 Average of the field experts' ratings for each asphalt pavement section

Pavement section ID	SA1	SA2	SA3	A1	A2	A3	A4	A5	A6	A7
Average of experts' ratings	4.25	6.13	6.38	3.25	4.25	5.63	4.88	2.38	3.38	5.38
Pavement section ID	A8	A9	A10	A11	A12	A13	A14	A15	A16	A17
Average of experts' ratings	4.75	4.38	2.57	4.25	7.00	4.75	3.50	3.13	7.50	3.13
Pavement section ID	A18	A19	A20	A21	A22	A23	A25	A29	A32	A34
Average of experts' ratings	5.00	4.75	3.63	4.78	3.88	5.67	6.00	5.00	3.14	3.86

4.2 Defects Influencing the Pavement Condition

The pavement surface defects that will be considered as the independent variables of the PCI model are first established using Pearson correlation analysis. Ideally, chosen independent variables are the parameters with a strong correlation to the dependent variable, given that the correlation is identified as significant. The strength of the correlation can be inferred from the correlation coefficient, while the significance of the correlation can be identified from the p-value [7].

The correlation between the field experts' rating and the surface defects considering severities is first identified. The calculated correlation coefficient between each variable and the corresponding p-value of the correlation are summarized in Table 5. As observed from the table, the majority of the calculated p-values are higher than the set threshold (0.200), implying that the established correlation is insignificant. As such, in this study, severities of the asphalt pavement defects are not considered in the development of the PCI for Philippine asphalt national roads.

Table 6 summarizes the Pearson correlation between the asphalt pavement defects and the field experts' ratings. Results indicate that the correlations between the field experts' rating and particular asphalt pavement defects (longitudinal cracks, transverse cracks, edge breaks, surface failures, rutting, and wearing surfaces) are significant, given that their calculated p-values are less than 0.200. Hence, the independent variables that will be considered for the development of the PCI for asphalt pavements are longitudinal cracks, transverse cracks, edge breaks, surface failures, rutting, and wearing surfaces.

Table 5 Pearson correlation between asphalt pavement defects (with severity) and field experts' ratings

Asphalt pavement defects (with severity)	Field experts' rating	
	Correlation coefficient	P-value
Crocodile crack—narrow (%)	−0.234	0.201
Crocodile crack percentage—Wide (%)	−0.051	0.787
Longitudinal crack percentage—narrow (%)	0.379	0.027
Longitudinal crack percentage—wide (%)	−0.103	0.588
Transverse crack percentage—narrow (%)	0.469	0.004
Transverse crack percentage—wide (%)	−0.214	0.246
Edge break percentage—small (%)	−0.245	0.179
Edge break percentage—medium (%)	0.119	0.527
Edge break percentage—large (%)	−0.221	0.229
Wearing surface percentage—minor (%)	−0.199	0.283
Wearing surface percentage—severe (%)	−0.375	0.028

Table 6 Pearson correlation between asphalt pavement defects and field experts' ratings

Asphalt pavement defects	Field experts' rating	
	Correlation coefficient	P-value
Crocodile crack percentage (%)	−0.170	0.371
Longitudinal crack percentage (%)	0.355	0.040
Transverse crack percentage (%)	0.451	0.006
Edge break percentage (%)	−0.260	0.152
Patching percentage (%)	0.008	0.967
Number of surface failures	−0.430	0.010
Mean rut depth (mm)	−0.290	0.111
Number of potholes	−0.100	0.582
Wearing surface percentage (%)	−0.510	0.001

4.3 *Expert-Based PCI for Philippine Asphalt National Roads*

With the independent variables identified in the previous section and the dependent variable as the field experts' ratings, the PCI estimation model for Philippine asphalt national roads is developed using multiple linear regression analysis. In the analysis, the intercept is set to 100 to represent the perfect condition of the pavement. The resulting equation (Eq. 8) obtained an adjusted coefficient of determination (R^2) of 0.58. The result implies that the developed PCI estimation model captures 58% of the variability of the expert-based dataset.

$$\begin{aligned} \text{PCI}_{\text{asphalt}} = \max(0, 100 &- 1.559x_{\text{LC}} - 0.354x_{\text{TC}} - 0.123x_{\text{EB}} - 0.101x_{\text{SF}} \\ &- 1.981x_{\text{RD}} - 0.838x_{\text{WS}}) \end{aligned} \tag{8}$$

The PCI for asphalt pavements ($\text{PCI}_{\text{asphalt}}$) ranges from 0 to 100, with 0 as the worst pavement condition and 100 as the perfect pavement condition without any defects. It is calculated based on the longitudinal cracking in percent (x_{LC}), the transverse cracking in percent (x_{TC}), the edge break in percent (x_{EB}), the number of surface failures (x_{SF}), the mean rut depth in millimeter (x_{RD}), and the wearing surface in percent (x_{WS}) of the pavement section.

Table 7 Comparison of asphalt pavement defects considered in expert-based PCI and VCI

Asphalt pavement defects	Expert-based PCI	VCI [2]
Crocodile cracks		O
Longitudinal cracks	O	
Transverse cracks	O	O
Edge breaks	O	O
Patching		O
Potholes		O
Surface failures	O	O
Rutting	O	O
Wearing surface	O	O

4.4 Comparisons with VCI

Asphalt Pavement Defects Considered. The asphalt pavement defects considered in the development of the PCI model in this study are identified using Pearson correlation analysis (Sect. 5). This set of defects is compared to the set of defects considered in the VCI currently used by the DPWH. Table 7 shows the comparison between the expert-based PCI model developed in this study and the VCI presently used by the DPWH based on the surface defects influencing the indices.

The VCI formula currently used for asphalt pavement condition assessments does not consider longitudinal cracking, as observed in Table 7. However, the results of this study have identified that the overall surface condition of asphalt pavements is significantly correlated to longitudinal cracks. In consequence, longitudinal cracking is included as one of the independent variables of the expert-based PCI developed in this study.

Conversely, crocodile cracks, patching, and potholes are not considered in the expert-based PCI but are included in the VCI. Based on the Pearson correlation analysis, the correlation of these particular defects to the field experts' ratings is not significant.

In this study, however, the confidence level considered in establishing the significance of a correlation is set to only 80%. This threshold is suggested to be increased to 90–95% once additional data is acquired.

Pavement Condition Evaluation of Historical Data. To further compare the expert-based PCI and VCI, the 2019–2020 data on surface defects of 44,698 asphalt pavement sections in the country is utilized. The corresponding expert-based PCI of the asphalt pavement sections is calculated using Eq. 8, given the information on the defect measurements. The calculated expert-based PCI is then compared to the corresponding VCI of the asphalt pavement sections. Since the motivation behind this study is to provide documentation on the localization of VCI, the expert-based PCI developed in this study should ideally be equal to the existing VCI. Therefore, fitting the two parameters in the equation $y = x$ is logical. When the two indices are fitted into this equation, the corresponding R^2 is 0.72, implying that the expert-based

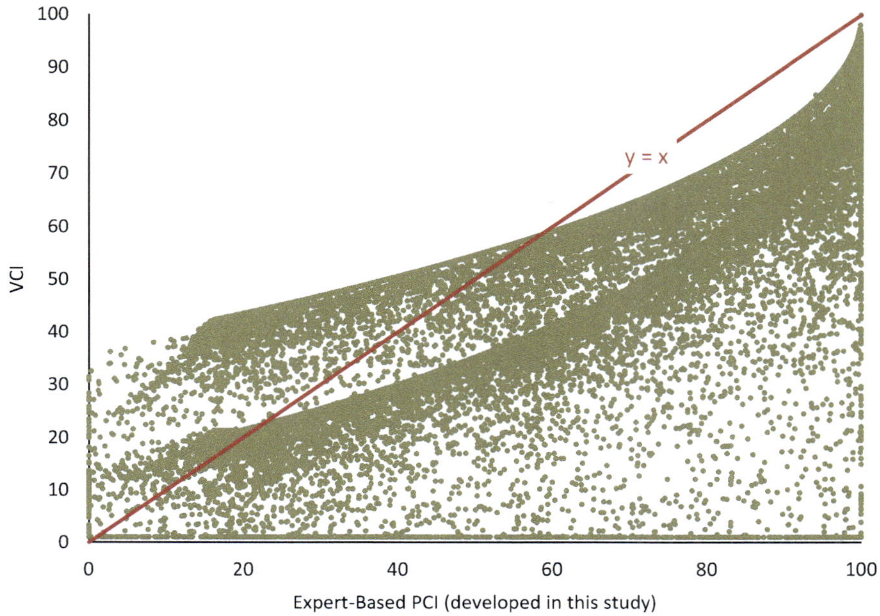

Fig. 3 Graphical representation of comparison between expert-based PCI and VCI

PCI developed in this study is a good representation of the VCI. The comparison can also be visually represented by plotting the VCI against the expert-based PCI (Fig. 3). From the figure, it is evident that the relationship between the two indices is not linear which is expected because the expert-based PCI developed in this study assumes a linear relationship between the pavement condition and the defects, while the existing VCI demonstrates exponential relationships between the pavement condition and the defects.

5 Conclusion

This study was able to establish a PCI estimation model for Philippine asphalt national roads based on the judgment and experience of field experts from different parts of the country. The resulting PCI model is a function of longitudinal crack percentage, transverse crack percentage, edge break percentage, number of surface failures, mean rut depth, and wearing surface percentage.

The expert-based PCI model developed in this study is also compared to the VCI currently used in the Philippines. Comparisons showed that the expert-based PCI is a good representation of the VCI. Therefore, this study can serve as a basis to support the claim that the VCI for asphalt pavements is localized to Philippine conditions, given that the expert-based PCI is developed based on the locally observed defects and local field expertise.

To develop this PCI model, a methodology that incurs zero cost and takes advantage of electronic survey platforms is formulated and utilized in this study. This methodology can be replicated and explored by other institutions with limited budget that also aim to develop an expert-based index of their own.

Similarly, this methodology is also highly recommended for developing the PCI model for Philippine concrete national roads.

The results of the correlation and multiple regression analysis can also be further improved by using more data. Therefore, the acquisition of additional data using the electronic survey on pavement condition assessments is highly suggested. Other analysis methods, such as artificial neural networks and genetic programming, are also recommended to be explored.

Acknowledgements This work is under the Project PAVE—Prototype Automated Visual Survey Equipment, supported by the Department of Science and Technology (DOST), and monitored by the Philippine Council for Industry, Energy, and Emerging Technology Research and Development (PCIEERD). The authors would like to acknowledge the collaborating agency, the Department of Public Works and Highways (DPWH), and their field experts for making this research possible.

References

1. Standard Practice for Roads and Parking Lots Pavement Condition Index Surveys, ASTM D6433–99 (1999)
2. Updating of the Road Network Definition and Inventory Update Manual and Visual Road Condition Assessment Manual Under the Road and Bridge Information Application (RBIA), DPWH DO 120 (2019)
3. Carey, W.N., Irick, P.E.: The pavement-serviceability performance concept. Highw. Res. Bull. **250**, 40–58 (1960)
4. Saitoh, M., Fukuda, T.: Modeling an asphalt pavement repair system considering fuzziness of budget constraints. Comput.-Aided Civ. Infrastructure Eng. **15**(1), 39–44 (2000)
5. Lee, J.-S., Suh, Y.-C., Kwon, S.-A., Kim, G.-Y., Lim, K.-S.: Development of pavement condition index for Korean asphalt national highway and decision criteria for resurfacing. Int. J. Pavement Res. Technol. **2**(3), 106–114 (2009)
6. Suh, Y.-C., Kwon, H.-J., Park, K.-S., Ohm, B.-S., Kim, B.-I.: Correlation analysis between pavement condition indices in Korean roads. KSCE J. Civ. Eng. **22**(4), 1162–1169 (2018)
7. Akoglu, H.: User's guide to correlation coefficients. Turk. J. Emergency Med. **18**(3), 91–93 (2018)

Hydraulic Engineering, Flood Control, and Bridge Engineering

Overview of Critical Vortex on Horizontal Jet Fluidization for Sediment Flushing Systems

Rudi Azis, Farouk Maricar, Muhammad Arsyad Thaha, and Bambang Bakri

Abstract The phenomenon of horizontal jets in the fluidization system can provide initial information on the fluidization phase of sediment in many fluidization experiments for the maintenance of the channel. The objective of this study was to investigate the hydraulic performance of horizontal jet fluidization in generating Critical Vortex until fluidization occurs. The research method is based on two-dimensional physical experience. The test results show that The discharge and Pressure requirement for achieving fluidization is achieved through the stages of reaching the critical vortex. Critical discharge and pressure cause fluidization of the horizontal jet when the critical vortex reaches 0.65 dB of the sediment layer. The addition of each layer of sediment increases the fluidization discharge to reach the critical vortex. The change in the dimensions of the vertical vortex on the horizontal jet is caused by friction between the fluid particles and the sediment layer, especially at the angle of the change in the direction of the jet holes from horizontal to vertical.

Keywords Critical vortex · Horizontal jet · Fluidization · Sediment flushing

1 Introduction

Sediment problems in open and estuary channels have greatly affected the quality of water resources. The quality of water resources is not only used directly (e.g. waterways, irrigation, etc.) but also used for sanitation and clean water [1].

One of the maintenance of channel methods that is easy and economical to apply is the fluidization method. The fluidization method is operated with hydraulic principles by utilizing high pressure in the fluidization pipe to produce powerful jets that can disturb sediment [2–4], and [5]. Therefore, it is important to understand in advance the hydraulic behavior acting on the fluidization pipe, both when the sediment is agitated by jets of fluid or fluidization under hydrostatic pressure.

R. Azis (✉) · F. Maricar · M. A. Thaha · B. Bakri
Department of Civil Engineering, Hasanuddin University, Makassar, Indonesia
e-mail: rudiazis.pwk015@gmail.com

E. Strauss (ed.), *Proceedings of the 7th International Conference on Civil Engineering*, Lecture Notes in Civil Engineering 371, https://doi.org/10.1007/978-981-99-4045-5_8

The water jet in the perforation hole which is operated in the fluidization system is one of the initial stages in the two-way fluidization system. The two-way fluidization mechanism or hybrid fluidization emphasizes the fluidization system as an important step before the sediment transfer process is carried out. The sediment flushing method in this system is carried out by the suction method with a flushing system or assisted by a slurry pump. Displacement of sediment with this system aims to facilitate the flushing of sediment to a deeper place.

The hydraulic system in sediment fluidization is a fluid movement mechanism that applies to fluid systems [6]. In the mechanism of fluid movement in the fluidization system, some parameters generally apply to measure hydraulic behavior from the beginning of fluidization to the full fluidization phase. Parameters of pressure gradient in the fluidization pipe, fluidization discharge, sediment thickness above the fluidization pipe (dB), and fluid parameters [7], are some of the parameters that can be measured to study the hydraulic behavior of the fluidization system. Of course, there are differences in the hydraulic behavior of the fluidized system, both in the direction of the fluid jet orientation and the type of material being fluidized.

Muhammad Arsyad Thaha [4] explains the hydraulic behavior in the vertical fluidization jet direction with sedimentary material as the fluidized material. Darcy's law is a mechanism that applies at the beginning of fluidization where its effect is to increase the hydraulic gradient (dh/z) on the pressure distribution in the sediment layer. So basically that there is a strong relationship between the hydraulic gradient and the formation of scour in the top layer of sediment to form a fluidization zone. The full fluidization will form a fluidization zone until the flow changes from darcy to non-darcy type [4].

This article discusses the result of the investigation about the hydraulic performance of horizontal jet fluidization in generating Critical Vortex until fluidization occurs to determine the hydraulic performance of the perforated pipe and also discuss the hydraulic mechanism for fluidization flow with a horizontal jet direction under the hydrostatic pressure. Experiments were carried out in a laboratory with a two-dimensional model.

2 Previously Published Experimental

Research on hydraulic behavior through a perforated pipe has been carried out by several previous studies such as Lennon P and Weisman [2, 8], Ledwith [9], and Thaha et al. [4].

Thaha et al. in a two-dimensional experimental study analyzed the hydraulic behavior of the process of achieving fluidization from the beginning to full fluidization. The increase in pressure in the sediment that occurred at the beginning of fluidization still showed a darcy type flow but after the fluidization zone was formed the flow type changed to non-darcy. Research on the behavior of hydraulics from Darcy to non-Darcy flow types was carried out using sediment in layers that were agitated by fluidized flow [4].

Lennon, G. P and Weisman, R. N do a three-dimensional experiment to determine the pressure required at the initial fluidization. The initial fluidization pressure requirement is generated by testing hydraulic parameters such as hydraulic gradient which is analyzed based on pipe parameters (hole size, hole spacing, and hole shape). In general, the results of testing with the finite-difference method using a theoretical hydraulic gradient with 3-dimensional and 2-dimensional simulations conclude that there is a loss of energy (head loss) when passing through a porous medium. Thus, the pressure required at the initial fluidization is strongly correlated with the loss of flow energy as the jet stream passes through the porous medium before the fluidization zone is formed [2].

3 Result and Discussion

3.1 Basic Knowledge of Horizontal Jet Above Hydrostatic Layer Without Sediment

The process of achieving fluidization in a system has been discussed in several previous studies. In the fluidization stage without sediment, the hydraulic behavior will be slightly different because the flow type is not Darcy flow as discussed by previous researchers. Therefore, the pressure parameters in the fluidization pipe (ΔP), fluidization discharge (Qf), and jet hole discharge (Qh) with various water levels above the perforated pipe (db) are presented.

The first step of the fluidization stage is the formation of bubbles form a turbulent flow zone which is relatively small with 14 cm of distance from the perforated pipe (see Fig. 1). After the first step, the bubbles start to spread and form a fluidization zone of relatively limited size in the circled zone. The horizontal distance of the water jet is 24 cm from the perforation pipe.

The bubbles spread horizontally and forming a fluidization zone of a relatively wide size. the horizontal distance of the water jet is 31 cm from the perforation pipe

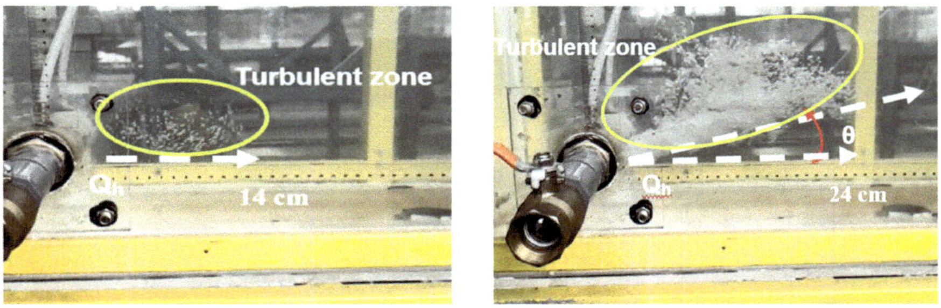

Fig. 1 Initial form of the turbulent zone of the fluidization without sediment layer

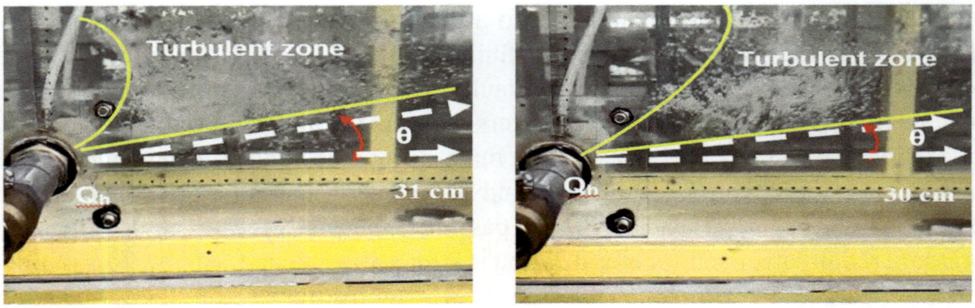

Fig. 2 The water jet spreads by expanding the fluidization zone

(see Fig. 2). The bubble jet distribution appears to lose pressure when it is above the perforation pipe. This is following the results of research by Thaha [4] and Lenon and Weisman [6]. The fluidization zone is formed and bounded by hydrostatic pressure in the horizontal direction where there is a reduction in the burst distance become to 30 cm. The direction of the burst moves upward following the zone bounded by the boundary line.

3.2 Hydraulic Performance to Create Vortex Dimension Above the Perforated Pipe

An understanding of the dynamics of the dimensional phenomena of the horizontal jet vortex has been clearly described by Thaha [3] in the analysis of the basic aspects of fluidization which was reviewed in the finite domain (1 Dimensional test) and infinite horizontal domain (2 Dimensional test). In a limited domain, the calculation in determining the sediment lift h is easy to determine which only involves the particle gravity (FG), the lifting force which is assumed to be a buoyant force equal to the weight of the displaced water (FB) and the drag force (FD), so that by following the equilibrium law then the height of energy (h) required is an Eq. 1.

$$\rho g h A \geq (\rho_s - \rho) g V) \tag{1}$$

The difference in velocity on the left and right of the sediment grains produces a force Fx so that a minimum speed is needed to be able to lift the grains where the condition is that the grains move up to follow the ideal line of vortex dimensions, the difference between v1 and v2 must be as small as possible and it is shown by an Eq. 2.

$$v^2 \geq \frac{4(\rho_s - \rho) g d}{3 C_D \rho} \tag{2}$$

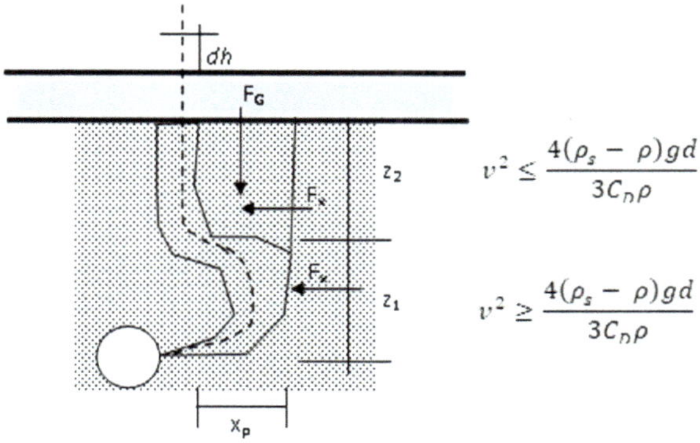

Fig. 3 Typical critical zone on the horizontal jet of fluidization

In the fluidization phenomenon that is generated in the infinite domain, the horizontal force (Fx) works as a consequence of the collective movement of sediments which tend to move following the combined gravity of the particles. The tendency of the combined particles to exert a horizontal force (x) in the opposite direction to the direction of the fluidizing jet produces a fluidization vortex phenomenon at full fluidization conditions whose magnitude is determined by the law of equilibrium. This phenomenon occurs in various ways where changes in fluidization flow occur when the minimum fluidization velocity is smaller than the combined weight of the sediment in the failure zone. Therefore, the test results describe the fluidization velocity conditions in each zone.

In layer z2, it can be seen that v2 is smaller than the combined particle settling velocity (ω) which means that the velocity difference between v1 and v2 is greater than that in layer z1 (see Fig. 3).

Following Fig. 4 at a sediment thickness of 40 cm the critical zone which was originally at a height of 30 cm to 40 cm (fluidized) again experienced a decrease in energy (12.5 cm) because in the boundary zone sediment collapse occurred in layers above 12 cm so that the fluidization pressure in this layer experienced a shift in the direction of the fluidization zone (see Fig. 4). The original horizontal (x) fluidization zone at a distance of 15.7 cm experienced a shift at a distance of 5 cm at a distance of 12.5 cm z (see Fig. 4).

The next failure phase is in the horizontal jet direction (x) up to a distance of 18 cm and the vertical direction (z) 12 cm. The deflection of the jet direction is caused by the loss of fluidization power in the sediment layer so that the collapse as high as he gives the effect of accelerating the fluidization of the next stage. The failure zone as high as hf experienced an acceleration equivalent to the settling velocity of fine sediment particles (0.030199 m/s). The sediment collapse zone and the fluidized jet burst zone are the forming zones for the fluidization zone. The failure zone is formed from a layer of sediment that has been agitated and has become slurry so that the cohesion between sediment grains = 0, this can be an advantage if there is surface runoff with a large discharge so that the slurry flow can be flushed.

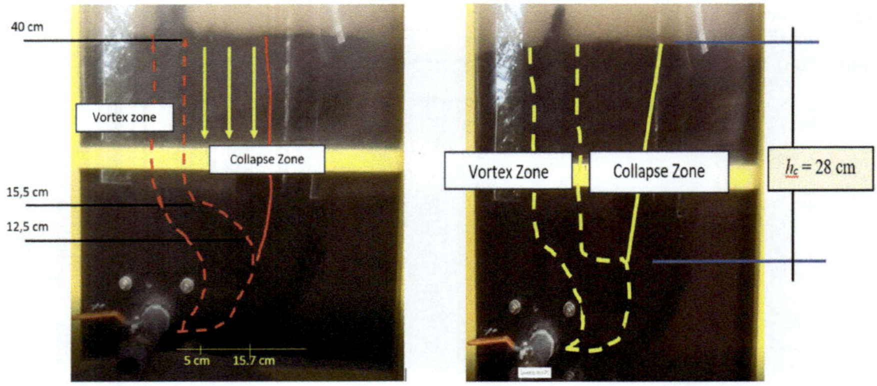

Fig. 4 Critical zone at the sediment thickness (db) 40 cm

The comparison of vortex height occurs at the minimum discharge requirement (Q0) and the critical zone between the vertical jet and horizontal jet experiments (see Fig. 5). The vertical jet experiment was carried out by Thaha [3] on a 1-dimensional finite domain while the horizontal jet experiment in the infinite domain by the current researcher (2022) has shown a hydraulic property of jet pressure which tends to depend on the porosity of the sediment. This has been discussed by previous researchers such as [2, 4, 8], and [3]. The critical vortex that occurred in the experiment on sediment with a thickness (db) of 45 cm above the perforated pipe on the horizontal jet was lower than the vortex that occurred in the vertical jet, namely at a height of Zp 32.5 cm with a minimum discharge (Q0) of 285.56 cm^3/sec. The phenomenon of the formation of eddies in horizontal jets can be understood through the mechanism of the formation of eddies by reviewing the zones forming the fluidization dimensions which consist of the fluidization/vortex zone and the failure zone as shown in Fig. 4.

The ability of the fluidized flow to penetrate the sediment layer at each thickness ranges from 11 to 23.5 cm which is the limit of the critical vortex zone (Zp). Table 1 presents some test results from various variations of sediment thickness. When Zp reaches about 0.65db, the scour at the top of the vortex appears to be growing dynamically increasing the vortex height at a constant pressure in the perforated pipe. This is following the theory that has been tested by [2, 3, 8, 10].

The discharge requirement for achieving fluidization is explained through the stages of reaching a critical vortex, namely where the discharge flowing through the perforation hole tries to penetrate the sediment surface layer. In the pre fluidization stage, the dimensions of the vertical jet vortex will increase with increasing discharge. However, in horizontal jet flow, the dimensions of the vortex enlarge in the horizontal direction, and as the direction of the jet flow changes toward the vertical surface, the dimensions of the vortex change. The change in the dimensions of the vertical vortex is caused by friction between the fluid particles and the sediment layer, especially at the angle of the change in the direction of the jet burst from horizontal to vertical.

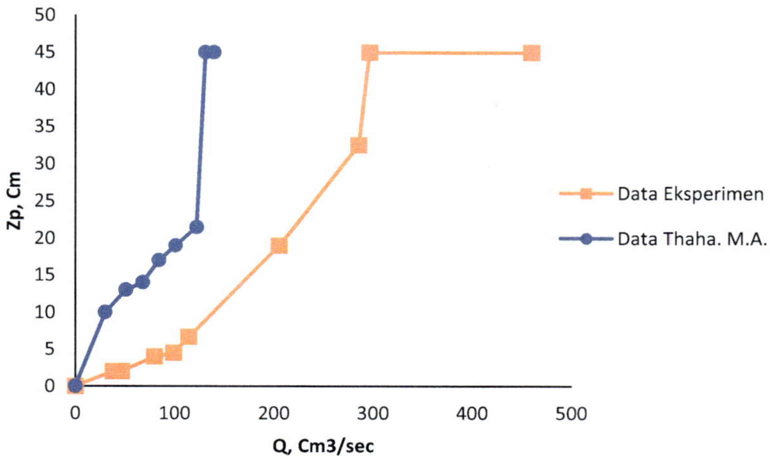

Fig. 5 Height comparison of single vertical and single horizontal jet vortex at sediment thickness 45 cm

Table 1 Achievement of critical vortex height (Zp) at sediment thickness (db)

Tes	db (cm)	Zc (cm)	Zc/db
Tes-1	20	11.0	0.55
Tes-2	20	16.0	0.80
Tes-3	20	16.0	0.80
Tes-4	25	15.5	0.62
Tes-3	30	21.5	0.72
Tes-4	30	21.0	0.70
Tes-2	35	22.5	0.64
Tes-3	35	22.0	0.63
Tes-1	40	23.5	0.59
Tes-2	40	22.0	0.55
Tes-1	45	23.5	0.52
Average			0.65

The achievement of fluidization is characterized by an increase in discharge for each additional layer of sediment (see Fig. 6).

The infinite domain limits the direction of the horizontal jet burst in certain segments, where the experimental results on fine sand sediments show that the increase in the sediment layer further limits the horizontal jet burst (see Fig. 7). The range of discharge that produces horizontal eddies after the critical vortex phase in the vertical direction is full fluidization discharge ranging from 200 to 600 cm^3/

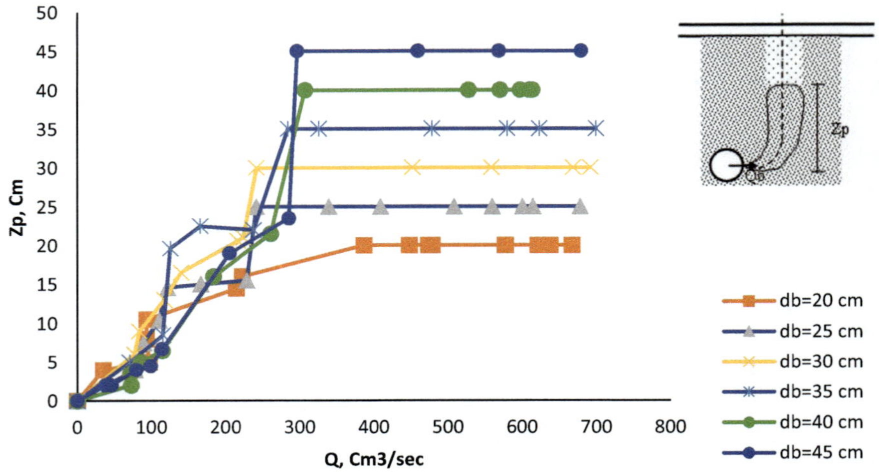

Fig. 6 Pre-post fluidization critical vortex (Zp) graph on a single horizontal jet

sec with average full fluidization occurring at a horizontal jet burst distance (Xp) of 10–13.3 cm. Although the dimension of the horizontal vortex does not develop linearly, the resulting geometric still forms a groove due to the combination of the vortex zone and the failure zone.

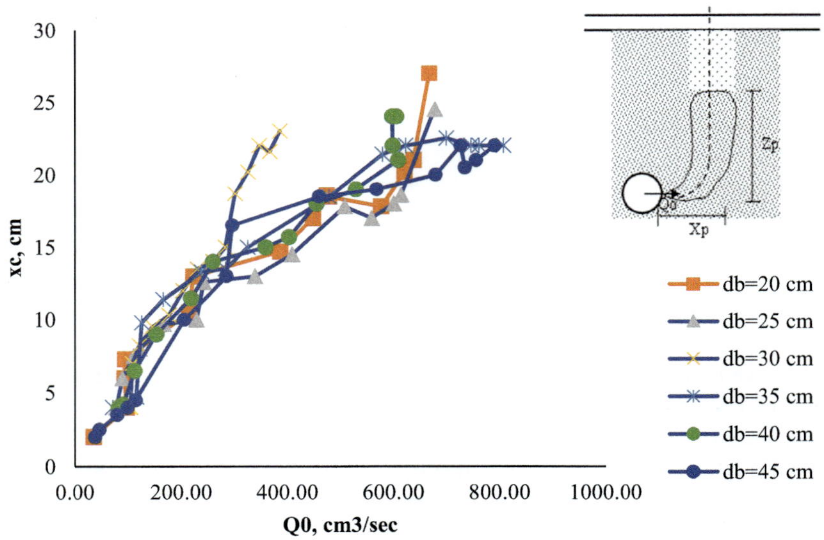

Fig. 7 Development of horizontal vortex dimensions (Xp) in single horizontal jet fluidization

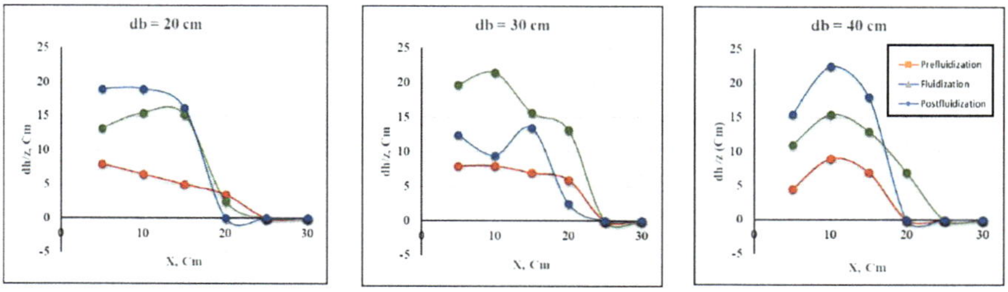

Fig. 8 Distribution of hydraulic head for sediment thickness of 20, 30, and 40 cm above the perforated pipe

3.3 Distribution of Hydraulic Head at the Sediment Layer

The figure (see Fig. 8) presents the pressure distribution in the sediment layers at the sediment height distance (z) above 5 cm from the perforation hole and the horizontal distance (x) from 5 to 30 cm. based on the experimental results on three variations of sediment thickness (20, 30, and 40 cm) showed that the high pressure (dh/z) in the post fluidization phase tended to be high because the flowrate (Q0) given was the maximum fluidization discharge while in the pre fluidization phase, the discharge was given slowly. Based on the point of 6 piezometer tubes placed at a horizontal distance, the sediment pressure is centered on the vortex zone, which occurs at a burst distance of 5–25 cm.

The distribution of pressure at a sediment thickness of 40 cm, the center of pressure occurs at a horizontal distance of 10–15 cm, this is following the visual image of the critical zone that forms the dimensions of the vortex. The farther the horizontal distance from the perforation hole, the greater the loss of power from the jet. The magnitude of the loss of burst energy is also influenced by the thickness of the sediment. Based on several variations of sediment thickness, it can be seen that the dimensions of the eddies that are formed at a thickness of 30 and 40 cm tend to decrease at a horizontal distance. These conditions have been explained above that in addition to hydrostatic pressure, the settling velocity of the combined particles also affects.

4 Conclusions

The main conclusion that can be drawn from the various studies above is as follows:

- The discharge and pressure requirement for achieving fluidization is achieved through the stages of reaching the critical vortex
- Critical discharge and pressure cause fluidization of the horizontal jet when the critical vortex reaches 0.65db

- The addition of each layer of sediment increases the fluidization discharge to reach the critical vortex
- The change in the dimensions of the vertical vortex on the horizontal jet is caused by friction between the fluid particles and the sediment layer, especially at the angle of the change in the direction of the jet burst from horizontal to vertical.

References

1. Bakri, B., Pallu, S., Lopa, R., Maricar, F., Sumakin, A., Maricar, M., et al.: Analysis of sediment distribution at the intake structure. IOP Conf. Ser. Mater. Sci. Eng. **875**(1) (2020)
2. Weismann, R.N., Lennon, G.P.: Head requirement for incipient fluidization of fine sands in unbounded domains. J. Hydraul. Eng. [Internet] **1990**, 838–841 (1995)
3. Thaha, M.A.: Sistem fluidisasi untuk rekayasa pemeliharaan alur. Dissertation. Gajah Mada University, Indonesia (2006)
4. Thaha, M.A., Triatmadja, R., Dwipuspita, A.I.: The hydraulic behavior of vertical jet sediment bed fluidization from the vortex growth point of view. Int. J. Hydraul. Eng. **2**(5), 85–92 (2013)
5. Thaha, M.A., Triatmadja, R., Yuwono, N.: Minimum jet velocity for unbounded domain fluidization as a new dredging method. Eng. J. **22**(5), 1–11 (2017)
6. Weisman, R.N., Lennon, G.P.: Design of fluidizer systems for coastal environment. J. Waterw. Port, Coastal, Ocean Eng. **120**(5), 468–487 (1995)
7. Azis, R., Thaha, M.A., Bakri, B.: Parameter affecting of slurry flow in perforated pipe on fluidization method to maintenance of channel. IOP Conf. Ser. Earth Environ. Sci. **841**(1) (2021)
8. Demchak, S.J., Weisman, R.N., Lennon, G.P.: Effecst of orifice spacing on the fluidization process. IMBT Hydraulics Laboratory Report. Lehigh University, Betlehem (1991)
9. Ledwith, C.A.: Effects of orifice size on the fluidization process. Theses and Dissertation. Lehigh University (1990)
10. Kelley, J.T.: Fluidization applied to sediment transport (f.a.s.t.), Theses, Lehigh University (1977)

Large-Scale in Situ Direct Shear Test in the Construction of Keureuto Dam, Indonesia

Abi Maulana Hakim, Samira Albati Kamaruddin, Andhika Sahadewa, Ramli Nazir, and Haris Eko Setyawan

Abstract A large-scale in situ direct shear test was developed to evaluate the shear strength behavior of random fill material for dam construction in Indonesia. A 70 × 70 × 30 cm large square soil inside a shear box was fabricated in situ. The testing mechanism followed a stress-controlled procedure. The term of random fill material is commonly used for dam construction to the major composition of earth-fill dams in Indonesia. The shear strength characteristic of random fill material may vary depending on the actual particle size distribution. In this paper, in situ shear strength testing on random fill was conducted in Keureuto Dam, Indonesia. The random fill material was majorly composed of coarse material: gravel and sand. A total of 4 samples were tested in situ with a variety of vertical stresses. The testing results showed that plastic deformation started to occur between shear strain of 1–2%. Dilative behavior with an increase in shear displacement was observed in all samples indicating dense condition of the material. Shear strength was evaluated using linear Mohr–Coulomb equation and exponential formula. The results show that both equations are preferable in this case. However, the linear curve is suggested to be intercepted at zero in order to represent non-cohesive behavior of the material. The shear strength of random fill material in this study may provide a new database as a contribution to the practice in dam construction in Indonesia.

A. M. Hakim (✉)
Department of Civil Engineering, Institut Teknologi Indonesia, 15314 South Tangerang, Banten, Indonesia
e-mail: abimaulanahakim@iti.ac.id

A. M. Hakim · S. A. Kamaruddin
Razak Faculty of Technology and Informatics, Universiti Teknologi Malaysia, 54100 Kuala Lumpur, Malaysia

A. Sahadewa · H. E. Setyawan
Department of Civil Engineering, Bandung Institute of Technology, 40132 Bandung, West Java, Indonesia

Indonesian Geotechnical Inztitute, 55165 Yogyakarta, Indonesia

R. Nazir
Faculty of Civil Engineering, Universiti Teknologi Malaysia, 81310 Johor Bahru, Malaysia

E. Strauss (ed.), *Proceedings of the 7th International Conference on Civil Engineering*, Lecture Notes in Civil Engineering 371, https://doi.org/10.1007/978-981-99-4045-5_9

Keywords Random fill material · In situ shear test · Large-scale shear test · Dam construction · Exponential shear strength

1 Introduction

Random fill material is commonly used for earth-fill dam construction in Indonesia. To the authors' knowledge, this type of fill term is not commonly used in other countries. Generally, materials such as rock fill or sand-gravel fill are commonly used to build dams [1]. Sahadewa et al. [2] stated one case of slope failure in dam construction using random fill material in Indonesia and suggested the importance of performing shear strength test activities on site as a control and verification purpose. Therefore, a series of testing to determine the shear strength behavior of random fill materials in situ becomes necessary.

Many researchers have developed large-scale in situ direct shear test apparatus. The most recent study has been conducted in Japan [3], Iran [4], United States [5], Europe [6], Turkey [7], and China [8–10]. Each researcher proposed different testing apparatus (i.e., shear box dimension) and testing method (i.e., load reaction, sample preparation). However, the testing mechanism remains similar to the conventional direct shear testing in laboratory (i.e., loading mechanism, determining failure, data recording, and post-processing).

An in situ shear strength testing is urgently needed to support dam development program in Indonesia. Building a large-scale in situ triaxial test equipment is very expensive. Additionally, it also demands expertise due to high complexity mechanism. Therefore, a large-scale in situ direct shear test is a very promising solution for this need. It is relatively inexpensive and only requires straightforward work process compared to other large-scale in situ testing methods.

In this paper, a large-scale in situ direct shear test at Keureuto Dam in Indonesia is presented. Regional geological conditions of the site are also discussed. The results of this study can be a supplement to the shear strength database of random fill for dam construction in Indonesia.

2 Theoretical Shear Strength of Fill Material

The shear strength of random fill in dam construction can generally be divided into 2 groups, depending on the composition of coarse and fine grains. The lesser the content of fine grains, the more the shear strength behavior closer to the rock fill material, which is cohesionless. If the content of fine particles is greater, then the treatment will be soil-rock material (SRM) or φ-c material, which has cohesive behavior.

An overview of various shear strength failure models is given by Okamoto [11]. In general, for soil materials, the Mohr–Coulomb (MC) model is widely used, which defines a straight failure line as described by the following equation:

$$\tau_f = c + \sigma . \tan\varphi \quad (1)$$

where τ_f is the shear strength, c is cohesion, σ is the normal stress, and φ is the internal friction angle.

For a material with coarser grains, such as rock fill, the power equation as shown below is commonly used to consider the stress-dependent behavior of the shear strength [12]:

$$\tau_f = A \cdot (\sigma)^b \quad (2)$$

where A and b is power equation parameter.

For random fill material, the less the content of fine grains, the more the shear strength can be considered cohesionless. Under these conditions, failure criteria may use either the exponential formula or the Mohr–Coulomb criterion. When the content of fine particles is greater, its behavior will be closer to Soil-Rock Material (SRM) or φ-c material, having cohesion. The Mohr–Coulomb is more appropriate for this situation. Meanwhile, the power equation may not be appropriate because it cannot predict the cohesive behavior. Generally, the Mohr–Coulomb can be applied to the entire grain size grading spectrum of random fill materials, provided that its use is complemented by a deep understanding and adequate experience.

3 Dam Information

3.1 Geological Condition

The research was conducted in Keureuto Dam during construction. The testing was located at the main dam. The satellite images are shown in Fig. 1.

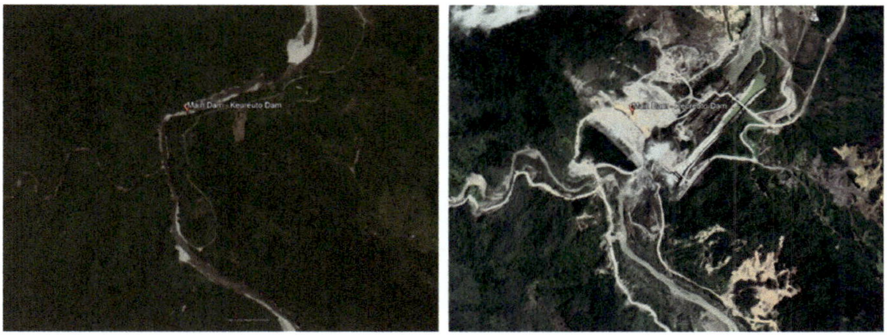

(a) (b)

Fig. 1 Satellite images of Keureuto dam

(a) (b)

Fig. 2 Testing location

Figure 1a provides the situation before construction, while Fig. 1b presents the current condition. Visual documentations of the location are presented in Fig. 2.

Geological conditions are presented here to explain the historical structure of subsurface material. The study location was close to Mount Krueng Pase and Mount Ulupeutu (Fig. 3).

A variety of geological conditions including Baong Formation (T_{mb}), Seureula Formation (T_{ps}), Keutapang Formation (T_{uk}), Geureudong Volcanic Central Rock (Q_{tvtu}), the Bampo Formation (T_{lb}), and the Peutu Formation (T_{mp}). According to Cameron et al. [13], T_{mb}, T_{ps}, and T_{uk} were grouped into the Lhoksukon Group. T_{mb} was a formation originally from Neogene period and Miocene epoch, T_{ps} originated from the Pleistocene epoch, while T_{uk} came from the transition between the Miocene and Pleistocene epochs. Q_{tvtu} was described as a volcanic rock that dated from the Miocene. T_{lb} and T_{mp} were grouped into the Jamboaye Group. T_{lb} was a quaternary rock from the Miocene epoch, while T_{lb} was from the Oligocene.

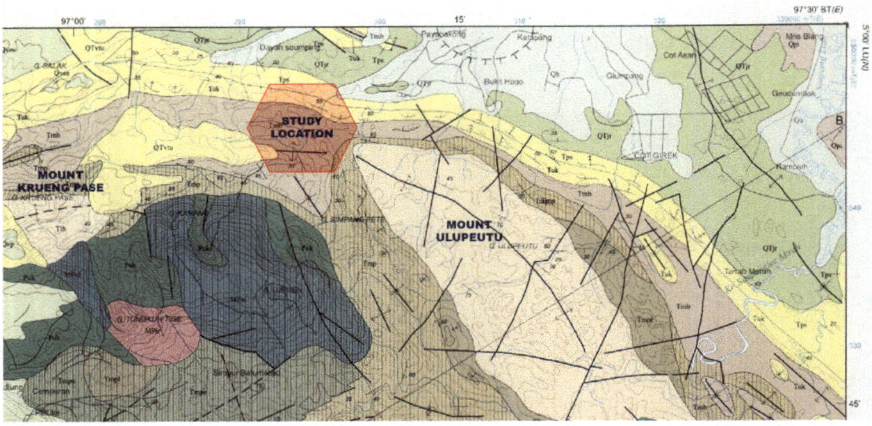

Fig. 3 Study location in geological map [13]

Keats et al. [14] described the Baong Formation as a calcareous mudstone. Additionally, the Seureula Formation was explained as a clastic volcanic sandstone and sublittoral calcareous mudstone. Then, the Keutapang Formation was defined as a sublittoral clastic volcanic sandstone and river delta, while the Geureudong Volcanic Central Rock was designated as a breccia and andesite rock formation. Finally, the Bampo Formation was depicted as dark mudstone.

3.2 General Design of Dam

The design height of the Keureuto Dam was approximately 70.5 m, with dam top elevation located at +107.5 m above sea level. At dam crest, the width and length are 12 m and 363 m, respectively. The upstream and downstream slopes are inclined at gradient of 1 (V): 2.5 (H). A detailed geometrical section of the dam is presented in Fig. 4.

The dam body was designed with diverse material and divided into 5 Zones as shown in Fig. 4. Zone 1 is a dam core utilizing silty clay material. Then, Zone 2 and Zone 3 are a fine filter and a coarse filter, respectively. Zone 4 is the major composition of the dam body, constructed using random fill material. Zone 5 is a protective surface using rock rip-rap material. Large-scale in situ direct shear testing was performed on random fill material in Zone 4.

Random fill material was collected from several quarry locations in the upstream area of the dam. This material was acquired in the form of a mixture of river sediment, such as soil-rock mixture or weathered rock, and was graded as required to meet design specifications. This material generally has two types of colors, namely brownish yellow and gray. The grain shape is commonly round and subrounded. Figure 5 presents random fill material on site. The sieve analysis has been performed on this material prior to construction (Fig. 6).

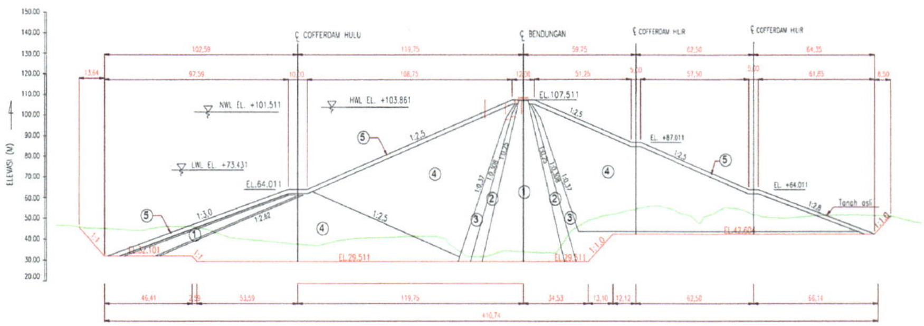

Fig. 4 Geometrical section of main dam

Fig. 5 Random fill material

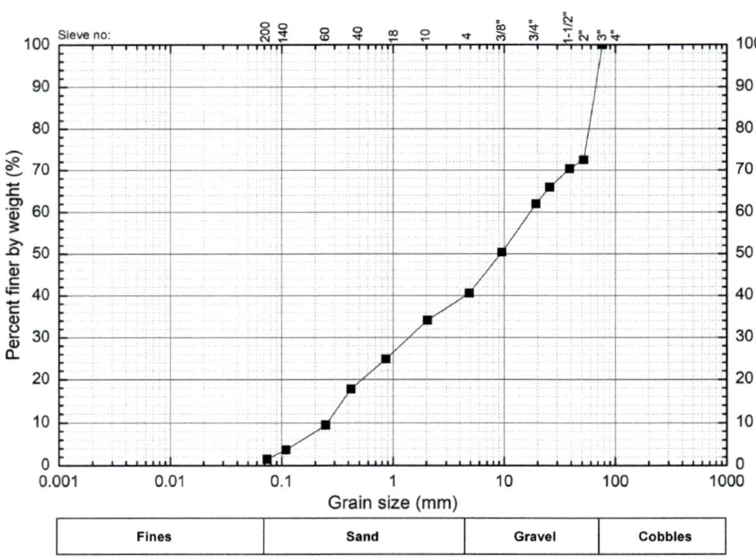

Fig. 6 Particle size distribution of the random fill material

4 Large-Scale in Situ Direct Shear Testing

4.1 Testing Method

A large 70 × 70 × 30 cm square shear box to contain random fill sample was fabricated on site. Each set of hydraulic jack, load-cell, and manual hand pump is provided in vertical and horizontal directions for applying respective load. Two dial gauges

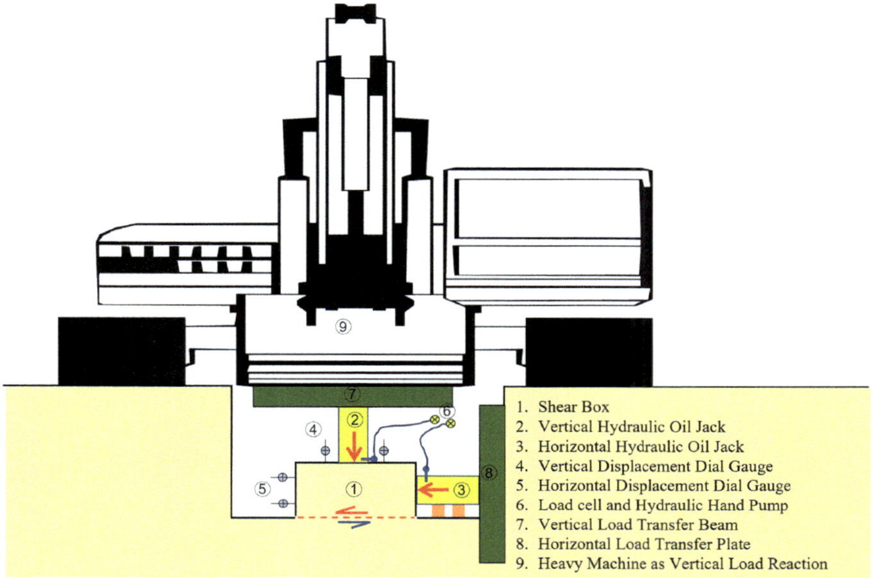

Fig. 7 Large-scale in situ direct shear test setup

were employed to measure displacement of the shear box for the respective direction. The testing mechanism was a manual system with a stress-controlled procedure. The vertical loading was applied to the shear box using heavyweight reaction via a steel load transfer platform. In the horizontal direction, the excavated side wall was used as a load reaction. Figure 7 shows the setup of large-scale in situ direct shear test. Test running is presented in Fig. 8.

4.2 *Testing Result*

Tests were carried out on random fill material at the upstream and the downstream of the dam. Two tests were conducted for different vertical stresses at each location (four testing samples). The test results are presented in Figs. 9, 10 and 11 and compiled in Table 1.

Stress-displacement Behavior. Figure 9 shows the relationship between shear stress and shear displacement. Based on the graph, plastic deformation started to occur in the displacement range of 10–20 mm, or approximately 1–3% shear strain. The shear failure was generally achieved between shear displacement of 60–80 mm, or about 8–11% shear strain. Because of the stress-controlled mechanism, the peak and residual stress behavior for relatively dense materials could not be determined at this rapport.

Fig. 8 In situ direct shear testing

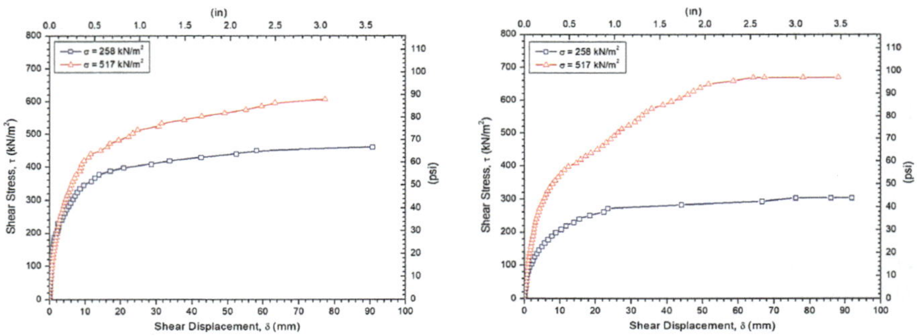

(a) **(b)**

Fig. 9 Shear stress vs shear displacement—**a** downstream and **b** upstream embankment

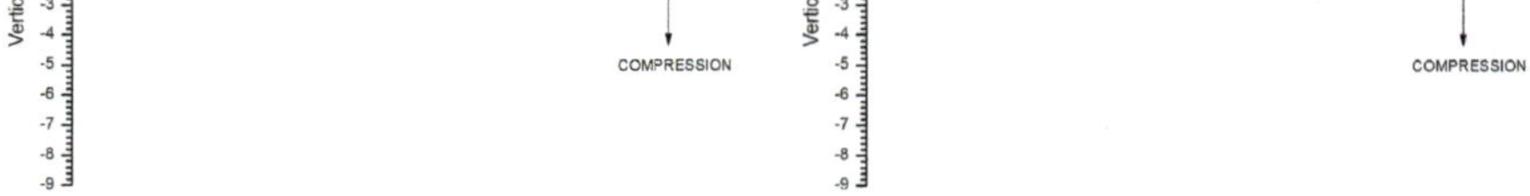

(a) **(b)**

Fig. 10 Vertical displacement vs shear displacement—**a** downstream and **b** upstream embankment

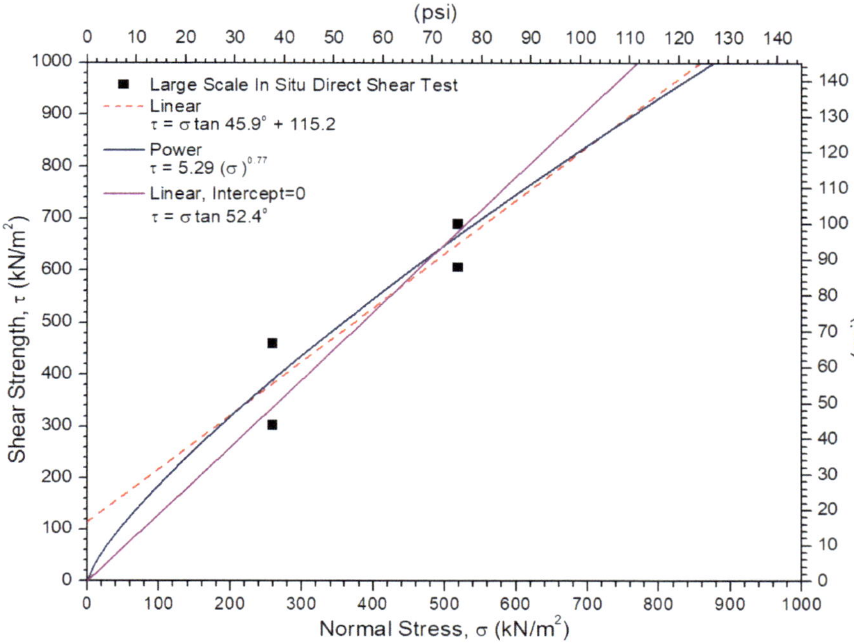

Fig. 11 Shear strength versus normal stress

Table 1 Shear strength parameter

Sample	Equation	Adj. R-square
Combined	$\tau = \sigma \tan 45.9° + 115.2$	0.727
	$\tau = 5.29\ (\sigma)^{0.77}$	0.992
	$\tau = \sigma \tan 52.4^{o}$	0.975

Figure 10 illustrates the relation of vertical and shear displacement. From the graph, it can be understood that the greater the normal stress applied, the greater the sample required shear deformation to come into the expansion phase. This dilative behavior of a non-cohesive granular material may indicate the relatively dense consistency.

Shear Strength. Shear strength was evaluated using linear approach (Eq. 1) and nonlinear power formula (Eq. 2) as presented in Fig. 11. Table 1 shows that the adjusted-R-square value for linear Mohr–Coulomb is smaller, while power curve and linear Mohr–Coulomb with zero intercept approaches has better adjusted-R-square number. Additionally, Mohr–Coulomb without zero intercept demonstrates unreasonably high cohesion, which is not realistic for granular material. Therefore, Mohr–Coulomb with zero intercept and power curve are well appropriate and more preferable for this study.

5 Conclusion

A series of large-scale in situ direct shear tests was performed on 4 samples at Keureuto Dam during construction. The condition of random fill material was closer to rockfill material based on the sieve analysis data. The shape of the materials was mostly rounded and subrounded. Schematic of testing method were presented in this paper. Several conclusions can be drawn according to the result of this study:

1. The stress-displacement behavior implied that the plastic deformation started to occur at shear strain ranging from 1–3%. Additionally, shear failures appeared at shear strain ranging from 8–11%. Furthermore, the material was indicated in relatively dense conditions due to the presence of dilative behavior.
2. The shear strength of random fill material might be defined by linear Mohr-Coulomb failure criteria and non-linear power curve equation. The Mohr-Coulomb approach should be intercepted at zero to obtain realistic value of the non-cohesive strength parameter. However, both methods were still appropriate to be used.

Acknowledgements This paper is written based upon a dam project owned by Indonesian Ministry of Public Works & Housing (PUPR). The authors would like to thank PUPR, PT. Brantas Abipraya, PT. Banyu Bumi Sangkara, Indonesian Geotechnical Institute, and Institut Teknologi Bandung for all supports and contributions. Authors are also grateful to Bagus Putra, Vallenozha, Al Baihaqy, and Dio for supporting site testing. Any opinions, findings, conclusions, and recommendations expressed in this paper are those of the authors and do not necessarily reflect the views of any supporting institutions.

References

1. Haselsteiner, R., Pamuk, R., Ersoy, B.: Aspects concerning the shear strength of rockfill material in rockfill dam engineering. Geotechnik **40**(3), 193–203 (2017)
2. Sahadewa, A., Setyawan, H., Tanjung, M., Pamumpuni, A., Hakim, A. M., Langit, J., Herry, P., Wibowo, A.: Construction failure of X dam: the importance of field monitoring. In: Proceedings on 11th International Symposium of Field Monitoring in Geomechanics, London, England. ISSMGE (2022)
3. Matsuoka, H., Liu, S.H., Sun, D., Nishikata, U.: Development of a new in-situ direct shear test. Geotech. Test. J. **24**(1), 92–102 (2001)
4. Fakhimi, A., Salehi, D., Mojtabai, N.: Numerical back analysis for estimation of soil parameters in the resalat tunnel project. Tunnel. Underground Space Technol. **19**(1), 57–67 (2004)
5. Fakhimi, A., Boakye, K., Sperling, D., McLemore, V.: Development of a modified in situ direct shear test technique to determine shear strength of mine rock piles. Geotech. Test. J. **31**(3), 269–273 (2008)
6. Oyanguren, P.R., Nicieza, C.G., Fernandez, M.I., Palacio, C.G.: Stability analysis of Llerin Rockfill Dam: an in situ direct shear test. Eng. Geol. **100**(3–4), 120–130 (2008)
7. Yunatci, A.A., Cetin, K.O.: Large scale direct shear box tests on gravels. Teknik Dergi **33**(1), 11617–11623 (2022)
8. Liu, S.H.: Application of in situ direct shear device to shear strength measurement of rockfill materials. Water Sci. Eng. **2**(3), 48–57 (2009)

9. Wen-jie, X., Qiang, X., Rui-lin, H.: Study on the shear strength of soil–rock mixture by large scale direct shear test. Int. J. Rock Mech. Mining Sci. **48**(8), 1235–1247 (2011)
10. Zou, Z., Zhang, Q., Xiong, C., Tang, H., Fan, L., Xie, F., Yan, J., Luo, Y.: In situ shear test for revealing the mechanical properties of the gravelly slip zone soil. Sensors **20**(22), 6531 (2020)
11. Okamoto, T.: Evaluation of in-situ strength of rockfill material taking into account of in-situ density and strength by laboratory test. In: Proceedings of the 4th International Conference on Dam Engineering—New Developments in Dam Engineering. New Developments in Dam Engineering, Taylor & Francis Group, London, pp. 693–704 (2004)
12. Marsal, J.R.: Mechanical properties of rockfill. In: Embankment Dam Engineering, Casagrande Volume, Hirschfeld and Poulos (Eds), J. Wiley and Sons, New York, pp. 109–200 (1973)
13. Cameron, N.R., Bennett, D.M.C., Clarke, M.C.G., Djunuddin, A., Ghazali, S.A., Harahap, H., Jeffery, D.H., Kartawa, W., Keats, W., Ngabito, H., Rocks, N.M.S., Thompson, S.J.: Geological map of the Takengon quadrangle, Sumatera. Geological Survey Institute, Indonesia (2007)
14. Keats, W., Cameron, N.R., Djunuddin, A., Ghazali, S.A., Harahap, H., Kartawa, W., Ngabito, H., Rock, N.M.S., Thompson, S.J., Whandoyo, R.: Geologic map of the Lhokseumawe quadrangle, Sumatra. Center for Geological Survey, Geological Agency, Indonesia (2011)

The Prediction of Lahar Flood Event Impact on the Inundation Areas in Gendol River, Indonesia

Jazaul Ikhsan, Elang Afif Hafizh Zhafran, Ani Hairani, and Mohd. Remy Zainol

Abstract One of the natural phenomena that inflict harm and fatalities is lahar flow. The Gendol River frequently experiences lahar flows from Mount Merapi, necessitating study. Predictions can be used to anticipate lahar flows and minimize casualties and material losses. The Nakayasu synthetic unit hydrograph is used in this study utilizing a simulation method based on the theory of Ashida, Takahashi, and Mizuyama, which modifies the hydrograph pattern and rainfall intensity values using the SIMLAR V2.1 application. Based on the simulation results, it was discovered that each simulated design had a different flow area, velocity, lahar volume, and flood height. Based on the result, it can be said that this study's hyetograph pattern significantly impacts how debris flows affect people. The more the hyetograph patterns, the higher the peak hydrograph value. Additionally, the increase in the hydrograph value is directly related to the increase in flow velocity. The rain intensity and pattern value also influence the number of areas affected by the lahar flow.

Keywords Lahar flow · SIMLAR · Mount Merapi

1 Introduction

Mount Merapi (2968 m asl.), one of Indonesia's active volcanoes, has attracted much public attention because of its activities and uniqueness from a scientific and cultural perspective. Many studies have been carried out related to monitoring for mitigation purposes as well as to increase understanding of the characteristics of Mount Merapi itself [1]. Mount Merapi is in Central Java at 7° 32.5′ latitude and 110° 26.5′ east longitude. This volcanic activity has been well recorded since 1768, or even earlier since 1006, associated with the history of Borobudur Temple. The

J. Ikhsan (✉) · E. A. H. Zhafran · A. Hairani
Universitas Muhammadiyah Yogyakarta, Bantul, Yogyakarta 55183, Indonesia
e-mail: jazaul.ikhsan@umy.ac.id

Mohd. R. Zainol
USM Engineering Campus, 14300 Nibong Tebal, Malaysia

E. Strauss (ed.), *Proceedings of the 7th International Conference on Civil Engineering*, Lecture Notes in Civil Engineering 371, https://doi.org/10.1007/978-981-99-4045-5_10

eruption of ancient Mount Merapi has left a horseshoe landscape, which includes the peaks of Selokopo, Batulawang, Pusung London, Kendit, and Plawangan. The activity of the Mount Merapi reviewer is centered on the dome of Mount Anyar, which is located on the plains of the Pasarbubar crater (+2500 m) [2].

Mount Merapi erupted from October 26, 2010, to November 6, 2010, and was the largest eruption in 100 years since 1870. Apart from causing hundreds of casualties, property, and infrastructure. The eruption also caused about 77% of forest ecosystems to be degraded, of which 12.5% or ±766.67 ha experienced severe degradation so that only volcanic sand was visible. Around 2,207.61 ha, or 36%, experienced moderate degradation, and 1,745.54 ha, or 28%, experienced mild degradation. Only 23% or ±1,425.22 ha were not affected by the eruption [3]. Based on data from the Health Office of Sleman Regency, on December 2, 2010, the eruption of Mount Merapi, which occurred in November 2010, resulted in 277 deaths. The Agency for Research and Development of Volcanic Technology provides a hazard limit of 10 km from the summit as a hazard area. However, the eruption on November 5, 2010, reached 14 km from the summit, with the main flow going to the Gendol River so most of the victims came from communities around the Gendol watershed [4].

Lahar flow is a collection of lava spewed by a volcano and reaches the lower surface with the help or encouragement of rainwater. The lava around the volcano will be carried down through the mountain's slopes when it rains heavily. As a result, rainwater that carries volcanic materials from this lava will hit the land below it or residential areas [5]. One way to research and mitigate lahar floods is to conduct a prediction study of the areas affected by lahar floods. Prediction of affected areas can be made through numerical modeling. Numerical modeling is one method that is often used to determine a process of hydrodynamic movements, such as the movement of waste and sediment.

SIMLAR is a GIS-based debris simulation application program integrating 3 (three) sub-programs, namely the flood hydrograph calculation sub-program, the hydrograph calculation due to natural bending collapse, and the 2D flood simulation program. Using a Graphical User's Interface (GUI) application that makes it easy to enter all model inputs, instruct to run and visualize the results in the form of topographic contours, animation of flood debris propagation, and threatened areas [6]. The presentation of the flood hydrograph can use the unit hydrograph reduction method from the measured flood hydrograph if data is available and use the empirical formula, namely Synthetic Unit Hydrograph (HSS), which is a hydrograph based on synthetic watershed parameters. One of the Synthetic Unit Hydrographs often used in calculating flood discharge in Indonesia is the Nakayasu Synthetic Unit Hydrograph (HSS) [7].

Digital Elevation Model (DEM) is a form of digital representation of the height of the earth's surface. Judging from the distribution of points that represent the shape of the earth's surface, it can be distinguished in regular, semi-regular, and random forms. When using DEM data, it should also pay attention to the level of DEM data resolution. The higher the resolution of the DEM data, the higher the accuracy in modeling. Good modeling simulation results can use DEM data in LiDAR (Light Detection Radar) with a resolution of 5 m. IFSAR (Interferometric Synthetic Aperture

Radar), SRTM (Shuttle Radar Topographic Mission), TerraSAR, and LiDAR (Light Detection and Ranging) methods are several methods that can be used to obtain DEM data.

Rainfall (mm) is the height of rainwater that falls on a flat surface with the assumption that it does not evaporate, seep, and does not flow. Rainfall of 1 (one) mm is rainwater as high as 1 (one) mm that falls (retained) on a flat area of 1 m^2 with the assumption that nothing evaporates, flows, and absorbs [8]. The average rainfall from the results of rain measurements at several measurement stations can be calculated using the Thiessen Polygon method. This method is considered quite good because it clarifies the depth of rain as a function of the area considered representative [9].

This study shows a linear relationship between rainfall and the height and velocity of the lava flood flow. The higher the rainfall, the higher the speed and height of the lava flood. However, this study still uses one rain event where there is only one peak of rain. Research on lava flooding due to two consecutive rain events has never been done. Lava floods will cause changes in river morphology which, of course, will also affect the occurrence of subsequent lava floods. River morphology will change due to deposits that occurred in previous lava floods. Therefore, this study will be conducted on the impact of lava flood events which are influenced by the intensity and pattern of rain, using SIMLAR assistance. Rain patterns with one rain series will be compared with two rain sets. The two series of rains referred to here are rains that experience two rain peaks. This study uses two different flood hyetograph patterns generated from maximum daily rainfall data using the Nakayasu synthetic unit hydrograph (HSS) model approach. The flood hydrograph generation is then used as input in SIMLAR lava flow modeling to predict the affected area. This study uses two different flood hyetograph patterns generated from maximum daily rainfall data using the Nakayasu synthetic unit hydrograph (HSS) model approach. The flood hydrograph generation is then used as input in SIMLAR lava flow modeling to predict the affected area. This study uses two different flood hyetograph patterns generated from maximum daily rainfall data using the Nakayasu synthetic unit hydrograph (HSS) model approach. The flood hydrograph generation is then used as input in SIMLAR lava flow modeling to predict the affected area.

2 Research Methods

2.1 Research Location

The study was carried out in the Yogyakarta Special Region Province's Gendol Watershed (DAS), which is situated in Sleman Regency. The study's location is depicted in Fig. 1.

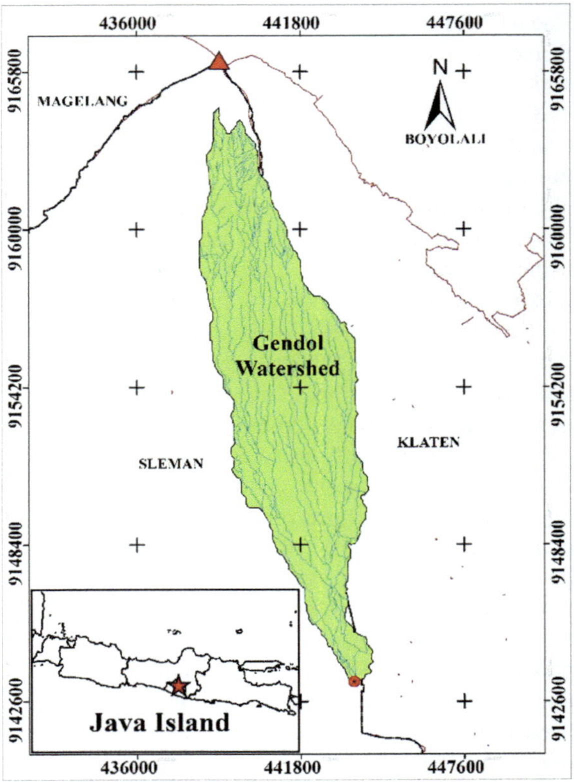

Fig. 1 Location of the Gendol watershed

2.2 Research Data

The Geospatial Information Agency (BIG) publishes DEM data nationwide. The name of this DEM is DEMNAS (National DEM). IFSAR data (5 m resolution), TERRASAR-X (5 m resolution), and ALOS PALSAR are all included in DEMNAS, which is an integration of altitude data (11.25 m). DEMNAS has a spatial resolution of 0.27 arc-second using these types of data. The Earth Gravitational Model 2008 (EGM 2008) is utilized as the datum or vertical reference [10].

The Yogyakarta SABO Center at the rain station used the Ngandong station, located at coordinates Latitude 7′35′43.80″ S Longitude 110′24′27.60″ E, which provided the rainfall data used in this study. The precipitation data utilized is from 2015. The Thiessen polygon method was then used to examine the rainfall data until 2019 to determine the maximum daily rainfall. The highest amount of rain for three hours was used for this study's rainfall. The duration of the dominant rain, which frequently happens in the Merapi area, determines three hours. According to data, it shows the annual maximum rainfall total for 3 h, where the most rain fell at the height of 162 mm.

The sediment samples utilized for the study were collected from several Gendol River crossings; at one cross river, three locations were collected: left, center, and right. There are numerous rock piles and a dry environment upstream. While the river has a flow downstream, its cross section has been normalized, and its right and left banks both have retaining walls. The samples used in this study were obtained, and their specific gravities were then determined in the Civil Engineering Laboratory at the Universitas Muhammadiyah Yogyakarta. This test was conducted based on SNI 1970–2008 on the specific gravity of the fine aggregate. According to the outcomes of the tests, sediment density has been determined for all samples. The average density is obtained in the upstream portion, the average density is obtained in the middle section, the average density is obtained in the downstream section, and so on. The results of this test determined the sediment's specific gravity at each place where a sample was taken. The average weight is 2.61 in the center and 2.64 downstream, with an average density of 2.64 upstream.

2.3 Simulation Scenarios

Six lahar flood simulation scenarios have been carried out to describe the impact of rain events. The rainfall intensity is the normal rainfall intensity, one and a half times the normal rainfall intensity and twice the normal rain intensity. The simulated rain events are one event for three hours and two events for six hours. On combining rainfall intensity and occurrence, the scenario is as follows:

a. Normal Rainfall Intensity and 1 Hyetograph Peak (1H1P)
b. Normal Rainfall Intensity and 2 Peaks of Hyetograph (1H2P)
c. 1.5 times Normal Rainfall Intensity and 1 Hyetograph Peak (1.5H1P)
d. 1.5 times Normal Rainfall Intensity and 2 Peaks of the Hyetograph (1.5H2P)
e. 2 times Normal Rainfall Intensity and 1 Peak of the Hyetograph (2H1P)
f. 2 times Normal Rainfall Intensity and 2 Peaks of the Hyetograph (2H2P).

3 Results and Discussion

3.1 Velocity

Figure 2 provides a comparison of the velocity value acquired from the simulation results that have been run.

Table 1 shows the simulation's generated maximum velocity value as follows.

Based on the results, it can be said that the velocity value follows the synthetic unit hydrograph pattern, rising and falling in step with the value of the synthetic unit hydrograph. The amount of rainfall is inversely correlated with the hydrograph

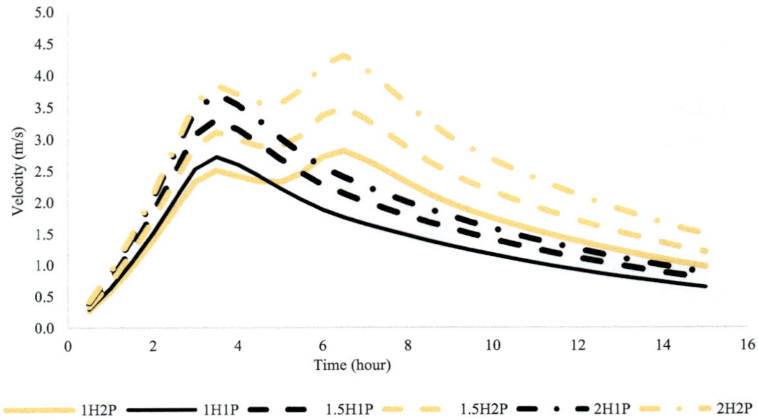

Fig. 2 Velocity of lahar flow for each scenario

Table 1 Maximum velocity value

Scenario	Velocity (m/s)
1H1P	2.72
1H2P	2.81
1.5H1P	3.30
1.5H2P	3.49
2H1P	3.71
2H2P	4.31

value. Rainfall and velocity are also inversely correlated, meaning that the heavier the rainfall, the higher the peak flow velocity.

The second peak of the two-peak hyetograph pattern has a more excellent discharge value, binding the velocity value. Therefore, it may be inferred that the hyetograph pattern influences the lava flood flow's high and low velocities. Table 2 shows the % increase in flow velocity compared to the baseline simulation.

Table 2 Percentage increase of velocity

Scenario	Velocity (m/s)	Velocity increase (m/s)	Percentage increase (%)
1H1P	2.72	0	0
1H2P	2.81	0.09	03.3
1.5H1P	3.30	0.58	21.4
1.5H2P	3.49	0.77	28.4
2H1P	3.71	0.99	36.4
2H2P	4.31	1.59	58.6

3.2 Volume

Figure 3 provides a comparison of the volume value acquired from the simulation results that have been run.

Table 3 shows the greatest volume produced by various simulation types as follows.

Based on the numbers acquired by the simulation results, it can be deduced that the peak value of the lahar flood volume will increase along with the intensity value and pattern of the hyetograph. This phenomenon is caused by an increase in the hydrograph's flow rate, which raises the velocity and, in turn, the scouring of the river's sediment and the volume of the lahar flood flow. Table 4 shows the percentage increase in flow volume compared to the simulation with typical rainfall and histography.

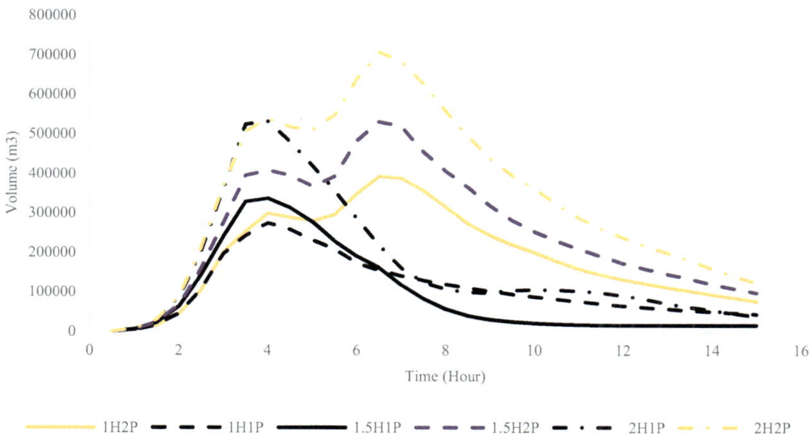

Fig. 3 Value of lahar flood volume

Table 3 Maximum value of lahar volume

Scenario	Volume (m^3)
1H1P	274,685.94
1H2P	393,162.41
1.5H1P	337,183.89
1.5H2P	530,748.70
2H1P	531,773.09
2H2P	707,060.15

Table 4 Percentage increase in lahar volume

Scenario	Volume (m^3)	Volume increase (m^3)	Percentage increase (%)
1H1P	274,685.94	0	0
1H2P	393,162.41	118,476.47	43.13
1.5H1P	337,183.89	62,497.95	22.75
1.5H2P	530,748.70	256,062.76	93.22
2H1P	531,773.09	257,087.15	93.59
2H2P	707,060.15	432,374.21	157.41

3.3 Affected Area and Height

Figure 4 shows the affected area value that was determined by the simulation results that were run as follows.

From the results above, it can be inferred that there is a direct correlation between rainfall and the lava flood's affected area: the more the rainfall, the larger the lahar flood-affected area. The size of the lahar flood is also influenced by the hydrograph pattern, with a 2-peak hydrograph producing a broader lahar flood.

Table 5 shows the percentage rise in the lava flood as well as the expansion of the area affected by the rainfall.

Figure 5 shows the high value of the lava flood as determined by the simulation results that were run as follows.

The flood height value can be seen in Table 6 as follows.

Based on the simulation's output, the maximum flood height varies based on the amount of rainfall and the hyetograph pattern. Most of the values demonstrate a

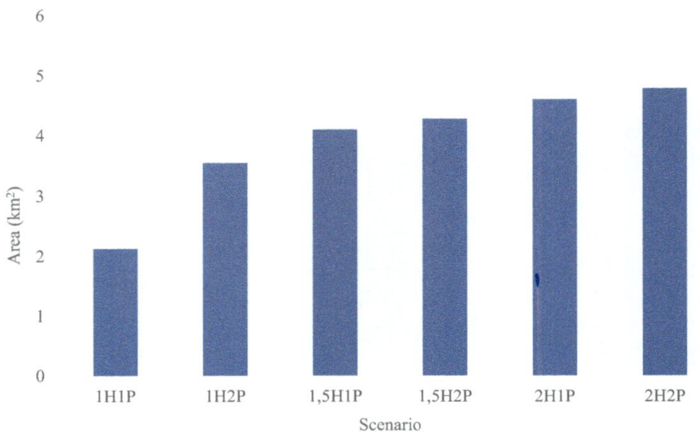

Fig. 4 Value of affected area

Table 5 Percentage increase affected area due to lahar

Scenario	Area (km^2)	Area increase (km^2)	Percentage (%)
1H1P	2.11	0	0
1H2P	4.27	2.16	102.37
1.5H1P	3.54	1.43	67.77
1.5H2P	4.59	2.48	117.54
2H1P	4.01	1.9	90.05
2H2P	4.77	2.66	126.01

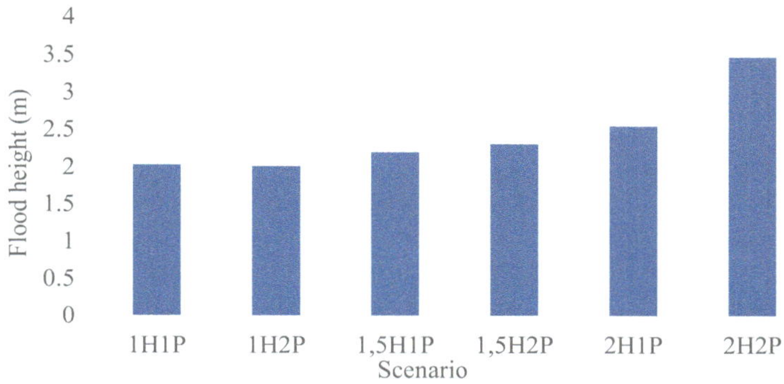

Fig. 5 Lahar flood height

Table 6 Lahar flood height

Scenario	Flood height (m)
1H1P	2
1H2P	2.01
1.5H1P	2.19
1.5H2P	2.3
2H1P	2.54
2H2P	3.46

direct comparison between flood height and rainfall and the impact of the hyetograph pattern, namely that the flood height rises when two hyetograph patterns are simulated. Table 7 shows the % increase in the lahar flood and the height of the lahar flood influenced by rainfall and the typical hyetograph (1H1P).

Table 7 Percentage increase in lahar flood

Scenario	Flood height (m)	Increased flood height (m)	Percentage (%)
1H1P	2	0	0
1H2P	2.01	0.01	0.5
1.5H1P	2.19	0.19	9.5
1.5H2P	2.3	0.3	15
2H1P	2.54	0.54	27
2H2P	3.46	1.46	73

4 Conclusion

This analysis concluded that the lava flood's values and velocity patterns matched the synthetic unit's hydrograph. As the intensity of the hydrograph pattern increases, so do the volume and area values. Additionally, the 1-peak and 2-peak hydrograph patterns yield precise results, with the 2-peak hyetograph producing higher volume and flood numbers.

Acknowledgements The author is grateful for the internal research grant 2022 from Research and Innovation Institute (LRI), Universitas Muhammadiyah Yogyakarta, Indonesia, and would like to thank the Ministry of Education and Culture of the Republic of Indonesia, which has funded this research activity through the 2022 Decentralization Scheme PDUPT Grant that has supported author to conduct research and publicize in this conference.

References

1. Andreastuti, S.D., Newhall, C., Dwiyanto, J.: Tracing the truth of mount Merapi eruption 1006. Indones. J. Geosci. **1**(4), 201–207 (2006). https://doi.org/10.17014/Ijog.Vol1no4.20064 (in Indonesian)
2. Pratomo, I.: Classification of Indonesian active volcanoes, case studies of several volcanic eruptions in history. Indones. J. Geosci. **1**(4), 209–227 (2006). https://doi.org/10.17014/Ijog.Vol1no4.20065(inIndonesian)
3. Gunawan, H., Heriyanto, N., Subiandono, E., Mas'ud, A., Krisnawati, H.: Invasion of exotic species in post-eruption degraded areas in mount Merapi national park **2011** (2015). https://doi.org/10.13057/Psnmbi/M010511erapi, https://doi.org/10.13057/Psnmbi/M010511 (in Indonesian)
4. Budiani, S.R., Lestariningsih, S.P., Gamayanti, P.: Coping capacity of das Gendol community in facing Merapi eruption disaster. J. Mns. Dan Lingkung **21**(1), 106–113 (2014) (in Indonesian)
5. Asmara, R.A., Prasetyo, A., Stevani, S., Hapsari, R.I.: Prediction of cold lava flood on the slope of merapi using rainfall data from satellite. J. Inform. Polinema **7**(2), 35–42 (2021). https://doi.org/10.33795/Jip.V7i2.494 (in Indonesian)
6. Hutahaean, E., Qomariah, S., Setiono, R.: Identification of lava-prone areas in Woro river with simlar V.1.0. application program. E-Jurnal Matriks Tek. Sipil **2**, 80 (2014) (in Indonesian)
7. Sutapa, I.W.: Nakayasu synthetic unit hydrograph study for calculation of design flood discharge in the Kodina river basin. Maj. Ilm. Mektek **7**, 35–40 (2005). (in Indonesian)

8. Mulyono, D.: Analysis of rainfall characteristics in south Garut regency. J. Konstr. **13**(1), 3 (2014). https://doi.org/10.33364/Konstruksi/V.12-1.274(inIndonesian)
9. Talumepa, M.Y., Tanudjaja, L., Sumarauw, J.S.F.: Analysis of flood discharge and water level of the Sangkub river, North Bolaang Mongondow regency. J. Sipil Statik **5**(12), 700 (2017) (in Indonesian)
10. Kasanah, N., Bashit, N., Hadi, F.: Analysis of flooded rice fields using change detection and Pppm (Phenology And Pixel Based Paddy Rice Mapping) Methods (Case Study: Demak Regency). Jurnal Geodesi Undip Januari 2021 (2021) (in Indonesian)

Captive Use Mini Hydropower Project for Pumping Station

Gautam Narula, Vijayinder Kumar Dogra, Rahul Sharma, Vaibhav Sapkal, Komal Bharadwaj, and Aradhyesh Sharma

Abstract The hydropower potential of India is one hundred and forty five Gigawatts, and the tapped potential is only twenty six percent. This potential is exploited in the form of large and small hydropower (SHP) projects. SHPs are classified as micro hydro for capacity less than 100 kW, mini hydro for 101 kW to 2 MW and small hydro for 2 to 25 MW. The electricity is vastly used for pumping water from dug wells at banks of Jhajhhar Nallah to Shri Mata Vaishno Devi University, Katra, Jammu & Kashmir, India. The monthly consumption of the pumping station is around 1lac units. The development of a mini hydropower project is proposed for the captive water pumping operation on the small stream, Jhajhhar Nallah, next to the pumping station. It was found that a Mini Hydropower Project of 200 kW shall be adequate to cater to the university's present and future pumping needs. The estimated cost of the project is $306 Thousand. The annual power expense of the pumping station is $88 Thousand. The cost recovery is estimated at 4.5 years, whereas the project life is 50 years. The pumping station is required to run for 14 to 16 h a day. For the rest of the day, electricity produced shall be used to fulfill the royalty paid to the Union Territory of Jammu and Kashmir. The pumping station must run on all days of the year, so the required discharge is 100% dependable, corresponding to the lean season discharge. In the absence of discharge data for the Jhajhhar Nallah, the discharge measurement was done using the float method for the day when there had been no rain for 10 days in the nearby catchment. The discharge measurements were done in the month of June 2022. The velocity was found to be 0.6 m/s. The area was measured at the upstream and downstream ends of the selected study site. The average area of the stream was noted as 1.2 m^2. Hence the average discharge was noted to be 0.72 m3/s. As per the calculations, the required head is 32 m.

Keywords Hydropower · Small Hydropower · Mini hydropower · Pumping station · Captive use · Float method

G. Narula (✉) · V. K. Dogra · R. Sharma · V. Sapkal · K. Bharadwaj · A. Sharma
School of Civil Engineering, Shri Mata Vaishno Devi University, Katra, India
e-mail: gnarula@ah.iitr.ac.in

E. Strauss (ed.), *Proceedings of the 7th International Conference on Civil Engineering*, Lecture Notes in Civil Engineering 371, https://doi.org/10.1007/978-981-99-4045-5_11

1 Introduction

Power as electricity is a basic need of human existence in the present world. With time the consumption of power is increasing and so one can either manage the production side or the demand side. The demand side management includes a reduction in demand by the use of efficient and smart technologies. The production side management includes revamping power by developing new power projects.

A nation needs to work on both aspects to get the best output from the available resources. Out of various power generation systems, hydropower is the most efficient, sustainable, and economic option.

1.1 Hydropower and Its Classification

Hydropower, which was initially used worldwide in form of Gharad, a mechanism used to convert the energy of flowing water into rotational motion, is now been used on a vast scale for power production. Every stream in the world has some hydropower potential. For optimum utilization of these resources, the resource from each end of every stream must be used judiciously. To harness the available energy, hydropower projects are developed at various scales carrying from large projects in Megawatts to small projects in Kilowatts.

Though it is been seen that hydropower projects of larger capacities are easy to manage hydropower projects of small capacities are a tough baby. Even after being a tough baby, hydropower projects of smaller capacities are more environmentally friendly and locally usable. Governments all around the globe tend to increase the use of hydropower potential available in every form.

Several subsidies and schemes are launched by the state to help the hydropower projects of small capacity in their commissioning and operation. To have better distribution and application of these schemes, the state classifies the hydropower project under various categories. In India, the hydropower projects are classified as shown in Fig. 1.

1.2 Small Hydropower Projects in India

The geography of India supports the development of small hydro projects to enhance energy generation [2]. The small hydropower projects are used in both ways in India, as grid-based and for captive use. The excellent grid network of India allows captive use power of hydropower projects for captive use at any location in India. The government itself is promoting small hydropower projects through subsidies and schemes because they are the hydropower projects that can save the Himalayas from various types of damage [3].

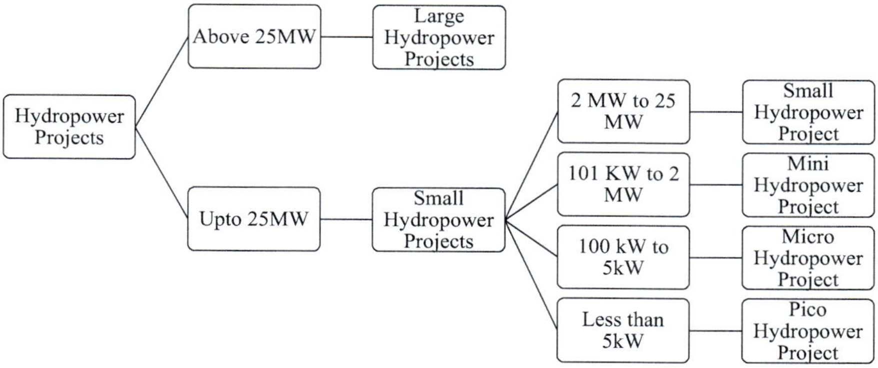

Fig. 1 Classification of small hydropower projects in India [1]

The small hydropower projects have multiple benefits above the conventional large hydropower projects in terms of less damage to the environment, less storage, and easy draw of potential from small streams and canals. As of 2013, 20% of India's electricity demands were fulfilled by small hydropower projects [4, 5].

1.3 Need of Project

Shri Mata Vaishno Devi University is situated in the beautiful hills of Katra City in the state of Jammu and Kashmir, famous for its Mata Vaishno Devi Shrine. The location of the university is shown in Fig. 2.

Fig. 2 Location of Shri Mata Vaishno Devi university, Latitude: 32.94168711360004, Longitude: 74.9529306471796 (*Source* Google Maps)

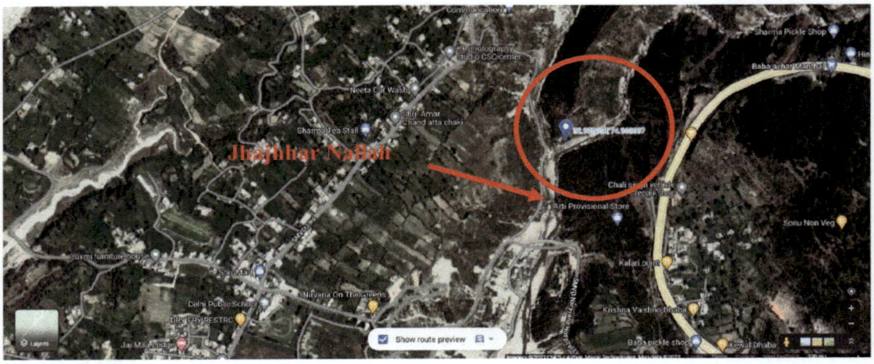

Fig. 3 Location of the proposed powerhouse site, Latitude: 32.9561272506164, Longitude: 74.96100596857538, (*Source* Google Maps)

The university is a residential university and hence has huge water requirement. The university has a pumping station installed which is running three pumps of 50 horsepower and two pumps of 60 horsepower. The pumps are run and maintained by the university itself. The combined operation of these pumps consumes 100 thousand units per month costing the University around $88 thousand per month. The pumping station is using dug wells. The pumping station is installed at the banks of Jhajjhar Nallah, a small stream flowing near the university as shown in Fig. 3. It is proposed to provide the pumping station with an autonomous operation through the installation of a mini hydropower project in Jhajjhar Nallah. The location of the proposed powerhouse site is shown in Fig. 3.

1.4 Projected Benefits

The pumping station, when powered by a mini hydropower project, shall allow multiple benefits such as independence from the external power source which is the grid in this case. The power which is around 100 thousand units can be distributed to the local villages and could power up more houses. This shall also reduce the load on the grid. The projected cost of the project is $306 thousand and the associated projected recovery time is 4.5 years. This project is having a life cycle of 25 years. Almost 20 years of free electricity shall be available for the pumping station. The project shall further enhance the goal of Sustainable India as the pumping station shall be dependent on power produced through renewable energy. This shall also work as an example for society to promote autonomous use of the available resources and reduce the burden on public resources.

2 Methodology

The first step of development was to analyze the power requirements. It was found by studying the electricity bills of the university that there is a consumption of n thousand units of electricity per month. The cost of pumping to the University is $88 thousand per month. Hence the power to be produced was noted to be 100 thousand units per month and a mini hydropower project of 200 kW shall cater to the required need. Further, the pumping station shall run 16–18 h per day. Since the nallah under consideration is a property of the state, the state of Jammu and Kashmir asks for a 12% royalty on the use of water for the power produced in the power plant. The electricity produced for the remaining 8–6 h can be used to pay the royalty to Jammu Kashmir Power Development Authority.

Hence the project size of 200 kW was finalized. The pre-feasibility analysis for the project is under process. The Jhajjahr Nallah is not been studied by any government organization and so the discharge data was not available. It was important to find the discharge. The discharge measurement is a tedious task and requires provision for space for mounting equipment [6]. The discharge measurement can be done using sophisticated instruments and techniques, but for pre-feasibility analysis of the project, basic and diverted methods can be used. Since the pumping station is a basic need and needs to be run every day of the year, hence the power station must be designed for 100% dependable discharge. The discharge measurement was planned for the lean period when there has been no rain in the catchment for 10 days. The discharge measurement was conducted in the mid of June 2022.

A very raw float method was used to study the velocity of the stream. A straight section, the best available, of 100 m was selected and three floats were dropped from a distance of 25 m upstream of the section considered for each experiment. Two observers were stationed at the upstream and downstream ends connected to cell phones. When observed at the upstream station sights the float entering the upstream section speaks up the color of the float and notes down the time. At the same call, the observer downstream also notes downs the time of the entrance of the float upstream. When the float exits the downstream section same process was repeated. The noting of readings by both observers reduced the errors drastically. The sections of the upstream and downstream were measured using tape and leveling staff. The depth of the water level was measured at a horizontal gap of 0.25 m using the trapezoidal method.

Since hydropower is a subject of social and economical development, the social surroundings of the proposed project were also observed. It was noted that some cremation sheds were developed at the river banks. But all the sheds were in a group and concentrated near a small temple. The strata of the mountains were also observed through the outcrops. It was noted that the strata were fragile and the stream has high boulder movements.

3 Result and Discussion

A number of experiments were run and the velocity was found to be 0.6 m/s. The area of the section was calculated upstream and downstream of the section. The average area was found to be 1.2 m^2. Hence the discharge was found to be 0.72 m^3/s. As per the plant requirements, the required head was found to be 32 m.

Other than the technical specifications of the project, it is required to meet the riparian rights of the stream. The cremation sheds that were observed at the site must be respected and given water for religious rights. It is proposed that the powerhouse and tailrace of the project must open almost 50 m above the temple where cremation sheds are concentrated. Further, the electricity produced can also be supplied to the temple for basic lighting. Since the strata are fragile, some soil stabilization techniques may also be required. Though the required head is 32 m, a slight increase in head to find a suitable location for the setup of the dam can be accommodated. Looking over to the boulder movement the rubber dam is proposed at the dam site. The inflation and deflation of dams in very less time can be useful in case of boulder movements and floods.

4 Conclusion

The data was collected from the site using raw methods and a diverted approach to the floating method to fit the need. No specialized pieces of equipment were used. At the level of the pre-feasibility checking, it is a very reliable result that the discharge was measured and found to be 0.72 m^3/s. Though the discharge calculations may not be precise a rough estimation of the project viability, proposing the location of the dam, powerhouse, penstock, and forebay can be finalized based on the required head calculations.

5 Future Scope

The discharge measurement for the project is only completed. The selection of the dam, dam site, and development of the project is the way to go. The measurements done at the pre-feasibility level can be cross-checked at the detailed analysis level using sophisticated instruments like current meters and acoustic Doppler current profiler. These results can be verified and correlated for checking the reliability of future operations.

The correlation shall also be studied in perspective of reliability of raw techniques for measurement of discharge. If the results are in good correlation, the method can be vastly used for pre-feasibility analysis of future mini and micro hydropower project site. The future studies shall also involve the stabilization of strata and the techniques used for the same.

References

1. AHEC/MNRE/SHP Standards/1.1-General: Small Hydropower Definitions and Glossary of Terms, List and Scope of Different Indian and International Standards/Guidelines/Manuals September 2013, pp22
2. Nautiyal, H., Singal, S.K., Sharma, A.: Small hydropower for sustainable energy development in India. Renew. Sustain. Energy Rev. **15**(4), 2021–2027 (2011)
3. Sharma, A., Dubey, V.K., Rawal, Y.K., Sivakumar, K., Johnson, J.A., Sathyakumar, S.: India must protect Himalayan headwaters. Science **376**(6594), 706–706 (2022)
4. Sharma, N.K., Tiwari, P.K., Sood, Y.R.: A comprehensive analysis of strategies, policies, and development of hydropower in India: special emphasis on small hydropower. Renew. Sustain. Energy Rev. **18**, 460–470 (2013)
5. Saxena, P., Kumar, A.: Hydropower development in India. In: Proceedings of the International Conference on Hydraulic Efficiency Measurement, Roorkee, India (2010).
6. Kumar, A., Saini, R.P., Gandhi, B.K., Srivastava, R.K., Chandra, P., Dubey, A.K.: Experiences in discharge measurements at small hydropower stations in India. Flow Meas. Instrum. **69**, 101605 (2019)

An OSINT-Driven Security Analysis of Intelligent Construction of Water Conservancy Projects in China

Yuanbo Qi, Ronghui Yang, and Chenghe Su

Abstract China's major plan to build a modernised economic and ecological system puts water resources at the top of major infrastructure networks, but traditional water resources construction has certain safety hazards. As science and technology continue to advance, intelligent building of water conservancy projects has emerged as a market trend that can significantly lower human error and ensure construction safety. Using the Yangqu hydroelectric power station as a case study, this paper suggests that the intelligent construction of water conservancy projects should prioritise ecology, technicals, and network security through the comparison of positive and negative network volume analysis, POI warning index, and word cloud graph. This study concludes future implications for the trajectory of intelligent development of water conservancy projects regarding carbon–neutral development, intelligent system initiatives, and safety management upgrade.

Keywords Hydroelectric power station · Safety risks · Intelligent construction · Public perception · OSINT

1 Introduction

Water conservation projects are an important part of building the nation's infrastructure and are essential for preventing floods, allocating water resources and providing both economic and social advantages [1–3]. Water conservation projects have an increasing impact on people's livelihoods and quality of life as a result of the nation's economy's fast growth [4]. Water projects are active in a variety of contexts, including urban living, agricultural irrigation, industrial production, and the natural environment [5]. The pace of development of new energy power systems has substantially increased as the energy revolution proceeds, notably with the introduction of the "dual carbon" objective [6]. A renewable energy source is hydropower, one of the finest energy sources for achieving carbon neutrality [7].

Y. Qi (✉) · R. Yang · C. Su
College of Humanities, Donghua University, Shanghai, China
e-mail: y.qi@dhu.edu.cn

E. Strauss (ed.), *Proceedings of the 7th International Conference on Civil Engineering*, Lecture Notes in Civil Engineering 371, https://doi.org/10.1007/978-981-99-4045-5_12

Meanwhile, hydropower plants encounter difficulties in construction [8]. The majority of hydropower projects are situated in high mountains and valleys, and the construction process contains safety risks [9–11]. The development of hydropower projects is a challenging and significant undertaking because of the geological complexity of the project, the construction environment, the substantial number of participating units, the labour intensity, and the machinery and equipment [2]. Waterworks development has led to the creation of intelligent construction due to ongoing advancements in AI and other technologies. Drones, artificial intelligence (AI), 3D printing, and other related technologies are used in intelligent construction, a way of building that can successfully eliminate human error and ensure worker safety while guaranteeing construction proceedings [12].

2 Literature Review

There are relatively few articles on the use of intelligent technology in engineering construction, as well as views on the use of public opinion frameworks to analyse the intelligent construction of water conservancy projects [13–16]. The literature both domestically and internationally primarily concentrates on the use of intelligent technology for monitoring and management [17].

A new method for optimising the functioning of chain reservoirs using artificial intelligence to manage water resources and generate hydroelectric energy has been developed [18–20]. The water conservation business has a large volume of data, high value, and time-sensitive water conservancy as a result of the deployment of sophisticated technologies such as big data, cloud computing, and simulation [21]. The use of information technology and the mining and application of big data for water conservation projects is an effective way to promote the intelligent construction and sustainable development of water conservation projects and therefore accelerate the construction and development of the water conservation industry [22].

Advanced technologies like artificial intelligence, cloud computing, big data, and the Internet of Things actively contribute to improving the management of water conservation projects, enhancing the effectiveness of the industry and its management, and fostering its sustainable development [23, 24]. For instance, the cost of intelligent management platform for water conservancy project makes use of computer-selected statements to achieve automatic statistics calculation and increase work and management efficiency [23]. The development and successful implementation of intelligent testing systems in water conservation projects have helped to standardise the industry's regulatory framework and advance the uniformity of water quality monitoring [25–27]. A foundation for the scientific management of the water conservation industry is provided by the intelligent platform for water conservation projects, which combines 7*24 h sensing data with computing resources [28, 29]. Artificial intelligence, big data, cloud computing, and other cutting-edge technologies have been applied to the comprehensive management, quality testing, efficiency assessment, simulation, and other aspects of the water conservancy industry [30–33].

This promotes the industry's sane and sustainable growth as well as the intelligent construction and orderly operation of the sector [34].

3 Intelligent Aided Water Conservancy Projects

On the Qinghai-Tibet Plateau, the Yellow River Yangqu Hydropower Station is an example of an intelligently constructed water conservation facility in China [34]. On the one hand, Yangqu Hydropower Station creates a construction power supply local area network using "energy storage + photovoltaic" green electricity to supply power during the station's construction. The building industry, on the other hand, uses a combination of artificial intelligence with 3D printing [35–38]. Under the careful computer supervision, the materials are piled in accordance with the design model, producing the appearance of 3D printing [39].

The use of 3D printing technology for water conservation projects enables the nearly flawless realisation of fine printing of the form and surface contours of special structural parts of water conservation projects in accordance with predetermined effects, while the strength of the finished concrete exceeds the general requirements in engineering construction, thereby avoiding the production and use of a large number of formwork and fully mechanised pouring and increasing the cost of the project [40, 41]. A standardised and intelligent construction process is made possible by 3D printing technology, which also reduces construction costs and eliminates many oversights and mistakes that are ineluctable in traditional human labour [14]. The use of robots in the construction of the Yangqu hydropower station, in addition to 3D printing technology, demonstrates the benefits of AI in water conservancy engineering [42–46]. Tens of thousands of sensors were installed at the Baihetan hydropower station in site construction, and access to the port allowed for quick understanding of the condition of the site. Artificial intelligence technologies, such as the Intelligent Water Passage 2.0 system, are also available to analyse temperature changes and temperature differences in real time [47–52].

The use of intelligent technology throughout the entire water conservation engineering construction process helps to thoroughly address the technical and management issues associated with the creation of large water conservation projects in terms of complex environments, resource flow, changing conditions, adjustment of nature, and structural transformation, as well as to advance the development and use of intelligent construction technology [53–56]. To support the national objective of "carbon neutrality" and "carbon peaking," several intelligent building technologies in the water conservancy industry are deserving of promotion to more pertinent projects [57].

4 Public Opinion, Data Mining, and Analysis

4.1 Open-Source Intelligence Mining Algorithms, Models and Tools

(1) The artificial intelligence models BERT, LSTM, and CRF are used to build large-scale machine learning datasets by manually annotating sample data. Through self-supervised learning, the accuracy and recall rate of existing language models are improved, and massive heterogeneous text data processing and risk information mining are achieved [58–61]
(2) Risk discovery model: The model incorporates various sub-model learning for modelling and may produce both the end results of categorization and distinct judgments on various chain risks. The temporal heat model, the Bert language model, the topic word heat model, and propagation sub-model.
(3) Analysis tools: Through the use of web crawlers, text semantic analysis, algorithms, and calculations from risk discovery model, the pre-established public opinion analysis system and the big data system, allowing us to realise visual analysis results, such as word cloud graphs.

4.2 Network Volume Analysis

The requests and expressions of network participants through network channels are referred to as network volume and indicate the overall amount of network data in a certain field. Positive and negative volume are the two categories. The greater the negative volume, the greater the potential for danger in the field. Two sets of network volume combinations are formed based on the identification of representative keywords for water conservancy project concealment and the addition of intelligence-related keywords. Afterward, through open-source intelligence mining, the positive and negative network volumes of water conservancy safety and water conservancy intelligence are compared (see Fig. 1).

Figure 1 shows that during the sampling periods, the negative volume of water conservancy project safety was 15.05 percent higher than that of intelligence, indicating that there are more unfavourable public perceptions of water conservancy project safety as well as risks, while the intelligence has somewhat decreased the negative volume. Water conservancy intelligence has a considerably greater favourable voice than water conservancy safety, which suggests that the market has a higher level of acceptance. The risk associated with traditional water conservation projects is substantial overall and there are numerous possible risk factors, whereas intelligent water conservation initiatives can successfully mitigate potential risks and generate favourable market response.

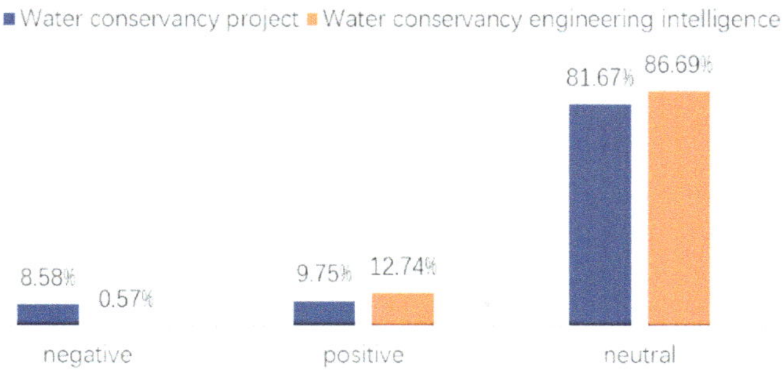

Fig. 1 Network voice analysis comparison of the discourse regarding water conservancy safety and water conservancy intelligence

4.3 POI Warning Index

The POI index, a thorough evaluation of the dispersion of unfavourable public opinion across platforms, has a significant risk prediction ability in addition to the total volume of unfavourable material. Figure 2 shows that the POI index of water conservation projects during the monitoring period is greater than 100 (grey warning), and that there have recently been more yellow warnings and even a new high of over 600 for orange warnings. This shows that the safety risks of water conservation projects are dispersing more quickly and widely, and the possibility of future risks is higher.

The POI feedback from the Water Resources Intelligent Monitoring Group is the opposite. Figure 3 depicts the monitoring period as having a POI index of less than

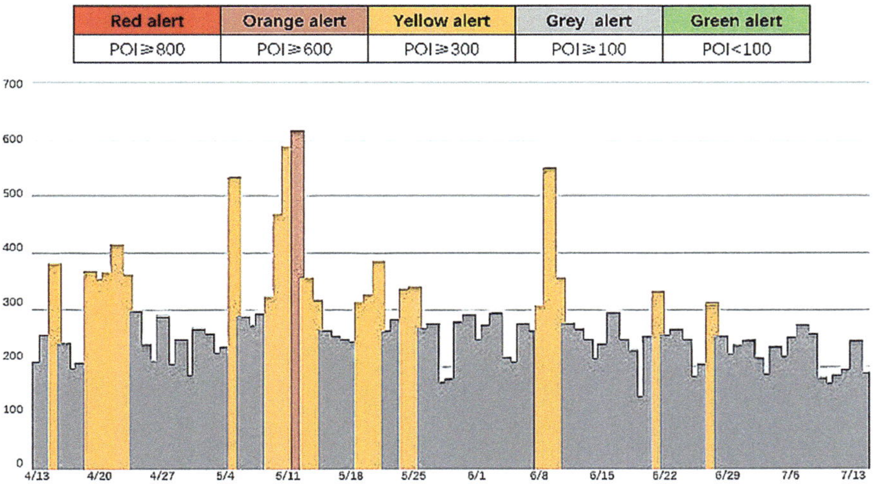

Fig. 2 Water resources engineering POI index

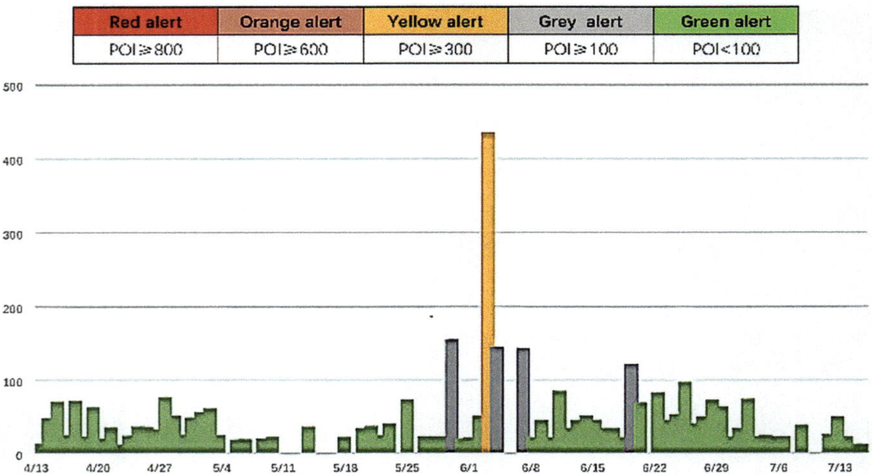

Fig. 3 Water conservancy engineering intelligence POI index

300 (Yellow warning). The majority of POI indices during the monitoring period were below 100 (green warning), showing that the negative effects of water engineering intelligence spread slowly, with a high overall safety factor and market acceptance, and a low probability of future risk.

4.4 Word Cloud Diagram

The word cloud diagram of the development of water conservancy intelligence is created and shown in Fig. 4. It is based on the real-life problems of water conservancy intelligence to determine the setting scheme (the main word is set as "water conservancy", and the analysis words are set as "intelligent computing", "drone, intelligence", "3D printing", "artificial intelligence", and "robot").

Fig. 4 Word cloud diagram of intelligent development of water conservancy projects

Figure 4 demonstrates that the development of intelligent water resources is centred on the concepts of intelligence, technological bottleneck, carbon neutrality, robotics, intelligent water resources, ecology, and security. In addition, the terms "drones," "region," "network security," "strategic," "technology," "3D printing," and "infrastructure construction" are also used more frequently. This suggests that the future development of intelligent water conservation projects will focus primarily on regional development, infrastructure building, drones, and 3D printing technology, while also paying attention to the internet.

5 Implications and Suggestions

Through data mining and Yangqu hydropower station, this article combines the focus on the direction of intelligent development of water conservancy projects and puts forward proposals for the future development of intelligent construction of water conservancy projects [60].

5.1 Development of a Carbon–Neutral Landscape and Water Eco-Energy System

Although hydropower is one of the best energy sources for achieving the double carbon goal, the carbon neutral goal must be accomplished by developing an ecological energy system, fully utilising hydropower's beneficial contributions to carbon peaking and carbon neutral, and assembling a trinity of wind, photovoltaic, and hydropower ecological energy systems [59]. Eco-energy systems can increase power generating efficiency by incorporating technology such as artificial intelligence, cloud computing, and big data with water engineering [53, 62–65]. These systems are also useful for promoting energy reform and hastening the achievement of carbon neutrality. It improves the "national energy and ecological map" and supports the sustainable growth of China's ecological environment, social economy, and other elements. It also advances the intelligence of power planning, engineering construction, power management, and social services [58].

5.2 Creation of Intelligent System Initiatives

Intelligent water conservation led to the intelligent development of planning and systematisation for water conservation and hydropower. Intelligent water conservation refers to the use of cloud computing, mobile terminals, IoT, AI, sensors, water model, and other new generation intelligent computing technologies to design water conservation projects that are informative and contemporary [65]. Applications of water wisdom mostly involve resources, ecological construction, disaster warning, engineering construction, oversight and management, and public services [66–68]. Using intelligent hydropower, water conservation projects integrate water resources, water environment, water ecology, and other information resources based on multiple intelligent perceptions. This dynamically monitors and assesses engineering construction and ecological conditions of rivers and lakes and achieves intelligent management of water conservation through forecasting and warning [69–71].

5.3 Upgrade of the Safety Management and Monitoring System

Construction-related IT technologies such as automation, the IoTs, 3D modelling, drones, as well as new equipment, provide comprehensive support for the development of intelligent water conservation construction projects [72–74]. Along with the advancement of technology, security management and a monitoring-based network security defence system might receive substantial attention and form a closed-loop security OS by achieving comprehensive network security visibility, monitoring, and troubleshooting capabilities [75–78]. This strengthens the core data protection in the water conservation industry as well as the infrastructure's information security.

6 Conclusion

This paper primarily monitors and analyses public opinion data based on text semantic analysis, artificial intelligence algorithm, risk discovery model, and intelligent application of water conservancy engineering technology, in contrast to the literature on intelligent construction of water conservation projects at domestic and international levels. This study addresses the direction that needs essential attention in the intelligent development trajectory of hydraulic engineering from three dimensions: network traffic, POI warning, and word cloud map, using Yangqu Hydroelectric Power Station as an intelligent construction scenario. It has been determined through the analysis of survey data on public opinion that intelligent engineering must be used to create water conservation plans in China. The reliance on information technology necessitates the employment of artificial intelligence, 3D printing, big data, and other technologies, henceforth enhancing the efficacy, quality, and safety of the water conservancy project and operation. The most significant of those are increasingly popular intelligence technology. Additionally, the major trajectories that must be given attention to in the intelligent construction of water conservation projects entail word cloud analysis, network security, ecological, and POI warning. Therefore, relevant practitioners must comprehend the actual situation of water conservancy engineering construction based on the automated inspection system and safety assessment data model to effectively improve the management quality of the project and to promote sustainable development of water conservancy engineering.

Funding Information The research for this article was supported by grants from Donghua University's Applied Linguistic Research Committee [Y2022-3] and the International Programme [N19].

References

1. Li, F., Wang, H.: Architectural landscape design in the water conservancy project of the coastal city. J. Coast. Res. 46–49 (2020)
2. Jiang, H.: Construction and reform of water conservancy undergraduate majors facing new water conservancy situations in new era: thoughts based on engineering education accreditation. Adv. Sci. Technol. Water Resour. **41**(1), 1–8,15 (2021)
3. Renli, W., Bo, C., Yang, Y., Wenjun, Z.: Application of 3S technology to digital water conservancy. Sci. Surv. Mapp. **33**(3), 210–212 (2008)
4. Zheng, X.: Application of spatial information technology in water conservancy industry: retrospect and prospect. J. Yangtze River Sci. Res. Inst. **38**(10), 167–173 (2021)
5. Li, L., Deng, X., Huo, W. (eds.): Analysis of Inspection and Testing Standards for Water Conservancy Projects in China, pp. 129–136.
6. Lin, W., Xu, H., Zhu, Y.: Assessment of water conservancy modernization level of Zhejiang province. J. Irrig. Drain. **35**(2), 40–43 (2016)
7. Fu, C., Yang, Z., Liu, C.: Connotation and evaluation index system for water convervancy modernization. Adv. Water Sci. **13**(4), 502–506 (2002)
8. Deng, M., Huang, Q., Chang, J., Huang, S., Guo, A.: Connotation, process and dimensionality of generalized ecological water conservancy. Adv. Water Sci. **31**(5), 775–792 (2020)
9. Liu, J., Zang, C., Tian, S., Liu, J., Yang, H., Jia, S., You, L., Liu, B., Zhang, M.: Water conservancy projects in China: achievements, challenges and way forward. Glob. Environ. Chang.-Hum. Policy Dimens. **23**(3), 633–643 (2013)
10. Lv, D., Li, G. (eds.): Water Conservancy in New Village Construction, pp. 242–244 (2010)
11. Che, P.: Water conservancy automation monitoring system based on VR image video and internet of things. Mob. Inf. Syst. (2022)
12. Yan, Y., Yang, L., Xu, H. (eds.): The Content Analysis of Rural Water Conservancy Construction Facing the Well-off Society, pp. 484–489 (2011)
13. Xu, S.: Deepening innovation of water conservancy technology and enhancing water conservancy management ability. Water Resour. Hydropower Eng. **45**(12), 20–21 (2014)
14. Liu, M., Liu, Z.: Design and implementation of water conservancy section drawing program. Sci. Surv. Mapp. **39**(2), pp. 142–145,149 (2014)
15. Song, J., Luo, H., Wu, L., Pan, B., Chen, Y.: Development trend of irrigation water conservancy construction in Karst mountain areas a case of Guizhou province. J. Agric. Sci. Technol. **18**(1), 112–119 (2016)
16. Yang, H., Zhao, Y., Song, H., Wen, H., Zhai, R. (eds.): Discuss on Management of Water Conservancy Planning in the Yellow River, pp. 140–145 (2012)
17. Ying, L., Huang, X., Mao, M.: Discussion of the distribution form of the water conservancy tourist area in Zhejiang Province. J. Zhejiang Univ. **29**(4), 470–475 (2002)

18. Ma, C., Cao, G., Ma, Y. (eds.): Discussion on the sustainable development of urban water conservancy infrastructure construction. 733 (2012)
19. Wang, L., Teng, H., Wang, K., Iop.: Discussion on Zhejiang Provincial Water Conservancy Local Standard System in the New Period (2019)
20. Wang, Y., Huang, Y., Tang, X.: Division of China water conservancy development stages: theoretical framework and evaluation. J. Nat. Resour. **28**(6), 922–930 (2013)
21. Liu, J., White, C.: Consuming missionary legacies in contemporary China: eric liddell and evolving interpretations of Chinese Christian history. China Inf. **33**(1), 46–65 (2019)
22. Zhu, J.W., Njing, X., Guo, P.L., Jiang, J.J.: Evaluate the Efficiency of Water Conservancy Investment in Shaanxi Province Based on Super-Efficient DEA and Malmquis Index (2019)
23. Wu, D.: Evaluation and strategic conception of China's water conservancy green modernization process. China Popul. Resour. Environ. **25**(9), 114–123 (2015)
24. Mao, M., Ying, L., Yang, X.: Evaluation of the development of water conservancy tourism resources in Zhejiang province. Bull. Sci. Technol. **18**(3), 213–218 (2002)
25. Yi, M., Xia, Q., Zhang, X.: Grey correlation analysis of farmers' demand priority for the rural water conservancy facilities based on entropy weight method. Guangdong Agric. Sci. **41**(8), 223–227 (2014)
26. Wang, C., Chen, R.: Haihe conservancy and temporal-spatial variation of the channel in modern times. Acta Sci. Nat. Univ. Pekin. **51**(6), 1141–1148 (2015)
27. Song, Y., Yuan, L., Wu, D., Ba, C. (eds.): Impact Analysis of Replacing Business Tax with Value-added Tax on Water Conservancy Project Cost, pp. 271–274 (2017)
28. Zhang, X.H., Kubota, K., Kubota, M., Li, K.: Designing blended learning environments with thinking tool strategies: examining a Chinese teacher's decision-making and beliefs. Interact. Learn. Environ. **29**(2), 301–314 (2021)
29. Wu, D.: Modeling an evaluation method of water conservancy development in Huaihe River Basin. Resour. Sci. **38**(7), 1323–1335 (2016)
30. Jingli, H.E., Ende, X., Yanping, L.I.U.: Water conservancy and water conservation control pattern in Keshi Keteng banner in inner Mongolia. Res. Soil Water Conserv. **14**(6), 143–145 (2007)
31. Liu, D.F.: The image impact of mega-sporting events perceived by international students and their behaviour intentions. Int. J. Sports Mark. Spons. **16**(2), 103–117 (2015)
32. Fan, C., Chen, Y., Zhu, Y., Zhang, L., Wu, W., Ling, B., Tang, S.: Using fuzzy comprehensive evaluation to assess the competency of full-time water conservancy emergency rescue teams. Mathematics **10**(12) (2022)
33. Rui, X., Liu, N., Zhu, J.: Understanding of some problems in development of water conservancy and water conservancy science. Adv. Sci. Technol. Water Resour. **31**(5), 1–4 (2011)
34. Zhang, H.-X., Wang, Y.-T., Liu, L., Lei, W., Zhou, Q.-H.: Modern water conservancy system in Guizhou, pp. 1100–1104 (2015)
35. Zhang, J., Liu, J., Jin, J.: Understanding and thinking of smart water conservancy. Hydro-Sci. Eng. **6**, 1–7 (2019)
36. Xie, N. (eds.): Three Dimensional Visualization Technologies for Water Conservancy Works, pp. 1809–1812 (2012)
37. Chen, J., Zhu, K., Li, Y.: Theoretical research on management mode for rural water conservancy. J. Hohai University. Nat. Sci. **32**(1), 95–99 (2004)
38. Wen, C. (ed.): The Study on Water Conservancy Project Control System, pp. 1996–1999 (2012)
39. Qian, Z.: New conception in Chinese water conservancy-harmonious coexistence of humanity with nature. J. Hohai Univ. Nat. Sci. **32**(3), 243–247 (2004)
40. Liu, B., Wang, X., Gao, L., Ieee. (eds.): Overview of the Application of Artificial Intelligence in Several Key Issues of Water Conservancy, pp. 7458–7465 (2021)
41. Jiang, C., Wang, J.: Recent advances of ecological water conservancy in China. Adv. Sci. Technol. Water Resour. **35**(5), 168–175 (2015)
42. Nan, T.: Study on Water Conservancy and Hydropower Engineering Risk, pp. 442–445 (2016)

43. Ma, Y., Tong, Y.: A study on the spatial structure features of water conservancy tourism resources and self-driving accessibilitytaking the national water conservancy scenic spots in city clusters along the middle reaches of the Yangtze river as an example. Resour. Environ. Yangtze Basin **25**(8), 1167–1175 (2016)
44. Ma, Z., Xie, J., Ieee (eds.): Study on the Mobile Water Conservancy Service Platform Based on Ontology, pp. 5043–5048 (2008)
45. Deng, X., Chen, J., Cui, S.: Study on the management performance evaluation model of water conservancy agent construction project based on SCP. South-to-North Water Transf.S Water Sci. & Technol. **12**(5), 183–186 (2014)
46. Li, J., Zheng, F., Feng, Z.: Study on the improvement path of water resource utilization efficiency from the perspective of public products. Resour. Sci. **41**(1), 98–112 (2019)
47. Yang, C., Iop (eds.): A Study on the Importance of Water Conservancy Engineering Design (2018)
48. Cuiling, B.A.I.: Study on the development of tourism resources of water conservancy. J. Anhui Agric. Sci. **36**(14), 6013–6015 (2008)
49. Chen, J., Wang, Q., Destech Publicat, I. (eds.): Study on the Countermeasures of Water Conservancy Talent Team's Construction under the New Normal Background in China, pp. 1909–1912 (2015)
50. Yi, X., Dongguo, S., Rui, D., Yunqing, W.U.: Study on support system for comprehensive evaluation of water conservancy projects. Eng. J. Wuhan Univ. Engineering Edition. **40**(4), 49–51,54 (2007)
51. Xing, M., Yang, L. (eds.): Study on Standard of Farmland Water Conservancy Engineering in Mine Reclamation-Area, pp. 993–997 (2010)
52. Song, Y., Tian, L., Han, X., Feng, S. (eds.): A Study on Optimizing the Water Conservancy Landscape Design by Applying AHP, p. 3328 (2011)
53. Huang, L., Zhu, J., Zhang, K.: Study on multi-stage risk assessment of construction quality of water conservancy project. Water Resour. Hydropower Eng. **48**(9), 117–125 (2017)
54. Zhang, Y., Luo, W., Yu, F.: Construction of Chinese smart water conservancy platform based on the blockchain: technology integration and innovation application. Sustain. **12**(20) (2020)
55. Zhang, R., Shi, G. (eds.): Study on Evaluation of Submersed Cultivated Land Resources Caused by Water Conservancy and Hydroelectric Engineering, pp. 1421–1426 (2008)
56. Zhou, X., Zhang, H., Xie, X., He, H.: Study on evaluation model for water conservancy guarantee capacity of Liaoning province. Water Resour. Hydropower Eng **46**(3), 12–16 (2015)
57. Wu, S., Xie, J., Ma, B., Ieee. (eds.): Research on Support Technologies of the Management Informationisation for Water Conservancy Project Construction, pp. 12352–12355 (2008)
58. Duan, H., Han, K., Zhao, H., Jiang, Y., Li, H., Mao, W.: Research on water conservancy comprehensive knowledge graph construction. J. Hydraul. Eng. **52**(8), 948–958 (2021)
59. Ji, C., Li, F., Xing, J., Cui, X., Ieee. (eds.): Research on the water conservancy and hydropower system engineering based on ADHALL theory (2010)
60. Li, S., Chen, X., Zhang, X., Xie, J., Wang, X., Zhang, B.: Research on the Overall Framework of Intelligent Water Conservancy Construction. In: Wenzheng, L., Jixing, Y. (eds.) Proceedings of 2021 Ieee 11th International Conference on Electronics Information and Emergency Communication, pp. 145–150 (2021)
61. Wang, Y., Fu, H., Zhang, S., Jiapaer, N.: Research on the comprehensive evaluation of the value of water conservancy heritage based on expert knowledge and public cognition: a case study of Red Flag Canal. Water Policy (2022)
62. Wei, N., Lu, K.M., Xie, J.C., Lin, M.K. (eds.): Study on Evaluation Index System for Ecological Effect of Water Conservancy Projects (2019)
63. Yan, C.: Study on economic assessment of water conservancy construction project based on sustainable development. J. Anhui Agric. Sci. **34**(20), 5159–5160,5164 (2006)
64. Zhen, F.: Study of Shanxi province rural water conservancy facilities based on factor analysis. Math. Practice Theory **41**(13), 40–44 (2011)
65. Chen, J.-F., Wang, H., Lin, W. (eds.): Risk Evaluation of Water Conservancy Investment Projects based on TOPSIS Model, pp. 281–283 (2008)

66. Chen, Y., Duan, X., Huang, C.: A Study of Relationship between Water Conservancy Facilities and Grain Yield Based on VAR Model, pp. 1935–1939 (2017)
67. Zhang, Y., Yu, S., Lu, M.: Study of index weight of evaluation index system for water conservancy modernization in northern Jiangsu province. J. Hohai University. Nat. Sci. **42**(2), 171–176 (2014)
68. Chang, C., Wen, Z., Shu, L., Lu, C., Gu, S., Su, J., He, M., Xing, K. (eds.): Study of an Evaluation Index System of Well-Off Water Conservancy in Yunnan Province, pp. 366–372 (2014)
69. Gao, C., Wang, A., Guo, X., Wang, S.: The structure and development of regional ecological water conservancy economic system. J. Coast. Res. pp. 65–69 (2020)
70. Feng, G.Z. (ed.): Strategies for Sustainable Water Resources Management in Water Scarce Regions in Developing Countries, pp. 107–112 (2000)
71. Chen, J.-J. (ed.): Some Thoughts on the Construction of Water Conservancy Project management, pp. 1873–1877 (2013)
72. Sun, J., Chen, C., Yan, Z., Wang, H., Zhu, Q., Zhang, M.: Rural water conservancy modernization based on environment-friendly agriculture. South-to-North Water Trans. Water Sci. & Technol. **11**(4), 161–165,205 (2013)
73. Xiaoping, S.H.I., Peng, W., Tianxing, Y.U.: A simulation system of water conservancy project. Comput. Simul. **25**(6), 270–273 (2008)
74. Qi, Y.: The language of terror: exploring speech acts in official english-language ISIS videos, 2014–2017. Small Wars & Insur. **31**(6), 1196–1241 (2020)
75. Qi, Y.: Illuminating terror: content analysis of official ISIS english-language videos from 2014 to 2017. Behav. Sci. Terror. Polit. Aggress. 1–29 (2020)
76. Qi, Y.: Unending capitalism. How consumerism negated China's communist revolution. Eur. Asia Stud. **74**(4), 703–704 (2022)
77. Qi, Y., You, K., Guo, S.: The battle for public opinion of "Japan's nuclear wastewater sea discharge." Open J. Polit. Sci. **12**(3), 363–372 (2022)
78. Zhao, Y., Shen, Y., Qi, Y. (eds.): A Security Analysis of Chinese Robot Supply Chain Based on Open-Source Intelligence, pp. 219–222. IEEE (2021)

Urban Planning, Construction, and Sustainable Development

Research on the Application of Comprehensive Geophysical Methods in Tunnel Investigation

Yong Hai He and Hong Qiang Zhang

Abstract In the construction of highway tunnel engineering, it is of great significance to do a good job in the preliminary investigation. With the geological conditions of highway tunnel becoming more and more complex, it is difficult to meet the requirements of design and construction only by drilling and one geophysical method. The shallow buried section of the highway tunnel portal adopts a high-density electrical method and the deep buried section adopts the EH4 geophysical method. The survey work is based on geophysical exploration, combined with geological mapping and drilling, which can provide accurate geological data, effectively reduce geological diseases, reduce project cost and ensure the safety of traffic after operation. This method is worth using for reference.

Keywords High density electrical method · EH4 · Highway tunnel · Geological diseases

1 Introduction

At present, the exploration of tunnels in China is still based on drilling methods, but the geophysical methods developed in recent years are playing an increasingly important role. Commonly used geophysical methods include seismic reflection method, electromagnetic sounding method, elastic wave CT method, high-density electrical method, electromagnetic field method, etc. Among them, the EH4 method and high-density electrical method have the advantages of high efficiency, stable and reliable data acquisition, and intuitive and clear imaging maps [1, 2].

Although the exploration depth of the high-density electrical method is limited, it has high resolution in the shallow layer and can accurately and quickly detect the karst caves, faults and other geological structures in the shallow buried section of the tunnel portal. The results play a great guiding role in the prevention and treatment

Y. H. He (✉) · H. Q. Zhang
Hebei Provincial Communications Planning, Design and Research Institute Co. LTD, Shi Jia Zhuang 050011, China
e-mail: zhang023541@alumni.tongji.edu.cn

E. Strauss (ed.), *Proceedings of the 7th International Conference on Civil Engineering*, Lecture Notes in Civil Engineering 371, https://doi.org/10.1007/978-981-99-4045-5_13

of geological disasters such as landslide, collapse and roof fall at the tunnel entrance [3, 4].

The measurement depth of EH4 equipment is about tens of meters to 3000 m. This method can be used to detect faults, water content and other geological conditions in the deep buried section of the tunnel. Combined with the drilling method, the engineering geological conditions of the whole tunnel can be accurately judged [5, 6].

Taking the Yuerya tunnel of national highway G508 as the engineering background, this paper explains the application of the high-density electrical method and the EH4 geophysical prospecting methods in tunnel investigation.

2 Techniques and Principles of the Geophysical Methods

2.1 High Density Resistivity Method

The high density resistivity method is to infer the formation change by measuring the resistivity change of the measured section and analyzing the geological drilling data. Different formation lithology has certain electrical differences, which can be inferred and interpreted according to electrical differences [7, 8].

The tunnel adopts the high-density resistivity method, the measurement and the observation system adopts Wenner device, and the interpretation section is an inverted trapezoid. The distance between the measuring points is 5 m, and the maximum distance between the power supply poles is 600 m. The field recorded data is replayed to the computer. After inspection and calibration, it is processed by the professional processing software. The processing work mainly includes section resistivity contour drawing, forward and inverse interpretation, drawing the forward and inverse interpretation color section map, and interpreting the resistivity section map according to the resistivity contour map and forward and inverse interpretation. We carried out geological interpretation in combination with drilling data and drew the geophysical interpretation geological profile as shown in Fig. 1.

2.2 EH4 Magnetotelluric Method (Acoustic Magnetotelluric Method)

The EH4 magnetotelluric system is applicable to various geological conditions and harsh field environment. The method principle is the same as the traditional MT method. It uses the natural electromagnetic field signals incident on the earth such as solar wind and lightning in the universe as the excitation field source, also known as the primary field. The primary field is a plane electromagnetic wave, which is vertically incident into the earth medium. According to the electromagnetic field

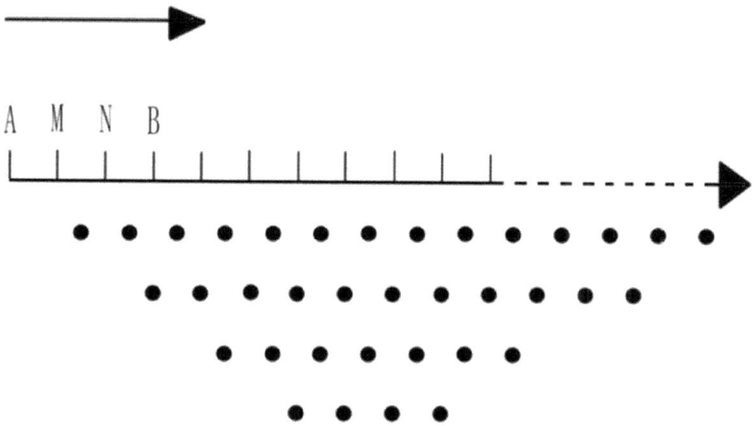

Fig. 1 Schematic diagram of high density resistivity method device

theory, the induced electromagnetic field will be generated in the earth medium, The induced electromagnetic field has the same frequency as the primary field, and the wave impedance Z is introduced. In the case of uniform earth and horizontal layered earth, the wave impedance is the ratio of the horizontal component of electric field E and magnetic field H. As shown in Eqs. (1)–(3) [9, 10].

$$Z = \left|\frac{E}{H}\right| e^{i(\varphi_E - \varphi_H)} \tag{1}$$

$$\rho_{xy} = \frac{1}{5f}\left|Z_{xy}\right|^2 = \frac{1}{5f}\left|\frac{E_x}{H_y}\right|^2 \tag{2}$$

$$\rho_{yx} = \frac{1}{5f}\left|Z_{yx}\right|^2 = \frac{1}{5f}\left|\frac{E_y}{H_x}\right|^2 \tag{3}$$

In Eqs. 1−3, f is the frequency, the unit is Hz, ρ is the resistivity ($\Omega \cdot M$), E is the electric field strength (mv / Km), H is the magnetic field strength (nT), φ_E is the electric field phase, φ_H is the magnetic field phase, and the unit is mrad. It must be pointed out that E and H at this time should be understood as the comprehensive field after the superposition of the spatial tensors of the primary field and the induced field, which is referred to as the total field for short. In electromagnetic theory, when the electromagnetic field (E, H) propagates the earth, the depth when its amplitude attenuates to the initial value of 1/e is defined as the penetration depth or skin depth (δ), as shown in Eq. (4).

$$\delta = 503\sqrt{\frac{\rho}{f}} \tag{4}$$

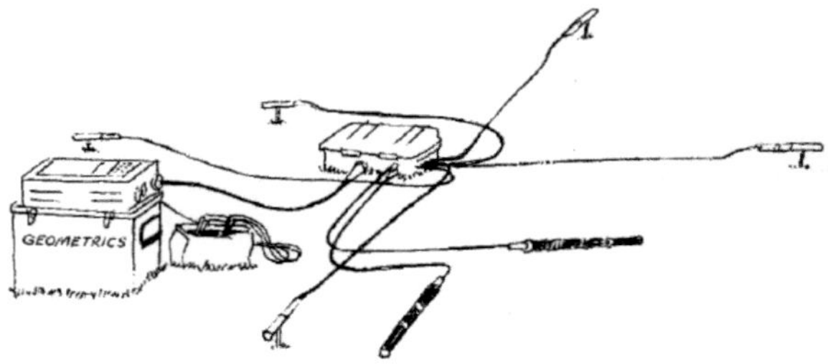

Fig. 2 Schematic diagram of EH-4 device for magnetotelluric sounding

According to Eq. (4), the skin depth (δ) will change with resistivity (ρ) and frequency (f), and the measurement is carried out in the frequency band corresponding to the underground research depth. Generally speaking, the data with higher frequency reflect the electrical characteristics of the shallow part, and the data with lower frequency reflect the characteristics of deep stratum. Therefore, the electric and magnetic field information is observed in a wide frequency band, and the apparent resistivity and phase are calculated. The geoelectric characteristics and underground structure of the earth can be determined, which is the simple method and principle of the EH4 observation system as shown in Fig. 2.

3 Project Overview

Yuerya tunnel of the reconstruction project from Yuerya to the large section of national highway G508 is located in the east of Yuerya Town, Kuancheng County, Chengde city. It is a separated long tunnel with right chainage of K1 + 365 ~ K3 + 596, length of 2231.0 m, design bottom elevation of 385.55 ~ 398.25 m and height difference of 12.70 m; Left section stake No.ZK1 + 383.0 ~ ZK3 + 635.0, 2252.0 m long, design tunnel bottom elevation 385.30 ~ 398.25 m, height difference 12.95 m; The cave body is curved on the plane and distributed in the southeast–southwest direction. The maximum buried depth of the tunnel is about 222.10 m.

The geophysical prospecting methods adopted this time are the high-density electrical method and magnetotelluric method. The instruments used in the high-density electrical method are the duk-2a high-density electrical method measurement systems produced by Chongqing Geological Instrument Factory of the China Decoration Group. The magnetotelluric method mainly uses stratagem EH-4 magnetotelluric sounder produced by the American laurel company.

3.1 Topography and Geomorphology

The proposed Yuerya tunnel is located in the east of Yuerya Town, Kuancheng County, Chengde city. It belongs to the middle and low mountainous area of Yanshan Mountains. The terrain of the tunnel area is undulating and the transportation is convenient. The ground elevation of the tunnel area is 376.1 ~ 640.0 m. The mountain slope at the entrance section is inclined to the west, with an overall slope of about 28°, and the terrain is steep; In the exit section, the mountain slope inclines to the south, with an overall slope of about 45°, the terrain is steep, the mountain vegetation is developed, and the intervisibility conditions in most sections are poor.

3.2 Geological Overview

The main stratum of the surrounding rock in the tunnel area is the Proterozoic great wall system Gaoyuzhuang Formation (CHG) dolomite. The rock stratum in the tunnel area has a monoclinic attitude, the attitude trend of the rock stratum is roughly W35° ~ 47° s, the dip angle is 38° ~ 48° NW, and tailings ponds are accumulated in some sections.

The groundwater is mainly bedrock fissure water, with less water content in rock mass, belonging to poor water area. There is no groundwater runoff.

3.3 Geophysical Characteristics of Survey Area

The tunnel area is covered with quaternary diluvium (Q4pl + dl) soil layer and gravel (including soil), which is thin. The tailings pond in the valley of K2 + 480.0 ~ K2 + 750.0 and K3 + 250.0 ~ K3 + 320.0 sections is accumulated, with a thickness of 5 ~ 50 m. The underlying bedrock is mainly dolomite of Gaoyuzhuang Formation of the great wall system.

According to the previous data, there are obvious electrical differences between the Quaternary loose layer and the underlying bedrock, and there are also electrical differences between different geological lithologies. If there are fault structures and rock fracture zones, the rocks within the fault structure zones are broken, and there are electrical differences with the intact rocks, which provides a geophysical premise for the application of the high-density resistivity method and the magnetotelluric method.

4 Data Interpretation and Result Analysis

4.1 Interpretation Method

The high-density resistivity method preprocesses the collected data by using the data processing function in the two-dimensional resistivity inversion imaging software, inputs the ground elevation data of the survey line for topographic correction, and then reasonably sets the parameters necessary for interpretation in the software according to experience. After these preparations are completed, run the software for inversion interpretation. During the interpretation process, adjust the parameters at any time to make the results true and reasonable. The calculation results are marked with the resistivity chromaticity diagram for comprehensive interpretation.

Firstly, the magnetotelluric method uses image and "HMT (EH4) data processing system" series software to process the data. In the processing process, the field data are first processed by non-value elimination and noise removal, and then one-dimensional and two-dimensional inversion imaging. Then, combined with the ground elevation data, the interpretation depth is converted to the absolute elevation, and then the surfer8 software is used for post-processing, to generate a color section view.

4.2 Result Analysis and Geological Interpretation

In combination with the site topographic conditions, vertical and horizontal high-density resistivity survey lines are arranged at the shallow buried section at the tunnel inlet, and vertical and horizontal EH4 survey lines are arranged at the large buried depth of the tunnel body.

Results of the high density electrical method K1 + 340.0 ~ k1 + 490.0 longitudinal section: the surface layer of this section is quaternary upper Pleistocene proluvial and Deluvial gravel, the overburden is thin, and the underlying bedrock is Proterozoic Changcheng System Gaoyuzhuang Formation dolomite; The low resistivity anomaly at the bottom of K1 + 340.0 ~ k1 + 430.0 is speculated to be fracture development.

Drilling is carried out at K1 + 400. According to the verification of drilling data, the rock stratum is strongly ~ moderately weathered dolomite, the rock mass is relatively broken, massive, and the joint fissures are relatively developed. The geophysical results are in good agreement with the drilling data.

Results of Magnetotelluric EH4 The lithology of K1 + 480.0 ~ k1 + 560.0, K2 + 230.0 ~ k2 + 290.0 and K2 + 480.0 ~ k2 + 780.0 sections is Proterozoic Changcheng System Gaoyuzhuang Formation dolomite with thin overburden; The resistivity of geophysical prospecting is well layered, and the high and low resistances are evenly distributed; The apparent resistivity value at the tunnel excavation is low, and it is speculated that the rock mass is broken ~ relatively broken.

The lithology of K1 + 560.0 ~ k1 + 660.0 and K1 + 880.0 ~ k2 + 060.0 sections is dolomite of thw Gaoyuzhuang Formation of Proterozoic Changcheng System, and the overburden is very thin; The resistivity of geophysical prospecting is well layered, and the high and low resistances are evenly distributed; The apparent resistivity at the tunnel excavation is relatively high, and it is speculated that the rock mass is relatively broken ~ complete.

The lithology of K1 + 660.0 ~ k1 + 880.0, K2 + 060.0 ~ k2 + 230.0 and K2 + 290.0 ~ k2 + 480.0 sections is dolomite of the Gaoyuzhuang Formation of Proterozoic Changcheng System, and the overburden is very thin; The resistivity of geophysical prospecting is well layered, and the high and low resistances are evenly distributed; The apparent resistivity value at the tunnel excavation is relatively high, and it is speculated that the rock mass is relatively complete ~ complete.

Drill holes at K2 + 240.0 and K2 + 950.0. The drilling data show that the rock stratum is strongly ~ moderately weathered dolomite, the rock mass is relatively broken, massive, and joint fissures are developed. The geophysical results are in good agreement with the drilling data.

5 Discussion

The tunnel adopts the high-density resistivity method and the magnetotelluric sounding method. The technical method is reasonable and the resistivity characteristics are obvious, which can better reflect the actual geological situation.

The Quaternary strata in the tunnel area are distributed on the surface layer in the form of alluvial proluvial and Deluvial proluvial. The lithology is mainly gravelly soil, with a thickness of about 0.5 ~ 3.0 m, and the thickness of the local slope toe is up to 5.0 m. Tailings pond in the valley of K2 + 480 ~ K2 + 750 and K3 + 250 ~ K3 + 320 sections is accumulated, with a thickness of 5 ~ 50 m. The underlying bedrock is mainly dolomite of Gaoyuzhuang Formation of the great wall system, mainly strongly ~ moderately weathered, The buried depth of the weathered layer basically fluctuates with the terrain, and the rock mass at some valleys is broken.

In this EH4 geophysical exploration, the apparent resistivity at K2 + 230 ~ K2 + 290 and K2 + 910 ~ K3 + 000 longitudinal faults is relatively low and steep, so it is speculated that the rock mass is broken or small faults are distributed; The apparent resistivity of K1 + 340 ~ K1 + 430 high density is low at the bottom, and it is speculated that the rock mass fractures are developed; After drilling verification, the rock stratum is strongly ~ moderately weathered dolomite, the rock mass is relatively broken and massive, and there is a risk of block falling and collapse during excavation. It is recommended to make advance geological prediction and advance support before excavation, control the excavation footage during excavation, timely anchor and shotcrete support, close the surrounding rock, strengthen monitoring and measurement, and ensure the safety of construction and operation.

6 Conclusion

Nowadays, with the rapid development of highway engineering construction in China, due to the increasingly complex geological conditions in the crossing area, highway tunnel survey has always been a technical difficulty. The traditional single drilling method is highly uncertain and the geological condition survey is unclear, which often leads to tunnel collapse and a large number of changes in the later stage, increases the project cost, and even has potential safety hazards after opening to traffic. Therefore, the combination of the geophysical prospecting method and the traditional drilling method in the highway tunnel survey can greatly improve the accuracy of survey data, and provide escort for future construction safety, cost reduction and operation safety after opening to traffic.

References

1. Hongyue, X., Xingjian, L., Wan, L.: Application of environmental geophysical technology in Karst exploration and its effect [J]. Sci. Disaster **22**(3), 58–62 (2007)
2. Sheng, Z., Xuheng, H., Bingqiang, L.: Application of high density resistivity method in Goaf survey [J]. Sci. Disaster **20**(4), 64–66 (2005)
3. Li, Y., Fei, L., Xiaonian, H.: Application and research of comprehensive geophysical exploration technology in permafrost survey of northeast highway engineering [J]. Sci. Disaster **33**(S1), 25–29 (2018)
4. Yongjin, C., Shanpu, T., Guanglong, P.: Application of comprehensive geophysical method in high altitude buried tunnel survey [J]. Subgrade Eng. **5**, 175–179 (2018)
5. Shu, L., Yong, P., Yangfei, L.: Applicability analysis and application of high density electrical method and audio frequency magnetotelluric method in highway tunnel investigation [J]. West-China Explor. Eng. **4**, 164–167 (2018)
6. Huang, J., W, B., Li, H., et al.: The application of EH4 high-frequency magnetotel-luric sounding method to highway tunnel exploration [J]. J. Sichuan Geol. **40**(3), 495–499 (2020).
7. Xie, K., Xing, L., Xie, Y., et al.: Application and comparative analysis of high-density electrical method and transient electromagnetic method in engineering investigation [J]. Yunnan Metall. (5), 89–92 (2018).
8. Su, P.: Research on the application of comprehensive geophysical prospecting in the detection of railway subgrade and tunnel diseases in karst areas [D]. Chengdu University of Technology (2016).
9. Xin, H.: Discussion on the application of high density electrical method in engineering geophysical prospecting [J]. West. Resour. (5), 095–097 (2021).
10. Huiming, L., Hui, W.: Application of comprehensive exploration technology in ExPressway along the Yangtze river [J]. Eng. Constr. **6**, 803–805 (2009)

Experimental Assessment of Leakage in Water Distribution Network

S. Atabay, T. A. Ali, Md. M. Mortula, and S. Sharifi

Abstract Water is one of the most important and scarce resources needed to sustain life. Wasting this invaluable resource will have negative impacts on humans and the environment. Producing potable water requires significant energy and hence a huge budget, therefore water leakage from distribution networks could lead to financial instability. To prevent this, water leakage must be detected, minimized, and controlled. This paper presents a dimensionless mathematical model that relates pressure drop to leakage in a water distribution system. The relationship was achieved through an experimental setup at the American University of Sharjah. The dimensionless relationship between pressure drops percentile and the flow ratio were obtained exponentially. The proposed dimensionless model correlated well with all experimental data and calculated the pressure drop with the amount of leakage within 10% accuracy in all pipe diameters. This model provides a basis for an easy and efficient method to detect leakage in pipe systems without the need for expensive equipment.

Keywords Water distribution network · Leakage and pressure drop

1 Introduction

Leakage in water distribution networks (WDNs) has become a progressively significant problem that all countries, including the most developed ones, are encountering. Leakage is causing big losses in water resources every year in addition to poor quality of service. Thus, it is vital to reduce water leakage to ensure sustainable water use. In highly developed and densely populated countries, water tends to be lost more because of the large consumption amounts associated with high usages and aging

S. Atabay (✉) · T. A. Ali · Md. M. Mortula
Department of Civil Engineering, American University of Sharjah, P.O. Box 26666, Sharjah, United Arab Emirates
e-mail: satabay@aus.edu

S. Sharifi
Department of Civil Engineering, School of Engineering, University of Birmingham, Birmingham B15 2TT, UK

E. Strauss (ed.), *Proceedings of the 7th International Conference on Civil Engineering*, Lecture Notes in Civil Engineering 371, https://doi.org/10.1007/978-981-99-4045-5_14

of infrastructures. Water leakage is not just the loss of a vital resource; it also has detrimental effects not only on the economy [1] but also on the environment since the location of the leak becomes a potential entry point for contaminants [2].

Leakage occurs due to faults in pipe fittings and aging of pipes, which are both inevitable causes [3]. Hence, the problem cannot be eliminated. However, a partial solution to this problem can be attained, by controlling the amount of water losses in the distribution networks [4]. The first step towards this aim is detecting leakage locations. The methods currently implemented for leakage detection involve the use of equipment that are relatively expensive and inefficient [4, 5]. Some analytical methods have been discussed in the literature; however, these proposed models are designed to estimate the amount of leakage occurring due to the existing pressure at a point before the leakage location [6]. Therefore, there is a need to achieve a mathematical relationship that relates leakage amount with pressure drop. This study therefore focused on the experimental study and development of a mathematical model in which the dimensionless relationship between pressure drops percentile and the flow ratio were obtained. This will help to estimate the significance of the water loss and identify the locations of water leaks in the water distribution network. Some of the previous studies related to the mathematical model are briefly summarized below.

The most prominent model used for leak detection is the Fixed and Variable Area Discharges model (FAVAD), which shows the relationship between the leak area and pressure as shown in Eq. 1 [7].

$$Q = C_d\sqrt{2g\left(A_o h^{0.5} + mh^{1.5}\right)} \quad (1)$$

As seen in the model, the main components contributing to leakage are the leakage flow (Q), initial leak area (A_o), pressure head (h), slope of pipeline (m), and discharge coefficient (C_d). The model shows that the flow rate varies with the square root of the pressure head. However, the unspoken assumption that C_d is constant might be invalid [8], since C_d can change depending on whether flow is laminar, transitional or turbulent which depends on Reynolds number ($R = V^*H_d/K_v$), where K_v is the kinematic viscosity and H_d is the diameter of the orifice [8]. The diameter of the orifice represents the leak diameter. A study conducted using finite element analysis to investigate the effect of pressure on the leak diameter showed that as pressure is applied on the holes, the diameter of the leak increases [9]. This implies that leakage amounts will increase overtime due to the increase in C_d as discussed earlier by Lambert [8]. It is also noted that this problem is more significant in uPVC pipes as the material is significantly affected by pressure and leak diameter increases exponentially [7].

Aburawe et al. [10] studied the changes in hydraulic behavior of the network as a result of changes in leakage volumes and locations. An analytical approach was suggested to identify leakage locations based on computer modeling. The first step entailed working with the network laboratory setup, which involved the identification of leakage locations as nodes in the network and assignment of leakage volumes as demands the measurement of the resulting pressure. The second step consisted of modeling the network with the volume of leakage, then the field and the model pressures were compared. Nodes of matching pressures are deemed as leakage nodes and the pipe after the node is the leakage pipe. To further improve the accuracy, an extra node was suggested in the leakage pipe and the comparison was again conducted to determine whether the leakage pipe is after the virtual node or before it. This was done repetitively until the accuracy of the location was satisfactory [10] and this procedure gives a good estimation of the location of leakage.

This paper presents experimental results between the amount of leakage and pressure drop in a network. It serves two purposes: the first of which is the detection of leakage locations in pipe systems. It is expected that points of significant pressure drops are at the potential leakage locations; thus, they can be further inspected. The second is to find the amount of leakage which gives an indication of the significance of water losses from the WDN. The ultimate aim behind this project was to achieve an empirical mathematical model (in dimensionless form) that will aid the process of leakage detection.

2 Experimental Apparatus and Procedure

To achieve the objectives, several tests were carried out using an experimental setup that was constructed using uPVC pipes as shown in Fig. 1.

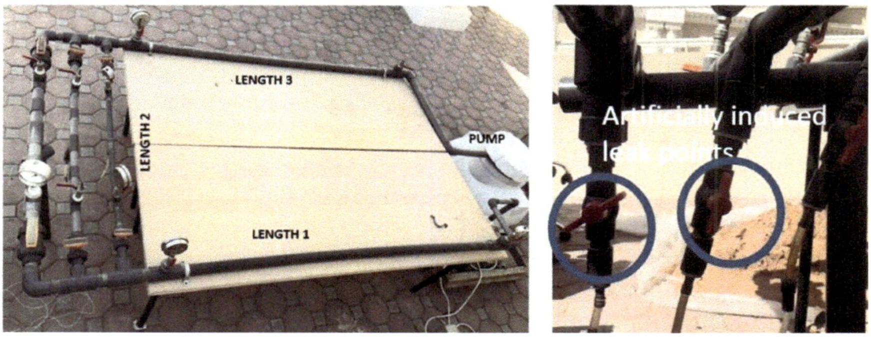

(a) (b)

Fig. 1 Experimental setup; **a** Pipe network with pressure gauges, **b** Artificially induced leakage points along Length 2

2.1 Experimental Program

The experimental setup consisted of a pump with a water tank to supply the flow to the network, ten digital pressure gauges, and six valves. Figure 1a and b show the pipe network with pressure gauges along the network and artificially induced leakage points along Length 2 respectively. Two pressure gauges were located along the Length 1 (upstream side of the system), two along the Length 3 (downstream side of the system), and two along Length 2. Three pipes with different diameters of 0.75″, 1.0″ and 1.5″, referred to D_1, D_2 and D_3, respectively, were used in Length 2 at which an artificially leakage point was constructed in each of these diameters (Fig. 1b). Two valves, located at the upstream and downstream of each different pipe diameter in Length 2, were used to control the flow direction. Two pressure gauges were used along each of the pipe diameters before and after the point where leakage is simulated in Length 2. Three different tubes with the size of 7, 6 and 5 mm were used to simulate the leakage and referred by L_1, L_2, and L_3, respectively, and the amount of leakage was controlled with these three different tube sizes.

2.2 Experimental Techniques and Procedures

The flow (F) into the system was provided from a water tank using a pump and it was measured digitally using a digital flow meter and manually. The experiments were run until the flow reached steady state flow conditions i.e. where pressure variation was not significant within the system. The pressures (P) were then measured using the digital pressure transducer at six different points as shown in Fig. 1 and the minimum three different pressure readings were recorded. All these data represent the pressure in the system without any leakage. Then, the valve at the point where the artificial leakage induced along Length 2 as seen in Fig. 1b was opened to simulate the leak in the system, the flow at the leak point (F_L) and the pressure was recorded especially at those points immediately before (P_B) and after leakage (P_A) points. The amount of leakage was measured manually, and meanwhile instantaneous pictures were taken of the pressure gauges.

The pressures, and flow before (P_B, F_B) and after the leakage (P_A, F_A) were then measured. The flow was measured before leakage three times for each combination of flow, diameter, and leakage size. The three flows were averaged to find the flow before leakage. Pressures were also read three times with each flow measurement, however, each reading entailed three readings to be jotted down off the pressure gauge; that was done for higher accuracy, since the pressure kept fluctuating within a certain pressure range, and it was not fixed at one reading. The three readings were averaged to find the pressures for each flow measurement and then the three averages were averaged again as those averages were the pressures used to analyze the data.

The second part of the experiment involved changing the amount of leak, as explained in the procedure, which was done along Length 2 in Fig. 1 for each pipe. Once the leakage was introduced in the system, leakage flow was measured once, while outflow was measured three times which were averaged. Three pressure readings were taken that were also averaged.

3 Results

It was observed that pressures within the entire system are affected by leakage, since the system is very small relative to a real water supply system. This means that the pressure drop is not only limited to the point just before the leakage but to all points surrounding the point of leakage. However, the largest drops occur in the vicinity of the leakage (P_3 & P_4) along Length 2 in Fig. 1.

This correlation or mathematical model was obtained in a dimensionless form (such as P_B/P_A and F_B/F_A) to enable the use of this model for prediction of leakage in real WDN. Several attempts were conducted to provide a dimensionless relationship between the amount of leakage and pressure drop. It should be noted that the largest drops occur in the vicinity of the leakage (P_3) just immediately upstream of the leakage point, and the analysis was focused on this pressure drop.

The best relation for all different pipe diameters was obtained between the pressure drop percentile in P_3 (P_{3B}-P_{3A}/P_{3B}) and relative flow ($Q_r = Q_L/Q_A$). This dimensionless relationship using the data along D1 with different flow and three different leak diameters (L_1, L_2, and L_3) is shown in Fig. 2. The best fit to the data points was concluded to be an exponential with all pipe diameters of D_1, D_2 and D_3 with determination coefficients of 0.89, 0.93 and 0.92 respectively.

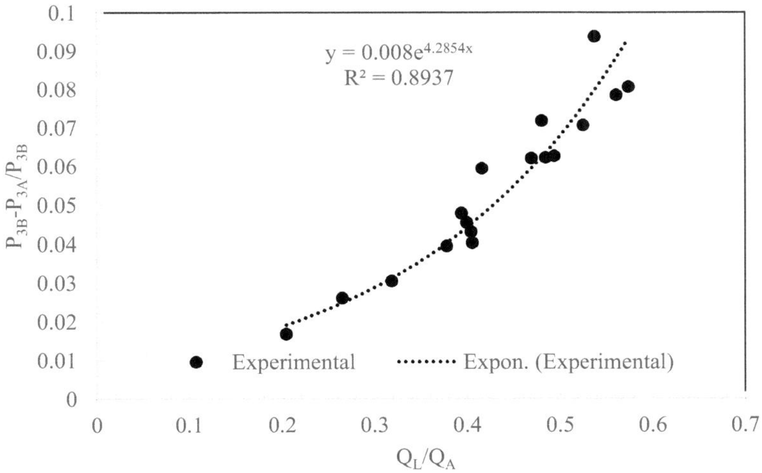

Fig. 2 Pressure drop percentile versus Q_r using all the data collected for D_1

4 Discussion of Results

After concluding the exponential relationship between pressure drop percentile and relative flow (Qr) all different pipe diameters, Eq. 2, as a single formulate that fits well with all data set, is proposed.

$$\% P_3 \text{ drop} = \frac{P_{3B} - P_{3A}}{P_{3B}} = 0.0077 \frac{D_1}{D_x} e^{4.113\left(\frac{Q_L}{Q_A}\right)} \quad (2)$$

where, $\%P_3$ is the percentile drop in pressure at just immediately upstream of the leakage point, D_1 is the diameter of the main pipeline, 1.5inc here, D_x is the diameter of pipe where the leakage occurred, Q_L is the amount of leakage and Q_A is the total discharge after leakage. The letters A and B refer to after and before leakage respectively.

The proposed mathematical model, as given in Eq. 2, was compared with all experimental data sets and they are shown in Figs. 3,4 and 5. Comparison between experimental results and the proposed mathematical model for D3 for the data collected along D_1, D_2 and D_3 respectively. All these Figures show very good correlations between the proposed model and all data sets regardless of pipe diameters and the amount of leakage. The proposed model calculated the pressure drop just immediately upstream of the location of the artificially induced leakage within 10% accuracy in all pipe diameters tested in this study. This model is very simple and efficient enough to detect leakage in pipe systems without the need for expensive equipment.

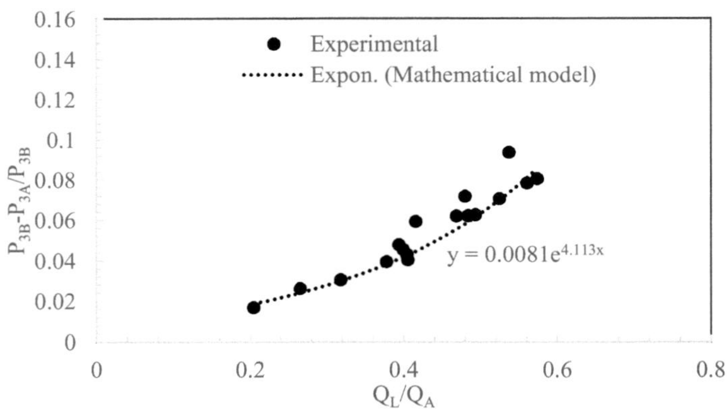

Fig. 3 Comparison between experimental results and the proposed mathematical model for D_1

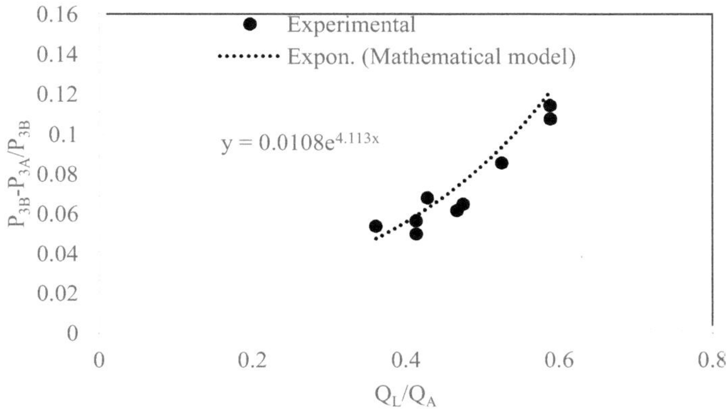

Fig. 4 Comparison between experimental results and the proposed mathematical model for D_2

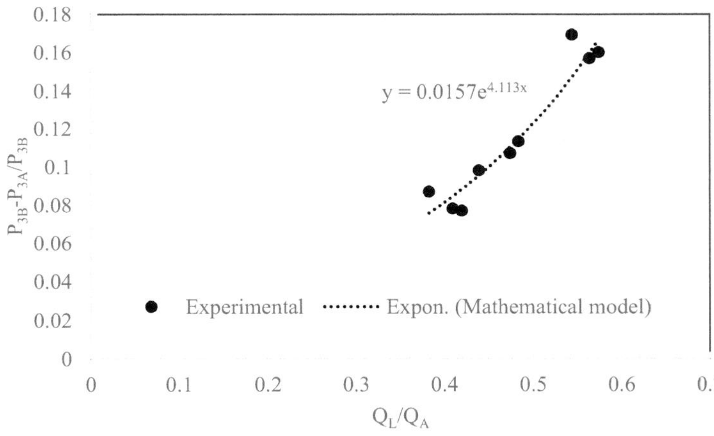

Fig. 5 Comparison between experimental results and the proposed mathematical model for D_3

5 Conclusion

This manuscript presents experimental results between the amount of artificially induced leakage and pressure drop in a simple water distribution network. A dimensionless relationship between amount of leakage and pressure drop was proposed. This proposed method is successful in quantifying the amount of leakage using the pressure drop percentile. The location of leakage could also be identified by the largest pressure drops. This research has shown that the relationship between the pressure drop percentile right before leakage and the relative leakage to total flow ratio is exponential. It should be noted that the proposed dimensionless model presented in this paper has proven to be quite accurate as the pressure drop percentile is calculated within the accuracy of 10% for different amounts of leakage and pipe diameters. The experiments were conducted using PVC pipes, clean water conditions, and certain pipe diameters.

Further research is therefore required for different pipe materials and dimensions to validate further the proposed model since the problem lies in determining the right coefficients. However, this proposed dimensionless relationship is very simple to use and will provide aid to people working with infrastructure water management to locate and quantify leakage based on the pressure drops.

References

1. Xu, Q., Liu, R., Chen, Q., Li, R.: Review on water leakage control in distribution networks and the associated environmental benefits. J. Environ. Sci. **26**, 955–961 (2014)
2. Amitangshu, P., Krishna, K.: Water flow driven sensor networks for leakage and contamination monitoring in distribution pipelines. ACM Trans. Sens. Netw. **15**(4), 1–43 (2019)
3. Hinaidi, O.: Detecting Leaks in Water-Distribution Pipes. National Research Council of Canada, Ottawa (2000)
4. Maninder P., Neil, D., Flint J.: Detecting & Locating Leaks in Water Distribution Polyethylene Pipes. In: Proceedings of the World Congress on Engineering, London (2010).
5. Aslam, H., Mortula, M.M., Yehia, S., Ali, T., Kaur, M.: Evaluation of the factors impacting the water pipe leak detection ability of GPR, infrared cameras, and spectrometers under controlled conditions. Appl. Sci. **12**, 1683 (2022).
6. Schwaller, J., van Zyl, J.E.: (2015). Modeling the pressure-leakage response of water distribution systems based on individual leak behaviour. J. Hydraul. Eng. **141**(5) (2015).
7. Schwaller, J., van Zyl, J.E.: Implications of the known pressure-response of individual leaks for whole distribution systems. Procedia Eng. **70**, 1513–1517 (2014)
8. Lambert, A.: What do we know about pressure leakage relationships in distribution systems?. In: Proceeding of IWA System Approach to Leakage Control and Water Distribution Systems Management (2001).
9. Oven, S.: Leak Detection in Pipelines by the use of State and Parameter Estimation, NTNU (2014).
10. Aburawe, S.M., Mahmud, A.R., Mohammad, T.A., Ahmed, N.: A laboratory based study of hydraulic simulation of leakage in water distribution networks. Civ. Environ. Res. **3**(12) (2013).

A Multi-task Oriented Optimization Method for Urban Rail Overhaul Workflow Based on Critical Chain Method

Shan Huang, Qin Luo, Jingjing Chen, and Tian Lei

Abstract With the expansion of the urban rail network scale, multi-tasking is becoming an inevitable trend. However, the Linear method of the workflow is still being adopted. The lack of time control leads to the low efficiency of project management, which further affects the daily operation of the metro. Regarding such a situation, this paper therefore aims to analyze the most unfavorable situation in the overhaul project arrangement of urban rail transit lines using the critical chain method. Taking the path with the longest processing time in the project as the critical path, the Critical Chain Project Management is established and calculated considering the resource constraint problem. Taking Shenzhen as an example to examine the effectiveness of the model, the results show that the critical chain method can significantly shorten the construction period and can make use of construction resources more effectively, which provides a guideline for multi-task scheduling of Urban Rail Overhaul projects.

Keywords Metro operation · Critical chain · Project scheduling

1 Introduction

With the rapid expansion of the urban rail transit system in big cities, the scale of the urban rail network has grown. As a result, the complexity of the Urban Rail Overhaul (URO) projects keeps increasing. The URO projects, as an essential part of the urban rail transit operation system, is critical to maintain the system's availability. With such situation, how to best arrange the URO workflow and optimize the resource usage becomes essential. For years, it is necessary to make appropriate plans with limited human resources, materials, and area resources so that the construction tasks

S. Huang · Q. Luo · J. Chen · T. Lei (✉)
College of Urban Transportation and Logistics, Shenzhen Technology University, Shenzhen, China
e-mail: leitian@sztu.edu.cn

E. Strauss (ed.), *Proceedings of the 7th International Conference on Civil Engineering*, Lecture Notes in Civil Engineering 371, https://doi.org/10.1007/978-981-99-4045-5_15

can be completed within the "Non-traffic Hour". Therefore, the URO projects can be classified as multi-tasking scheduling problems.

Multi-tasking problems aim at processing with efficiency and increasing the completion rate, and multi-tasking problems are now widely used in process industries such as cloud computing, industry as well as construction, etc. [1–5]. Multi-tasking problems can minimize the maximum completion time. However, their study focused on the completion time and rate, without considering the resource allocation required for URO projects. Therefore, as a multi-tasking problem, URO projects need to realize the resource allocation while optimizing the completion rate.

Aimed at achieving resource allocation in multi-tasking problems, the existing methods mainly use the Linear method [6–13]. The study of MTR [7] tried to schedule the overhaul plan through the Linear method. Hyun et al. [8] established a project information management system and analyzed the complexity of data from a data mining perspective to use Linear method for overhaul projects. Argyropoulou et al. [9] established an integer programming model to optimize the scheduling of maintenance tasks from the perspective of the passenger service level. The main problem of using the Linear method to realize multi-task scheduling is that it is inflexible as it is difficult to achieve efficient use of time, space and human resources. Thus the Linear method may not fully apply in URO projects.

As to this situation, network analysis techniques have been explored in recent years to solve multi-tasking problems, including Gantt Chart, Critical Path method (CPM) and Plan Review Techniques (PERT) [14–16]. For example, Muthamil et al. [17] used the CPM to solve the overhaul problem with time and operating space constraints. Delias et al. [18] proposed a workflow management system under resource constraints. Still, their proposed Resource Conflict Joint Optimization method, which is only applicable to highly repetitive operations, is partially applicable to URO projects. Araszkiewicz et al. [19] indeed claimed that the Critical Chain Project Management (CCPM) is better than CPM and PERT and can solve the Resource Constrained Project Scheduling Problem (RCPSP). Goto et al. [20] sought a method of buffer setting using mixed-integer linear programming. Zarghami et al. [21] analyzed many buffer setting methods and pointed out that the method based on activity attributes supports cooperation mode and can integrate many projects influencing factors. However, in solving URO management problem, these methods have net been applied and the construction efficiency of URO projects is still low.

Based on these considerations, this paper therefore aims to optimize multi-task workflow scheduling for URO projects using CCPM. Specifically, we apply Work Breakdown Structure (WBS) for the URO project and identify the critical path to complete the project. Based on the output, each subtask is then re-arranged. Then factors affecting the buffer arrangement were decomposed to allocate the buffer settings reasonably through principal component analysis. To evaluate the performance of the proposed method, the optimization results using CCPM are compared to those of the Linear method and CPM.

The rest of the paper is organized as follows, and we begin with the problem statement and a brief introduction of the theory and basic assumptions of the CCPM in solving the URO management problem. We then detailed the steps of the Critical Chain Method for overhauling management in metro operation, using Shenzhen metro overhauling management as an example. The subsequent section selects the most unfavorable situation in the subway system, optimizes the URO projects arrangement, and compares the optimization result with the traditional method. Finally, the paper draws the conclusions from the current work and points out the shortcomings of the research and the future research directions.

2 Materials and Methods

2.1 Problem Statement

Due to the characteristics of URO projects with many subtasks, large variation in duration and high extension cost, the often applied schedule management methods in the field of multi-tasking scheduling are the Linear methods. Since lack of consideration of resource factors, there often exits detachment of the plan from the site situation and a large amount of safety time is kept in the process while adopting these two methods, which further causes project delay due to the existence of Student Syndrome, Parkinson's syndrome and Murphy's law [22]. Through applying CCPM, the amount of project safety time can be used uniformly, and sufficient resources can be maneuvered to cope with the uncertainties and risks that may arise in the project. As a result, project efficiency can be improved and the risk of project failure can be reduced.

Multi-task workflow project scheduling problem is solved based on CCPM, which can be described as follows:

Given:

- S, the set of tasks, the URO projects consists of a combination of tasks. Each of these tasks can be clearly divided into subtasks. Each subtask has its own immediately preceding and immediately following process, which strictly follows the workflow of the directed acyclic diagram.
- R, the set of resources, each subtask has a specific, strictly delineated required human resources, materials, and area resources for the segment. A subtask can only proceed if the resource requirements of the subtask are met in the current segment area.

Determine:

- Process arrangement: to arrange the start point and end point of each subtask in order and arrange the buffer size and position based on CCPM output.

- Task allocation: to allocate the set of resources R to the set of tasks S.

Assuming:

- There is the phenomenon of not reporting the completion of work. Due to the unpredictable schedule of the operation, some construction directors may not report to their superiors after completion for fear of rejection by other employees, resulting in the phenomenon of delay.
- All projects are represented by directed acyclic graphs. Figure 1 illustrates the multi-task scheduling process in metro overhaul projects.

The URO projects have strict execution conditions and immediate pre-process arrangements, S_i is the start point of the ith $(i = 1, 2, 3, \ldots, I)$ task, without any directed edges pointing to it; D_i is the end point of the ith task and does not point to any vertex; E_m^j is the jth $(j = 1, 2, 3, \ldots, J)$ subtask of the mth $(m = 1, 2, 3, \ldots, M)$ level. In order to ensure the safety of people, vehicles and electricity, the approval

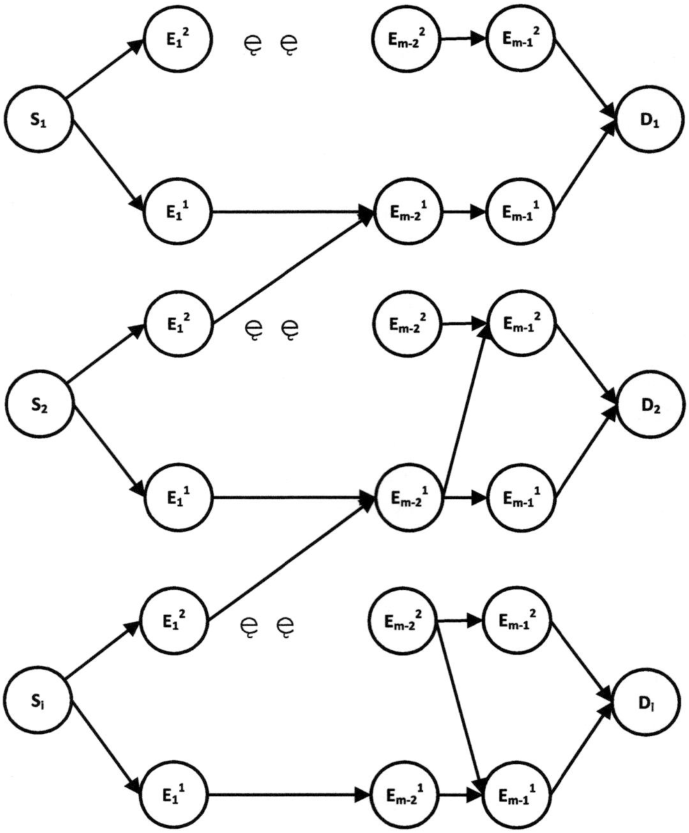

Fig. 1 Directed acyclic graph for the URO project

process can be parallel, while the execution process can only continue the subsequent work if the previous process is completed, for subtask E_m^j, there is at least one immediately preceding process E_{m-1}^j, E_{m-1}^{j+1},… , and the immediately following subtask E_m^j can start only when all immediately preceding processes E_{m-1}^j,… , are completed.

- It can monitor the project completion and resource utilization. The system can monitor the construction process and keep track of it in real time.

Objective:

Based on the above discussion, the objective of the given problem is to minimize the time to complete the set of tasks S within the specified construction time frame, i.e. the "Non-traffic Hour", to maximize productivity or minimize completion time.

Evaluation:

- The probability of completion; the probability that the combination of the set of tasks S can be completed within the duration given by the program. For the URO projects, it is usually considered that in the set of tasks S, each task is independent of each other and obeys the central limit theorem. The activity duration of each task independently and identically distributed approximately follow a normal distribution, i.e.:

$$P_s = P(t \leq T_s) = \int_0^{T_s} \frac{1}{\sigma_n\sqrt{2\pi}} e^{-\frac{1}{2}\left(\frac{t-T_n}{\sigma_n}\right)^2} dt \tag{1}$$

where, P_s is the probability of completion; T_s is the actual duration in the task set S; and T_n is the planned duration of the task.

2.2 Mathematical Formulation

The CCPM for the URO projects is proposed based on the above theories and assumptions. The algorithm is step by step as follows.

We first need to get the time of each subdivision. From established overhaul data the three-point estimation method [23] is adopted to obtain the most optimistic time a, namely the shortest time, the most pessimistic time b, the longest time, and the most likely time m, namely 50%-bit time and the immediate process. The expected time t is taken as the CCPM process time, namely:

$$e = (a + 4m + b)/6 \tag{2}$$

According to the CCPM theory, the difference between the expected time T and the most optimistic time a, Δt, is set as the safe time, i.e.

$$\Delta t = e - a \tag{3}$$

Identifying the critical chain is similar to critical path identification, but the difference lies in the optimal allocation of resources.

$$ES_i = \max\{ES_h + T_h\} \tag{4}$$

$$EF_i = ES_i + T_i \tag{5}$$

$$LS_i = \min\{LS_j - T_i\} \tag{6}$$

$$LF_i = LS_i + T_i \tag{7}$$

$$TF_i = LS_i - ES_i \tag{8}$$

$$TF_i = LF_i - EF_i \tag{9}$$

where, ES_i is the earliest start time of the working procedure i, ES_h is the earliest start time of the working procedure h, T_h is the construction period of the predecessor h; EF_i is the earliest end time of the process i; LS_i is the latest start time of the process i; LF_i is the latest end time of the process i; TF_i is the total time difference. When $TF_i = 0$, this process is denoted as a critical process, and critical chain is formed by many critical processes connected together.

Buffer setup is the key to CCPM [24] and is designed to reduce the overall project duration with the same completion risk. For operating lines with only precious 3–4 h of "Non-traffic Hour", the most widely used cut-and-paste method and root-variance method, although simpler in calculation, are too arbitrary, lacking in science and taking too few factors into account [21], and are not applicable to metro overhaul management projects.

In response to the shortcomings of the above mentioned traditional methods and the timeline requirements of URO projects, this paper adopts a PCA-based method for setting Feeding Buffers. Since the time of each subdivision, the process is affected by resources to different degrees. It is necessary to quantify the influencing factors first and then allocate the buffer. The following equation obtains each influencing factor.

$$\alpha = \sum_{j=1}^{n} r_j / R \tag{10}$$

$$\beta = N_p / N_t \tag{11}$$

$$k = \frac{m - a}{b - a} \tag{12}$$

$$P_i = \frac{D_i}{\sum_{i=1}^{n} D_i} \tag{13}$$

where, α is the degree of resource constraint, r_j is the number of spatial and tunnel resources utilized for the jth subdivision, R is the total supply of tunnel resources, and n is the number of subdivision; β is the degree of process complexity, differentiation by type of assignment, N_p is the number of immediately preceding processes for that subdivision, and N_t is the number of all subtasks for the overhaul of the tunnel area, P_i is the process duration ratio, and D_i is the duration of the segment process i.

The above is denoted as the original data of the sample $x_{ij}^{n \times p}$, i.e.

$$X = x_{ij}^{n \times p} = \begin{bmatrix} x_{11} & \cdots & x_{1p} \\ \vdots & \ddots & \vdots \\ x_{n1} & \cdots & x_{np} \end{bmatrix} \tag{14}$$

$$Y = AX \tag{15}$$

Among them,

$$y_j = a_j^T x, \quad a_j^T = \left(a_{j1}, a_{j2}, \ldots, a_{jp}\right) \tag{16}$$

If y_j is required to reflect $\left(x_{j1}, x_{j2}, \ldots, x_{jp}\right)$, so that y_j has the maximum variance, that is:

$$D\left[y_j\right] = D\left(a_j^T x\right) \tag{17}$$

$$\begin{cases} v a_j = \lambda a_j \\ a_j^T a_j = 1 \end{cases} \tag{18}$$

$$D y_j = \lambda, \; |v - \lambda E| = 0 \tag{19}$$

After obtaining the score of each component y and eigenvalues λ_i, the principal components are determined, the composite score is obtained according to the following formula, and the buffer distribution M_j is determined:

$$y_j = 1 \Big/ \sum_{i=1}^{m} \lambda_i (\lambda_1 y_1 + \lambda_2 y_2 + \ldots + \lambda_m y_m) \tag{20}$$

$$M_j = y_j \Big/ \sum_{j=1}^{n} y_j \tag{21}$$

The principal component with a cumulative contribution rate of more than 80% is generally taken in the application.

3 Application of Model in Shenzhen Metro Overhauling Management Project

In the present work, we take the URO project of Shenzhen Metro as an example to examine the feasibility and effectiveness of the proposed method. Based on the human resources, materials, and area resources for different segments, tasks in the URO project of Shenzhen Metro are divided into ten types. Each type of task is specified by combining the segment type (including "Z" for main line, "C" for yard line) and resource situation (including "A" for the power supply equipment, "B" for the train equipment, "C" for the tunnel maintenance, "D" for the station equipment, "E" for the signal equipment, "F" for all other situations). For instance, task "ZA" represents the situation that the operation of main line needs to cut off power and hang the ground wire. In this section, we take the URO project with three tasks, "ZA", "ZB" and "ZE" as examples. More specifically, except for the explanation for task "ZA", task "ZB" represents the situation that the power supply is required for the main line and trains are running and "ZE" stands for the situation that no power outage operation is needed for the main line.

In the following interpretation, "ZA" is recorded as task A, which includes a series of subtasks: Submit Task, Approval Task, Issue Task, Submit Work Ticket, Approval Work Ticket, Power Cut, Request Work Area, Execution of Works, Power Transmission, and Cancel Work Area. To simplify the expression, each subtask is then represented in the table with its acronym. For example, "Submit Task" is represented as "S.T.". Correspondingly, "ZB" and "ZE" are denoted as task B and C, respectively. For these two tasks, subtasks include Submit Tasks, Approving Tasks, Issuing Tasks, Executing Works, and Canceling Work Areas.

In this paper, Shenzhen Metro's annual operation and construction data are obtained, and the subdivision operation time of task A, B, and C are extracted, respectively. Substituting in Eqs. (2) and (3) shows the obtained operation time in Table 1.

We use the generally accepted "Non-traffic Hour" as 630 units of time.

According to the most unfavorable situation described in the previous section, at least two Train Dispatchers, noted as Train Dispatcher A and Train Dispatcher B, one Power Dispatcher and one Construction Director, are needed as a set of shifts to complete the three tasks A, B and C. According to the requirements of the Shenzhen Metro URO project management system, the directed acyclic diagram formed by the resource requirements of each subtask is shown in Fig. 2.

The essence of determining critical chain and critical path is to identify the path with the longest process time in a task. The difference lies in whether resource constraints are considered. Firstly, the critical chain is found from Eqs. (4), (5), (6),

Table 1 Work breakdown structure

Subtask	Predecessor	a	b	m	T	Δt
S.T. A		4	23	11	11.83	7.83
A.T. A	1	3	25	9	10.67	7.67
I.T. A	2, 16, 22	1	28	13	13.50	12.50
S.W.T. A	3	1	25	12	12.33	11.33
A.W.T. A	4	19	76	48	47.83	28.83
P.C. A	3	31	198	145	134.83	103.83
R.W.A. A	5, 6	4	12	7	7.33	3.33
E.W. A	7	131	187	162	161.00	30.00
P.T. A	8	4	27	10	11.83	7.83
C.W.A. A	9	1	28	5	8.17	7.17
S.T. B		4	27	11	12.50	8.50
A.T. B	11	3	35	9	12.33	9.33
I.T. B	12	1	8	3	3.50	2.50
R.W.A. B	13	4	51	11	16.50	12.50
E.W. B	14	48	224	196	176.00	128.00
C.W.A. B	15	1	60	11	17.50	16.50
S.T. C		3	27	14	14.33	11.33
A.T. C	17	2	49	8	13.83	11.83
I.T. C	18	1	37	7	11.00	10.00
R.W.A. C	19	3	43	9	13.67	10.67
E.W. C	20	28	225	148	140.83	112.83
C.W.A. C	21	1	41	3	9.00	8.00

(7), (8) and (9), as shown in Fig. 4. The task identifier starting with "+" indicates that the subtask is identified as a critical chain.

According to Eqs. (10), (11), (12) and (13), Tables 1 and 2, the values of influence factors for each subtask are calculated as shown in the following table.

The values of the above four influencing factors were taken as input variables x_1, x_2, x_3, x_4, from Eqs. (14) to (19). The principal components were obtained as follows

$$\begin{cases} y_1 = 0.17x_1 + 0.17x_2 + 0.72x_3 - 0.64x_4 \\ y_2 = 0.32x_1 + 0.83x_2 + 0.10x_3 + 0.43x_4 \\ y_3 = -0.92x_1 + 0.35x_2 + 0.13x_3 - 0.007x_4 \\ y_4 = -0.05x_1 - 0.38x_2 + 0.67x_3 + 0.62x_4 \end{cases}$$

The corresponding eigenvalues are $\lambda_1 = 1.9787$, $\lambda_2 = 1.0482$, $\lambda_3 = 0.8203$, $\lambda_4 = 0.1527$. Obviously, the contribution rate of the first three principal components has exceeded 80%, so the first three principal components, namely resource constraint

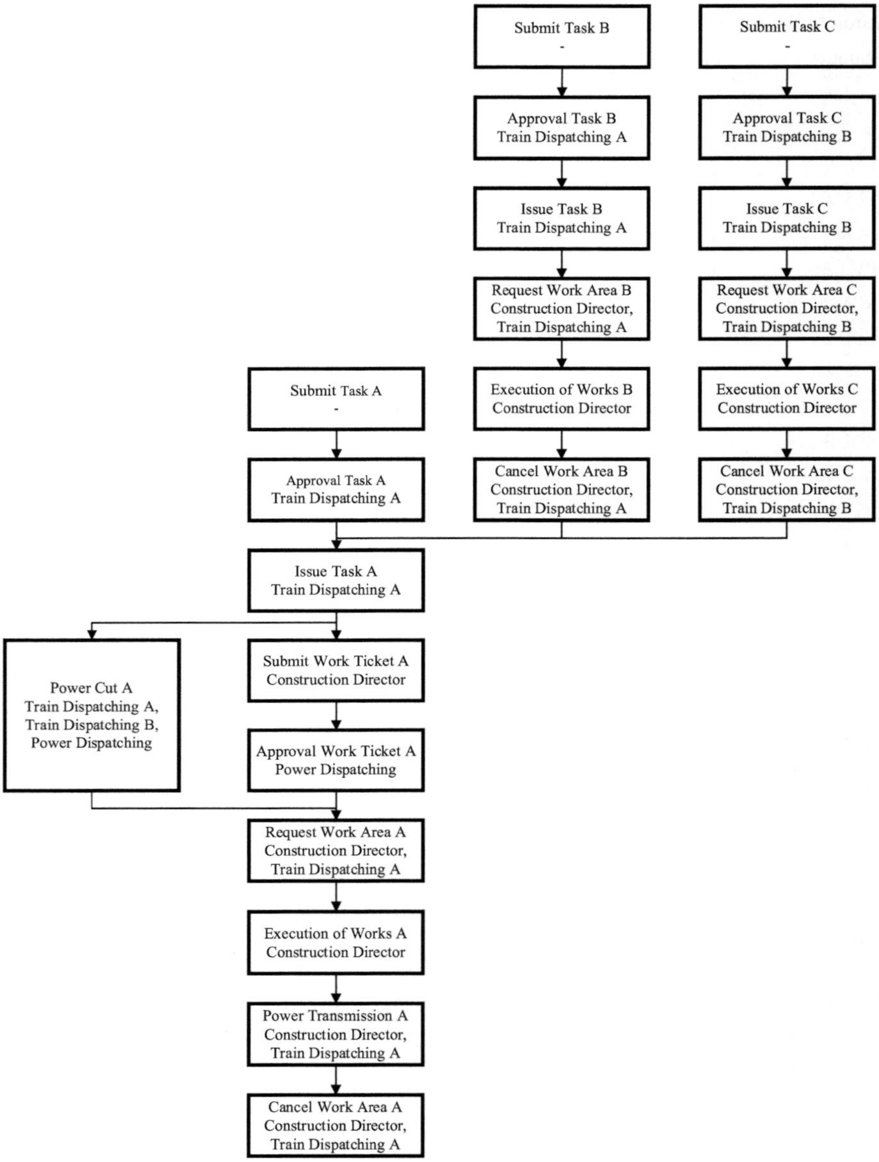

Fig. 2 Identify resource conflict

Table 2 Principal component analysis

Subtask	α	β	k	P
S.T. A	0.00	0.00	0.37	0.02
A.T. A	2.50	1.00	0.27	0.02
I.T. A	2.50	0.21	0.44	0.02
S.W.T. A	2.50	0.07	0.46	0.02
A.W.T. A	0.50	0.06	0.51	0.08
P.C. A	4.25	0.07	0.68	0.23
R.W.A. A	5.00	0.11	0.38	0.01
E.W. A	2.50	0.05	0.55	0.28
P.T. A	5.00	0.05	0.26	0.02
C.W.A. A	5.00	0.05	0.15	0.01
S.T. B	0.00	0.00	0.30	0.02
A.T. B	2.50	1.00	0.19	0.02
I.T. B	2.50	0.50	0.29	0.01
R.W.A. B	5.00	0.33	0.15	0.03
E.W. B	2.50	0.25	0.84	0.31
C.W.A. B	5.00	0.20	0.17	0.03
S.T. C	0.00	0.00	0.46	0.02
A.T. C	1.25	1.00	0.13	0.02
I.T. C	1.25	0.50	0.17	0.02
R.W.A. C	1.25	0.33	0.15	0.02
E.W. C	2.50	0.25	0.61	0.24
C.W.A. C	3.75	0.20	0.05	0.02

degree, process complexity degree, and process uncertainty, are taken as the main influencing factors.

To obtain the proportion of buffer size allocated to each Feeding Buffer pre-process, the principal component score and the composite score need to be calculated. Combining with Fig. 2 and Table 2, find the Feeding Buffer pre-process Approval Task A, Cancel Work Area C, and Approval Work Ticket A.

$$\text{Approval Task A:} \begin{cases} y_1 = 2.50 \\ y_2 = 1.00 \\ y_3 = 0.27 \end{cases}$$

$$\text{Cancel Work Area C:} \begin{cases} y_1 = 3.75 \\ y_2 = 0.20 \\ y_3 = 0.05 \end{cases}$$

$$\text{Approval Work Ticket A:} \begin{cases} y_1 = 0.50 \\ y_2 = 0.06 \\ y_3 = 0.51 \end{cases}$$

Substitute scores of Approval Task A, Cancel Work Area C and Approval Work Ticket A into Eqs. (19) and (20).

The buffer ratio and buffer time obtained are shown in Fig. 3.

In summary, the Gantt chart after adding the buffer is shown in Fig. 4.

In the CCPM method, the total process time is 574.9 units, and the buffer time is 208.07 units. Projects that exceed the whole process can be completed in "Non-traffic Hour".

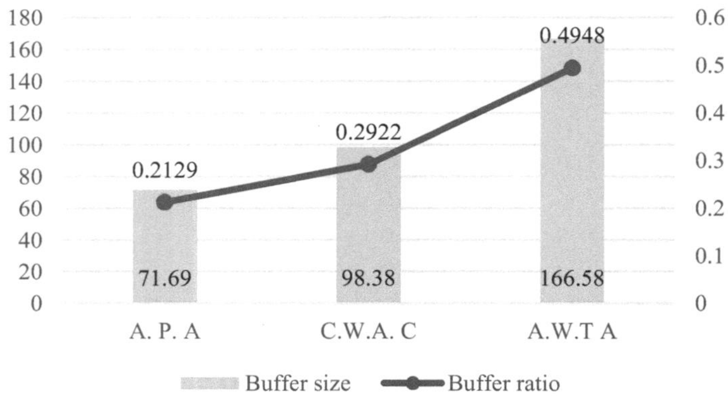

Fig. 3 Buffer size and buffer ratio

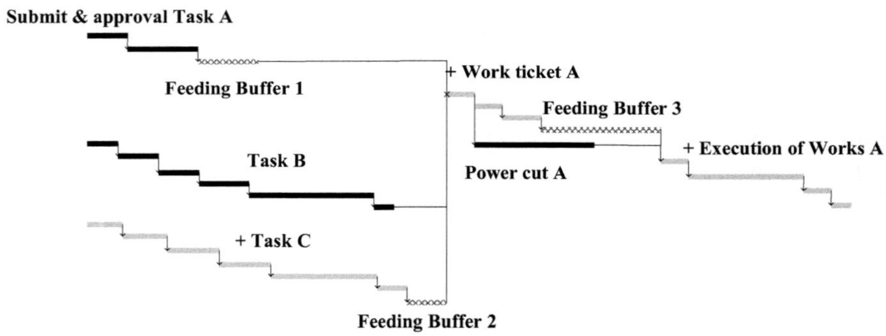

Fig. 4 CCPM

4 Result and Discussion

For the linear method, the total time is the sum of the sub-task times to complete tasks A, B, and C in 1197 units of time. Obviously, due to the "Non-traffic Hour" limitation, the Linear Method cannot complete three tasks.

For CPM, 90% of the time is used as the process time, and the process arrangement and Gantt chart are shown in Fig. 5 and Table 3.

Under the CPM method, the total time to complete tasks A, B, and C is 760 units of time. For the construction data obtained from Shenzhen Metro, all constructions can be completed on time for all possible construction combinations. However, the time required for CPM still exceeds the "Non-traffic hour".

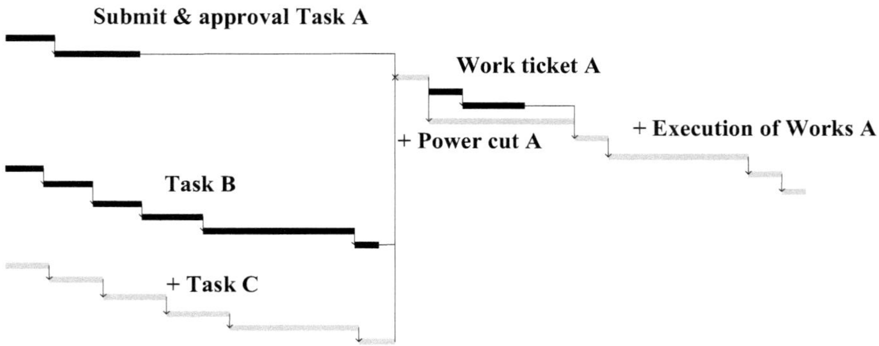

Fig. 5 The Gantt chart of CPM

Table 3 CPM

Subtask	Construction period	Predecessor	Resource
S.T. A	22		
A.T. A	21	1	T.D. A
I.T. A	20	2, 16, 22	T.D. A
S.W.T. A	21	3	C. D
A.W.T. A	71	4	P. D
P.C. A	183	3	T.D. A, T.D. B, P. D
R.W.A. A	12	5, 6	C. D., T.D. A
E.W. A	183	7	C. D
P.T. A	19	8	C. D., T.D. A
C.W.A. A	7	9	C. D., T.D. A

(continued)

Table 3 (continued)

Subtask	Construction period	Predecessor	Resource
S.T. B	19		
A.T. B	13	11	T.D. A
I.T. B	6	12	T.D. A
R.W.A. B	33	13	C. D., T.D. A
E.W. B	215	14	C. D
C.W.A. B	16	15	C. D., T.D. A
S.T. C	27		
A.T. C	29	17	T.D. B
I.T. C	13	18	T.D. B
R.W.A. C	29	19	T.D. B
E.W. C	223	20	C. D
C.W.A. C	15	21	C. D., T.D. B

Fig. 6 Methods contrast

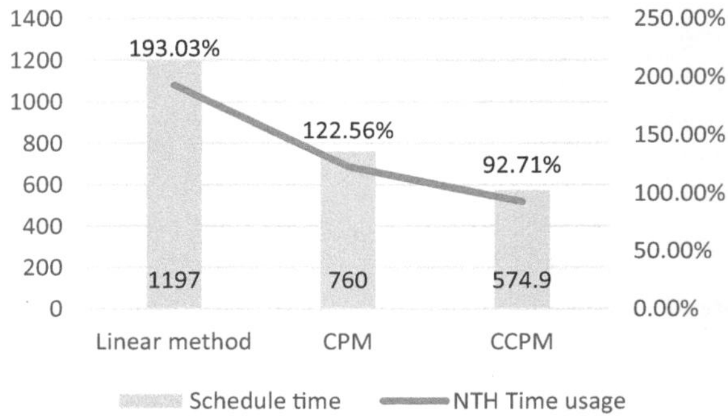

According to the evaluation method described in the previous section, as shown in Fig. 6, for Linear Method and CPM Method, the duration is 1197 and 760 units of time, respectively. The maximum allowable time for the problem is exceeded, i.e., "Non-traffic Hour". For the CCPM Method, the probability of completing the combination of task sets S within the given duration of the solution is 93.32%. The maximum allowable time utilization is 92.17%, as shown in Eq. (1).

5 Conclusions

This paper solves the construction allocation problem subject to resource constraints to rationally schedule URO projects using CCPM, taking the Linear Method, the CPM as comparisons. The multi-task scheduling problem allows multiple tasks to be processed parallel at the same timer model. We use the optimization model of critical chains to identify critical chains and use Approval Task A, Cancel Work Area C and Approval Work Ticket A as the sink points of non-critical chains. The degree of resource constraint, process complexity, and uncertainty coefficient is selected as the principal components. The buffer area size is determined using the PCA method to rearrange the URO projects and realize the workflow control. Meanwhile, the calculation results show that the CCPM can significantly shorten the construction period with a guaranteed completion probability compared with the Linear Method. The subsequent work can be considered in the actual construction, where there are multiple resource constraint situations, and the CCPM can be further improved based on the feedback. It should be noted that although the current work focuses on the multi-task scheduling problem of URO, the proposed method is also applicable to most multi-objective and multi-task workflow process industries. It is also instructive to schedule multiple tasks in the same processing unit.

Acknowledgements This paper is supported by the Stability support fund of Shenzhen colleges and universities, Shenzhen Technology University under Grant No. SZWD2021014, Basic research grants of Shenzhen Natural Science Foundation under Grant No. JCYJ20210324121203008 (Value Discovery Model and Method of Metro Operation Big Data in Spatiotemporal Complex Scenarios), School-enterprise cooperation R&D project of SZTU under Grant No. HT20221061030008 (Intelligent Metro Construction Management Technology Application Research and Intelligent Management Platform Development) and Natural Science Foundation of Top Talent of SZTU under Grant No. GDRC202130.

References

1. Patnaik, H.K., Patra, M.R., Kumar, R.: A workflow based approach for task scheduling in cloud environment. Mater. Today: Proc. (2021)
2. Shen, Y., Qi, D., Zhou, N., Wang, X.: Workflow task scheduling algorithm for marine environment data processing under multi-core computing environment. J. Coast. Res. 646–650 (2018)
3. Lappas, N.H., Ricardez-Sandoval, L.A., Fukasawa, R., Gounaris, C.E.: Adjustable robust optimization for multi-tasking scheduling with reprocessing due to imperfect tasks. Optim. Eng. **20**, 1117–1159 (2019)
4. Rakovitis, N., Zhang, N., Li, J.: A novel unit-specific event-based formulation for short-term scheduling of multitasking processes in scientific service facilities. Comput. Chem. Eng. **133**, 106626 (2020)
5. Arashpour, M., Wakefield, R., Blismas, N., Maqsood, T.: Autonomous production tracking for augmenting output in off-site construction. Autom. Constr. **53**, 13–21 (2015)
6. Chun, H., Yeung, W.M.: Proceeding of the International Conference on Computing, Communications and Control Technologies, pp. 14–17 (2004)
7. Chun, A.H.W., Yeung, D.W.M., Lam, G.P.S., Lai, D., Keefe, R., Lam, J., Chan, H.: AAAI, pp. 1467–1474 (2005)
8. Hyun, J.-H., Yang, B.-S., Moon, S.-W.: The development of urban metro maintenance facility system using construction classification system management. Korean J. Constr. Eng. Manag. **13**, 69–77 (2012)

9. Argyropoulou, K., Iliopoulou, C., Kepaptsoglou, K.: Model for corrective maintenance scheduling of rail transit networks: application to Athens metro. J. Infrastruct. Syst. **25**, 04018035 (2019)
10. Zhang, T., Andrews, J., Wang, R.: Optimal scheduling of track maintenance on a railway network. Qual. Reliab. Eng. Int. **29**, 285–297 (2013)
11. Chun, A., Suen, T.: Proceedings of the AAAI Conference on Artificial Intelligence, vol. 28, pp. 2890–2897 (2014)
12. Shim, C.-S., Kim, S.-W., Song, H.-H., Yun, N.-R.: Development of BIM for a maintenance system of subway infrastructures. J. KIBIM **1**, 6–12 (2011)
13. Sedghi, M., Kauppila, O., Bergquist, B., Vanhatalo, E., Kulahci, M.: A taxonomy of railway track maintenance planning and scheduling: a review and research trends. Reliab. Eng. Syst. Saf. **215**, 107827 (2021)
14. Palombarini, J.A., Martínez, E.C.: End-to-end on-line rescheduling from Gantt chart images using deep reinforcement learning. Int. J. Prod. Res. **60**, 4434–4463 (2022)
15. Olivieri, H., Seppänen, O., Alves, T.d.C., Scala, N.M., Schiavone, V., Liu, M., Granja, A.D.: Survey comparing critical path method, last planner system, and location-based techniques. J. Constr. Eng. Manag. **145**, 04019077 (2019)
16. Roos, E., den Hertog, D.: A distributionally robust analysis of the program evaluation and review technique. Eur. J. Oper. Res. **291**, 918–928 (2021)
17. Selvan, M.M, Ahuja, S., Parray, R.A., Rajaiah, P., Rangasamy, K.: ICAR (2015)
18. Delias, P., Doulamis, A., Doulamis, N., Matsatsinis, N.: Optimizing resource conflicts in workflow management systems. IEEE Trans. Knowl. Data Eng. **23**, 417–432 (2010)
19. Araszkiewicz, K.: Application of critical chain management in construction projects schedules in a multi-project environment: a case study. Procedia Eng. **182**, 33–41 (2017)
20. Goto, H., Murray, A.T.: Exact and flexible solution approach to a critical chain project management problem. Constraints **25**, 280–297 (2020)
21. Zarghami, S.A., Gunawan, I., Corral de Zubielqui, G., Baroudi, B.: Incorporation of resource reliability into critical chain project management buffer sizing. Int. J. Prod. Res. **58**, 6130–6144 (2020)
22. von Danwitz, S.: Managing inter-firm projects: a systematic review and directions for future research. Int. J. Proj. Manag. **36**, 525–541 (2018)
23. Sackey, S., Kim, B.-S.: Schedule risk analysis using a proposed modified variance and mean of the original program evaluation and review technique model. KSCE J. Civ. Eng. **23**, 1484–1492 (2019)
24. She, B., Chen, B., Hall, N.G.: Buffer sizing in critical chain project management by network decomposition. Omega **102**, 102382 (2021)

The Role of Public–Private Partnership on Preservation-Led Projects in Urban China—A Comparative Perspective

Ting Zhang, Fangqian He, Zhouquan Li, and Ran Xu

Abstract Driven by public–private partnerships, urban historic settlements in Chinese cities are increasingly being targeted for redevelopment, especially if they are located in urban centers. Previous scholars have featured state policy-led or market-led to understand the driving force of the transformation process of the urban heritage area. However, this understanding glosses over the role of actors during transformation process, which refers to the public–private partnerships. With examples of Huishan ancient town in Wuxi and Xintiandi in Shanghai, this study proposes an analytical framework to compare how different networks of transformation process are caused the similar heritage space among heritage, tourism and commerce. Finally, by comparing the two cases, subtle differences of the two spaces also reveal the different interests and needs of different public–private partnerships. This study calls for further attention to study urban China from a comparative perspective.

Keywords Public–private partnerships · Preservation-led projects · Urban China

1 Introduction

Since the twentieth century, heritage has played an important role in constructing and understanding memories. As a representation of history, heritage is a value-laden concept [1], which defined as what needs to be selected to inherit and pass on in contemporary society [2, 3]. Smith proposed that the heritage site is defined by policy and linked to material buildings, institutions, and local practices. This idea has influenced heritage management, interpretation and presentation. The formation of the policy is related to the urban regime, the members of one regime share the objectives of the policy and profit from their participation, such as the urban

T. Zhang (✉) · F. He · R. Xu
Wuxi Taihu University, Wuxi, China
e-mail: zhangt_wxu_edu@163.com

Z. Li
Jiangsu Chenggui Designing Institute Co., Ltd, Wuxi, China

E. Strauss (ed.), *Proceedings of the 7th International Conference on Civil Engineering*, Lecture Notes in Civil Engineering 371, https://doi.org/10.1007/978-981-99-4045-5_16

transformation process [4, 5]. A large number of urban transformation projects have been produced by various Public–Private coalitions, which means the co-operation relationship between different participants. In urban studies, it also refers to the Public–Private Partnerships (PPPs). PPPs is defined as ‘co-operation between public and private actors with a continuity feature, and the actors develop the production or serves and share the risk and benefits’. The actors focus on the additional benefits that will be more than the cost of co-operation [6]. This co-operation relationship can be explored by the urban governance through network perspective, which argues that the policy is framed and implemented in the network governance [7].

With the increasing number of preservation-led projects in Chinese cities, critical voices tend to question the homogeneity of these preservation-led projects, which are led by a different network governance. However, few studies explored if they are really the same in the different network governance. From comparative perspective to analyze two cases in different urban contexts, this paper discusses how different PPPs produce similar urban heritage space among heritage, tourism, and commerce, and it also reveals that the interests and demands of different PPPs that can make subtle differences to urban space.

This paper comprises six sections. Section 1 contains the introduction part; Sect. 2 develop an analysis framework among heritage, tourism and commerce; Sect. 3 analysis the role of PPPs in the trans-formation process of two cases under different urban contexts; Sect. 4 provides a comparative analysis of cases and reveals the differences under the different network governance; finally, Sect. 5 offers conclusion and calls for more interpretation of Chinese cities from a comparative perspective.

2 Methodology

To structure the comparative analysis, this study proposes a framework that consists of three key aspects of the transformation process: heritage, commerce, and tourism. This study firstly focuses on heritage space because it provides the basic element for the other two aspects. This part examines the formation of historical buildings in two cases through their own historical lines. In the transformation processes of two cases, how historical buildings are selected and implanted into the new space through different public–private partnerships will be analyzed. Second, the heritage has been transformed into consumption symbolic, commercial elements are implanted by developers into the new space. By analyzing different shop types, the degree of commercialization caused by different public–private partnerships can be examined. Third, driven by globalization, heritage is increasingly transformed into tourist production through “heritage industry” [8]. This process redefines and reinterprets the heritage while also enhancing the attractiveness and competitiveness [9]. Thus, the tourism aspect focuses on reproduction of the tourists’ performance spaces and their relationship with historical and commercial buildings. The next section proposed a comparative examination of how diverse public–private partnerships produce the new heritage space among heritage, tourism and commerce in different urban context. The two cases of Huishan ancient town and the Xintiandi are analyzed.

3 Case Study

3.1 Huishan Ancient Town, Wuxi, Jiangsu Province

The case area located in the inner city of Wuxi, which was the ancient capital and birthplace of the Wu Culture (eleventh century BC-473 BC). It covers an area of 300,000 m^2, and originally had almost 450 households with a population of more than 800 before renovation. In order to restore the cultural images of the historical city, the local government initiated the restoration project in 2008 and established the Wuxi Huishan Ancient Town Preservation and Construction Working Group to restore the historical character of the Huishan Ancient Town.

Heritage

From Ming dynasty (1368–1644) to contemporary (2000s), the Huishan Ancient Town has experienced more than six hundred years. The Ancestral shrine architecture (祠堂-Ci Tang) is the main historical architectural type in this area, which refers to temples where people worship their ancestors, thus, they are also called ancestral temples. In history, this building was an important place where the elders instilled family values to the clan members and enforced family law. Most of the ancient ancestral shrine architectures were built near mountains and rivers, which shows the ancient Chinese concept of ecological environment and return to nature. Comparing the ancestral shrine buildings in the north and south of China, it reflects the differences caused by geography. The ancestral shrine architecture in Huishan Ancient Town conforms to the characteristics of the southern ancestral shrine architecture: the courtyard style.

The renovation plan was started in 2006, the whole project was led by the local government, and some social investments are involved. The government formed the Wuxi Huishan Ancient Town Preservation and Construction Working Group to control the process of renovation project, at the same time, Wuxi Huishan Ancient Town Cultural Tourism Development Co., Ltd. was established as the main developer, both formed the network governance (PPPs) to redevelop this historical area.

During the transformation process, first, the local government set a workgroup to recover land property and provide various relocation plans for residents. The relocation plan comprised the directional commercial houses and the property rights replacement houses in other areas. Second, the main idea of the restoration project was ‘restoring the old as it was’, which followed the concept of authenticity in Chinese preservation [10]. According to the historical sources, the morphological characteristics of the main historical periods were selected and recycled to build the new space. The main gate of the area was rebuilt in its original location that means this

Fig. 1 The historical architecture in Huishan ancient town. *Source* By author

area was circled and protected (Fig. 1). The developer collected traditional materials from other cities to decorate the historical area, such as the streets, the roofs, the windows and the doors.

Commerce

Before renovation, the Huishan ancient town was a residential area with low-quality living environment. During the transformation process, the PPPs recycled historical elements and inserted the commercial elements into the newly created space. The developer adopts the international advanced management model of the shopping mall: 'only rent and not sell'. There are certain requirements for the commercial types, which are mainly divided into local food, cultural and creative crafts, garden and flowers, cultural health care, tea shop etc. Under the same condition, shops with local cultural characteristics are preferred, standards are also made in terms of store decoration, such as conforming to the feature of historical area. Table 1 summarizes the shop types of case area, where the traditional shops account for 62%, and there is no involvement of international shops. The basic consumption is far below the local consumption level through fieldwork. It shows that the developer pays more attention to preserve the cultural value.

Tourism

The whole area has become a natural museum of Ancestral shrine architecture, which has applied the Chinese National Scenic Area with 5A level. The developer designed the various activities of the new space that can be used to improve tourism attractiveness. According to the fieldwork, the dynamic activities are divided into

Table 1 The commercial types of Huishan Ancient Town

Types	Number	Proportion (%)
General shops	26	38
Traditional shops	42	62
International shops	0	0

Table 2 The dynamic types of Huishan Ancient Town

Types	Number	Proportion (%)
Shops	68	51
Hotel	1	1
Museum	6	5
The former celebrity residence	55	41
Chinese medicine clinic	2	2

five main types. Table 2 summarizes the main activity types of the Huishan Ancient Town. The data demonstrate that commerce is still the most important type with the ratio of 51%.

3.2 Xintiandi, Shanghai

In the early 1990s, Shanghai began a large-scale urban renovation and development. A large number of old houses have been demolished and replaced by modern high-rise buildings. In this context, many protectionist voices appeared to call the idea of urban preservation, which caused local people to discover that the old houses with nostalgia had gradually disappeared. Thus, the local government passed a series laws of urban preservation, such as making the historical building list. The Shikumen architecture with colonial characteristics has undoubtedly become the historical area. The Xintiandi renovation project is one part of the Taiping Bridge project, which was the most important project of the Luwan district in Shanghai. After 2002, the economic success of Xintiandi proposed the formula for other preservation-led projects [11].

Heritage

After Opium War (1840–1842), some foreign concessions were built in Shanghai, such as French concession and British settlement. Each settlement had its own management model. Because the Taiping Army (1851–1864) launched a war and invaded the surrounding cities in 1860, a large number of refugees poured into the Shanghai concession area to seek asylum. The concession administrators built Shikumen buildings to arrange for these refugees. In order to make full use of the land, most of these houses were built as row-style houses with narrow lanes. By considering the traditional Chinese family lifestyle, the layout of Shikumen houses was similar to residential buildings in the south of the Yangtze River, and the Western features are added to the building type (Fig. 2). Until 1980s, Shikumen architecture was the main building type in Shanghai at that time.

In order to renovate the Taiping Bridge area, the local government and the Shui On Group of Hong Kong signed an official document. They formed a network to develop the area. The local government provided political support to reclaim land property and resettle residents. After dealing with local residents and land issues, the land property was sold to the Shui On group for 50 years, and the Shui On group invested

Fig. 2 The Shikumen architecture. *Source* Shikumen architecture in Shanghai, *Author* Xueqiang Ma

to redevelop this area. The Shui On Group is a private developer that focuses on real estate, so it lacked experience to preserve historical buildings. Due to the National landmark—the First Communist Party (CCP) hall locates here, and the developer could not demolish it. Thus, the Shui On invited international designers to provide a plan proposal, and finally decided to retain the overall Shikumen architectural style and the CCP hall building. The developer recycled the original materials that can be used to build the two blocks of Xintiandi, the north block (北里) and the south block (南里). After renovation, most of the historical buildings were kept in the north block, and historical architectures in the south block were displaced by a modern shopping mall and fewer historical buildings. The PPPs of Xintiandi used historical elements to create a modern historical space, which also became a symbol case area of Chinese urban preservation.

Commerce

Since this study focuses on heritage space, this part analyzes the north block that has historical buildings. The Shui On provides the historical environment to attract international brands and stores to settle here, which made the area gradually become a high consumption area. In order to control the development, the Shui On Group introduced the shopping mall model to manage commerce. Table 3 summarizes the commercial types of North block, it is obvious that among the shop types, the international shops account for 63%. According to the survey, although traditional shops account for 16%, they are almost Chinese restaurants that exceed the local consumption level.

Table 3 The commercial types of the north block, Xintiandi

Types	Number	Proportion (%)
General shops	8	21
Traditional shops	6	16
International shops	24	63

Table 4 The dynamic types of Xintiandi

Types	Number	Proportion (%)
Shops	38	97
Hotel	0	0
Museum	1	3
The former celebrity residence	0	0

Tourism

In 2019, Xintiandi was selected as one of the top ten landmarks in Shanghai. Xintiandi has become a symbol of Shanghai that represents the colonial era. Although it is not a National-Level scenic area, it has become one of the must-visit spots for tourists after transformation. In addition to the CCP hall, the rest of other historical spaces are almost the commercial elements with the ratio of 97% (Table 4).

4 Comparative Analysis and Discussion

By analyzing two cases, we conclude that the different PPPs produce similar new heritage spaces among heritage, commerce and tourism, but these new spaces are slightly different, which reflect the different interests and concerns of different PPPs in the transformation process (Table 5).

Table 5 Summary of the PPPs' role in two cases

Cases	Huishan ancient town	Xintiandi
The organization of PPPs	Local government	Local government
	State-owned developer	Private real estate developer
Heritage	Reclaim the land from residents and sell to developer	Reclaim the land from residents and sell to developer
	Restore the historical form of area	Restore the historical building typology
Commerce	Traditional shops as main	International shops as main
Tourism	5A level of national scenic area	One of the 10 landmarks of Shanghai

In the heritage sector, the building typology of the two cases are different, however, the common is that the two PPPs used the same method to reclaim the land property and restore the historical buildings by remaining and rebuilding their historic condition. That is to say, the transformation logic of heritage in two cases is similar. Both the local government provided the policy support to relocate the residents, and the developers investment to redevelop the historical area. Significantly, both cases have chosen to erase local residents and create a space where has more attraction for the middle class.

With regard to dynamic activities, commerce and tourism are the main types. Comparing commercial types in two cases, although two cases adopted the shopping mall model to manage commerce, it is obvious that the concern of the two cases is quite different. The commercial types of Huishan Ancient Town focus on local characteristics, such as local handicrafts and local food. In contrast, Xintiandi is more interested in international brands, with very few shops that can represent the relationship with local culture. The main commercial types lead to the difference between two heritage spaces, which reflect the different interests and objectives of two PPPs. The developer of Huishan Ancient Town, a state-owned company, is more interested in the socio-cultural impact of the new space. On the contrary, the developer of Xintiandi is a private real estate company, which pays more attention to economic benefits.

In terms of tourism, the cultural value of historical buildings lies in historical buildings, which have strong social attractions. In both cases, historical elements are used to build the physical environment. The Huishan Ancient town allows visitors to enjoy the unique landscape through the ancestral shrine architecture. Visitors can experience the local life in the historical period through diverse activities infused by the PPPs. Similarly, the western colonial architecture of Xintiandi has the memory of the historical time. The Chinese Communist Hall, a national landmark, also provides the fuel for the educational significance of tourism at Xintiandi. Through the operation of PPPs, the entire area has been transformed from a shabby residential area to a modern entertainment area with high consumption.

5 Conclusion

This paper examines the role of different public–private partnerships in preservation-led projects and reveals the subtle differences in similar urban heritage spaces. PPPs for preservation-led projects consist of local governments and developers who have become the main drivers of the transformation of these historic areas. A comparison of the Huishan Monument and Xintiandi reveals that, despite their different PPPs, both cases utilize historic buildings as a means of redevelopment. In the transformation process, the two preservation-led projects adopted the same transformation logic, that is, the local government provided policies to reclaim the land property and solve the relocation problem of local residents, and the developers buy the land use right and invest to redevelop the historical area. After the reconstruction of the physical environment, the developer uses the shopping mall model to manage commerce, and has the right to select the commercial types in the newly built environment.

The comparison also argues the different interests and goals of different developers are reflected in the built environment. In Huishan ancient town, the state-owned developer pays more attention to social-cultural by controlling the commercial types. In Xintiandi, the private developer has more interest in international brands that can get high economic returns. Despite Xintiandi's economic success provides a formula to other preservation-led projects, the comparative study can help reveal the subtle differences of these urban heritage spaces.

Funding This study is supported by Philosophy and Social Sciences of Jiangsu Higher Education, the project number is 2021SJA0913; and is also supported by the Natural Science foundation of Jiangsu Higher Education, the project number is 20KJB560011, the Soft Science Research Project of Wuxi Association for Science and Technology, 2023 (Research on the Implementation Mechanism of Urban Old Community Renewal under the Multiple Partnerships in Wuxi, China).

References

1. Lowenthal, D.: The Past is a Foreign Country—Revisited. Cambridge University Press (2015)
2. Graham, B., Ashworth, G.J., Tunbridge, J.E.: A Geography of Heritage: Power, Culture, and Economy. Arnold, London (2000)
3. Smith, L.: Uses of Heritage. Routledge, London (2006)
4. Stone, C.N.: Regime Politics: Governing Atlanta, 1946–1988. University Press of Kansas, Lawrence, Kan (1989)
5. Ren, X.F.: Forward to the past: historical preservation in globalizing Shanghai. City Community **7**(1), 23–43 (2008)
6. Klijn, E.H., Teisman, G.R.: Governing public–private partnerships: analysing and managing the processes and institutional characteristics of public–private partnerships. In: Osborne, S.P. (ed.) Public–Private Partnerships. Routledge, London (2000)
7. Klijn, E.H., Koppenjan, J.F.M.: Public management and policy networks: foundations of a network approach to governance. Public Manag. **2**(2), 135–158 (2000)
8. Hewison, R.: The Heritage Industry: Britain in a Climate of Decline. Methuen Publishing Ltd (1987)
9. Nasser, N.: Planning for urban heritage places: reconciling conservation, tourism, and sustainable development. J. Plan. Lit. **17**, 467–479 (2003)
10. Zhu, Y.J.: Performing heritage: rethinking authenticity in tourism. Ann. Tour. Res. **39**(3), 1495–1513 (2012)
11. Luo, X.W.: Shanghai Xintiandi. Dongnan University Press, Shanghai (2002)

Gamification to Stimulate Green Behaviors in Cities

Joyce Ngo, Emmanuel Fragnière, Blaise Larpin, and Jean-Michel Sahut

Abstract Gamification is a new trend that has gained increasing importance in climate protection. In recent years, researchers have begun to integrate game elements into non-gaming contexts to encourage green behaviors. The city of Sierre jumped on the bandwagon and initiated a project aiming to use gamification to encourage its population to engage in green behaviors. The research focuses on identifying the barriers and motivators for people to engage in green actions in order to diminish or enhance them using games. Followed by a series of focus groups, 10 in-depth semi-structured interviews were conducted. The main result shows a significant gap between attitude and behavior despite a high level of awareness and positive sustainable intentions. Based on research propositions that have been confronted with the scientific literature, Sierre has therefore collaborated with the HES-SO Valais-Wallis to realize a gaming app on the theme of sustainable development for the population called ECOTREE.

Keywords Sustainability · Gamification · Behavior change

1 Introduction

Climate change is a current recurring issue, and the situation is increasingly deteriorating by the day. Although caused by human activities, the alarming issue poses an existential threat to human life. To counter the aggravating situation, interventions increasingly appeared, focusing primarily on raising awareness on the matter. Furthermore, attempts at changing environmentally harmful behaviors faced many hurdles, making the task a tedious one. However, in the last few years, gamification and serious games were discovered to be effective methods for climate adaptation.

J. Ngo · E. Fragnière (✉) · B. Larpin
HES-SO Valais-Wallis, Institute for Tourism (ITO), 3960 Sierre, Switzerland
e-mail: emmanuel.fragniere@hevs.ch

J.-M. Sahut
IDRAC Business School, Lyon, France

E. Strauss (ed.), *Proceedings of the 7th International Conference on Civil Engineering*, Lecture Notes in Civil Engineering 371, https://doi.org/10.1007/978-981-99-4045-5_17

As a result, researchers conducted many studies to assess the opportunity games present to increase sustainable behaviors and induce enduring behavior change. The city of Sierre (Canton Wallis in Switzerland) has decided to take advantage of this trend and to act on a state scale in favor of a sustainable city.

This paper aims at understanding how to use gamification to stimulate pro-environmental behaviors for the city of Sierre and to ultimately conceptualize a game on the theme of sustainable development for the population. The research objective is to obtain insights into people's experiences, desires, motivations, and needs and then go from insights to implementation. Directing the interest of the research toward people's needs, behaviors and motivations leads to a deeper understanding of human emotions and behaviors, enabling the development of a game based on what people really want.

In 2002, the city of Sierre has been awarded the Cité de l'Energie label (Ville de Sierre, n.d.a). This Swiss label rewards the performance of cities or municipalities that practice and implement an exemplary municipal energy policy in terms of sustainable development. Those Energy Cities promote renewable energies, and environmentally friendly mobility and are committed to the efficient use of resources. In October 2019, Sierre was awarded the Cité de l'Energie Gold label, the most prestigious and most demanding energy distinction at the European level (Ville de Sierre, n.d.a). Additionally, the Charter of Cities and Towns for Climate and Energy, issued by the Climate Alliance in Switzerland, has been ratified by several cities and municipalities in Switzerland, including the city of Sierre in June 2020. The charter unites the signatory cities and towns in a joint commitment to resolve effective climate protection. They acknowledge their responsibility for climate protection and are willing to support the federal government in its climate and energy policy. Consequently, signatories endorse the objectives of the Paris Agreement and support the Federal Council in its goal of reducing greenhouse gas emissions to zero by 2050.

The purpose of the research is to analyze people's attitudes and behavior toward climate change in order to identify barriers and motivations that influence them. Subsequently, the objective is to remedy or increase them by implementing game mechanics in their daily life to shift people's behavior towards a more sustainable one. Therefore, the research question is the following: How to use gamification to stimulate pro-environmental behaviors in Sierre?

The paper is organized as follows. In Sect. 2, we present a focused literature review carried on the topic of games and pro-environmental behaviors. Section 3 briefly explains the methodology we have used to analyze the data collected through semi-structured interviews to ultimately generate new research propositions on the topic. It also includes an overview of the sampling process and an explanation of the questionnaire. In Sect. 4, we provide a synthesis of results along with a discussion. In Sect. 5, we provide some snapshots of the (MI-FI) app that has been realized following this research and that will be developed soon for the citizen of Sierre City. In Sect. 6, we conclude and indicate directions for further research.

2 Literature Review

Computer games have been controversial for a long time due to their negative image related to violent entertainment games spurring aggressive behaviors. Nonetheless, more people have shown interest in their positive impacts instead of focusing on the negative ones. This interest later developed into the use of games for serious purposes to foster learning and the development of useful skills. It then resulted in the emergence of the terms serious games and gamification. This method of learning has led to optimism for inducing engagement and motivation to achieve a desired result [1]. Although they pursue the same goal, serious games, and gamification slightly differ in their approach. Gamification is a method that engages users in real-world activities by applying game design principles to a non-gaming context [2].

The principles of gamification use game design features such as clear progression paths with achievable goals, a point system, competition, leader boards, feedback, levels, rewards, and cooperation between players and badges. Gamification is perceived as an effective tool to change behaviors as it can intervene in users' behaviors in their daily life [3]. Therefore, it has gained popularity in recent years as a method for encouraging the development of specific behaviors and for stimulating motivation even in tedious activities, thus creating behavior change in several domains [4, 5].

Gamification contains gaming elements but may not operate as a full game, unlike serious games that are set in game-like contexts to offer a learning experience. Whilst gamification uses game design elements in non-playful environments and contexts, serious games place gaming at the core of the experience and as a central and primary medium [1]. The approach of serious gaming revolves around the stimulation of users through experimental learning by gaming. Serious games aim at providing an engaging environment, virtual or real, favorable to education that enables users to better understand complex problems in a fun [3].

In short, serious games work as simulations with educational goals that include mechanics such as rules, choices, and challenges. They can be effectively used to prepare people for real-world tasks by facing challenges in-game [1]. Researchers have increasingly become more interested in the topic to be used for sustainable development. In recent years, studies have proven that games are effective tools to educate individuals about sustainability. Gamification, as opposed to other approaches to behavior change such as nudging, can result in longer-term psychological involvement. It can teach people which pro-environmental behaviors to adopt and give instructions on how to perform certain sustainable behaviors [2]. Gamification can also stimulate individuals' motivation when performing tiresome and unrewarding tasks to guide their behavior toward a desired result [5].

Serious games are a good way for teachers to support student's education for sustainability [6]. In order to induce behavioral change in favor of sustainable development with games, it is necessary to understand how to do so. It was assumed that for behavior change to occur, changing the attitude of consumers was necessary in the first place. However, many researchers observed that a positive attitude toward pro-environmental behaviors does not always equate to sustainable behavior. Therefore, they came to the same conclusion that changing the attitude of consumers does not ensure behavior change [7]. This was further supported by the observation that people are inconsistent with their sustainable behavior [8]. This means that engaging in one green behavior does not always imply engaging in others. Someone who is good at recycling, for example, may not consider the environment while selecting a mode of transportation [1]. In addition, negative spillover effects can occur when pro-environmental technologies and resources are available after the observation of consumers using more resources when they knew that a recycling option was available [8].

These issues raise many questions on how to stimulate behavior change. The approach used to address this subject is to first identify the elements preventing behavior change and how to deal with them. There are many psychological barriers that prevent behavior change. Although people are invested in climate change issues, personal and social issues take precedence, for example for stakeholders in the tourism sector [9]. People are often present-focused, which makes the distant threat of climate change intangible and therefore difficult for people to grasp the emergency.

Many people struggle to conceive the benefits of long-term solutions and tend to focus on the short term. As the threat seems distant with stronger impacts in the future, subsequently the consequences of their sustainable actions will be achieved in the future which makes it difficult for people to act in the present. This is reinforced by the fact that not only do they not perceive a positive outcome immediately, but their current sustainable actions will benefit other people that might not even be them [4], making the beneficial impact of sustainable behavior more certain in the present that will increase the possibility of repeating it in the future.

Authors in [8] also developed the SHIFT framework based on the identification of five psychological factors to shift consumer behaviors to be more sustainable: Social influence, Habit formation, Individual self, Feelings and cognition, and Tangibility. One of the most powerful factors for sustainable consumer behavior change is the social factor. For instance, it was observed that another hindrance for people to act is the necessity of collective action for the benefits to be fully realized. Most individuals are prone to perceive their sustainable behaviors as insignificant, which leads them to believe that individual efforts do not matter overall [4].

3 Methodology

The research draws upon the practice of service design and UX design [10] and uses a qualitative research method. This research has an inductive approach; therefore, qualitative research is most suitable as it can lead to the development of hypotheses. The research aims to eventually deliver a game proposition that stimulates sustainable behaviors. In order to elaborate on the concept, it is primarily necessary to carry out research on people to understand their needs and motivations to play and engage in the game. This qualitative research uses the purposeful sampling technique, which relies on personal judgment and knowledge when choosing participants. The research aims at searching for patterns from observations to develop broad generalizations.

The objective of the qualitative data collection is to gather diverse insights, hence diverse samples, to reach a general observation of the relationship between people and sustainable behaviors. A total of 10 semi-structured interviews were conducted. The sampling includes foreigners and Swiss residents as well as students and full-time employees.

The interview guide is made of 8 open-ended questions with some completed by sub-questions to get a deeper insight.

Question 1: Tell me what you think about when we talk about climate change.

Question 2: Why is it important (or not) to act against climate change according to you?

Question 3: Could you please list pro-environmental actions that you know?

Question 4: Would you be more motivated to have pro-environmental behavior if you were given more information on the impacts of climate change and how to act upon it?

Question 5: Would you be more motivated to adopt a pro-environmental behavior if there was a reward?

Question 6: How would you change your behavior if you could measure your environmental impact in real-time and receive direct feedback to improve it?

Question 7: What is your attitude towards games in general? what do you like about playing games?

Question 8: How would you react if you could play a game and at the same time act against climate change just by playing it?

The method of thematic analysis was used (manually) to present the ideas and perceptions related to climate change and the pro-environmental actions of the interviewees. As one of the most common methods of analysis for qualitative research, the thematic analysis focuses on identifying patterns of meaning that come up frequently to establish common themes.

4 Results

This section provides an overview of the responses obtained from the interviews. To summarize, the results are presented by themes issued from the thematic analysis conducted on the 10 interviewees. The five themes identified are as followed:

- **High Level of Awareness**

Firstly, when asked about their thoughts on climate change, all the respondents acknowledged the threat posed by climate change. All of them mentioned the effects of climate change on the environment, which showed a high level of awareness about the consequences of climate change.

– **Gap between Attitude and Behaviour**

Another theme that came out from the analysis was the overall green attitude that all the interviewees seemed to share. When asked about the importance of taking action to fight climate change, the answer was unanimous. All the respondents agreed that the matter is important and that something has to be done. However, when they explained the concrete actions, they were undertaking to meet the above agreement, few of them actually had a matching behaviour.

– **Barriers and Motivators for Green Behaviours**

A sense of emergency was felt in general by the respondents. The majority of answers pointed towards the immediate necessity to save the planet and care for the environment in order to ensure the survival of humankind. Furthermore, half of the respondents experienced a sense of guilt regarding the consequences on the planet caused by human actions. They feel responsible for it and therefore feel the need to make up for it.

– **Influence of Rewards on Green Behaviours**

Although all the respondents acknowledged the threat of climate change and the urge to act, many of them openly admitted not doing much. The feelings ranged from indifference to not doing enough.

– **Games Benefits**

The respondents shared the reasons behind their motivation for playing games. The most recurring one was the social aspect as most respondents enjoy playing with or against other people. Other motivations were named such as playing games as a distraction or for relaxation.

This led to the generation of the following four research propositions that were confronted with the scientific literature.

Proposition 1: Feedback to Motivate Green Behaviours

A striking phenomenon identified in the results summary was the lack of perception of concrete consequences of sustainable choices. Moreover, this lack of perception creates the impression of having unlimited resources and therefore no need for restrictions or limitations on people's lifestyles. Overall, a lack of concrete perception of consequences for negative and positive behaviours prevents people from engaging in green behaviours. This issue leads to the following hypothesis: *Providing feedback that shows concrete effects of sustainable choices increases people's awareness and engagement in green behaviors.*

Proposition 2: Social Comparison to Decrease Discouragement

The second phenomenon raised from the interviews was social comparison. Respondents tended to mention other people when discussing the climate issue. For instance, there was a general understanding that helping with climate change is a collective effort to make. Thus, the hypothesis formed from it is the following: *Displaying others' green behaviors will consequently decrease the discouragement of the necessity for collective efforts and motivate people to engage in sustainable behaviors.*

Proposition 3: Acknowledgement and Sustainable Goals Contribution as Rewards

Another phenomenon discerned during the interviews was the need for incentives to stimulate pro-environmental behaviours. Green efforts are often regarded as a chore and as a cost to comfort. People's habits are strongly embedded, and they believe that engaging in green behaviour has consequences for their lifestyle. All in all, people need incentives to engage in green behaviours, but they do not necessarily expect direct or monetary rewards. Instead, satisfaction from the acknowledgment of concrete contribution and real changes induction might be enough. Therefore, the hypothesis proposed is: *Guaranteeing people that their sustainable actions are acknowledged and contribute to sustainable goals is an effective alternative to monetary rewards.*

Proposition 4: Intervention of the State to Ease the Process and Set Examples

The last phenomenon noted in the interviews was related to the expectation of higher institutions to intervene in the climate issue. Firstly, there were suggestions for governmental intervention to establish laws and policies to force people to be more sustainable. Also, there was a general understanding that certain people carried out sustainable actions such as recycling due to them being mandatory through fines and penalties. Secondly, there was a significant reproach issued towards the passivity of governments and big companies. As the intervention of governments or states could nudge the population to engage in pro-environmental behaviours, the following hypothesis is proposed: *The intervention of the state gives credibility to the population's sustainable behavior and consequently motivates them to perform green actions by providing proper infrastructures and setting a trustworthy example.*

5 Discussion and a New Mi-Fi App Design

There is an evident need for people to realize that their actions do have an impact on the planet and human life. Each action contributes to worsening or improving the situation related to climate change. Moreover, as resources become scarcer, there is an emergency for people to realize that resources are not unlimited, and one should be careful with their consumption. Providing people with feedback on their consumption enables them to have a concrete visualization of their impact, which is a necessary element to implement in the game. This can be done through the personification of a character to simulate the consumption of resources or by displaying the corresponding consequences in a virtual scenario.

Additionally, the social aspect is one of the most important to be implemented in the game. People seem to seek validation and reassurance. The game could act as a bridge to prove to people that they are not alone in making efforts and that their individual actions join collective efforts. Displaying other players' efforts and activity could be a way to convince them that everyone is engaged in the matter. Not only do people perceive fairness and being part of a group, but they also have the opportunity to show that they are contributing as well. They are provided a means to showcase their efforts and gain social recognition for that. The game mechanics to be used would be a leaderboard, competition, or the creation of a community.

Furthermore, it appears that incentives are indispensable for people to initiate sustainable actions. People seem to expect rewards for their contributions as they perceive it as a tedious chore. However, following the second hypothesis, tangible rewards are not necessarily expected but an acknowledgment of contribution seems to suffice to be effective. Therefore, the game could combine feedback and acknowledgment. Using feedback to effectively point out the consequences of people's actions and then to quantify their impacts. For instance, telling people how much of something they contributed to save or what difference they made by performing a sustainable action. Acknowledging their contribution to sustainable goals might encourage them to maintain their green behaviours.

Finally, state or government intervention can help in many ways. It can provide and finance the infrastructures needed to ease people's tasks to help with climate change, like recycling. It can act as a referee that monitors people's activity and rewards or punishes those who do not follow the rules. By doing that, they provide fairness and trust so that people carry on their sustainable duties. Moreover, showing that the state is involved in the climate change issue sets the right example for people to follow the initiative and make it more credible. In this case, the game being proposed and developed by the state can build trust with the citizens as it shows that the city cares about the environment and is willing to step up and act. This can create momentum and lead citizens to do the same. Moreover, it contributes to ensuring that people cannot cheat as it is monitored by a higher institution. Following the above research, the area that needs more focus and that has not been covered in the past projects submitted by the students is energy management in households. Therefore, the recommendation proposes ECOTREE, a simulation game where energy reduction in the household is at the core of the gaming and sustainable experience. The game aims at inducing people to save as much energy as possible at home. The efforts will be directed towards electricity, water, and heating consumption reduction. Energy is unknowingly consumed every day through small actions. Daily habits such as turning on lights, the heater, or the preparation of a meal consume energy without people noticing it. People seem to take it for granted and perceive the provision of energy as unlimited and running freely. Therefore, ECOTREE aims at raising awareness of energy consumption with the objective to reduce it by changing people's habits. Saving energy does not only benefit sustainable goals but also people themselves as the reduction of energy consumption will be reflected on their bills.

The goal of the game will be to take care of virtual little trees (see Fig. 1), called *Treenergy*. The game is similar to the game Tamagotchi, a keychain-sized virtual pet

Fig. 1 A few snapshots of the ECOTREE app

simulation game. However, the game will run on a mobile application that can be easily downloaded on smartphones or tablets. The main objective of the game will be to maintain a high level of health and happiness for the little trees, which are directly influenced by people's energy consumption at home. Therefore, the better they are at managing their consumption, the happier and healthier their *Treenergy* will be and be able to grow and will gain eco-badges (see Fig. 2).

Fig. 2 App pages related to the obtention of eco-badges

6 Conclusion

The city of Sierre has instigated considerable initiatives to commit to climate adaptation and to face one of the biggest threats of the current time. The climate emergency is worsening and one way to contribute to its improvement would be for individuals and large organizations to align. The idea of stimulating pro-environmental behaviors among the population with games has emerged in recent years, generating many studies and attempts on the topic. Sierre has decided to take advantage of that trend and apply it to its own population in order to achieve the climate objectives requiring the participation of the citizens. Using a qualitative research method, this research carried out in-depth semi-structured interviews with 10 respondents. The research identified the motivators and barriers that encourage or prevent individuals from engaging in green behaviors. Supported by a review of the theory and application of gamification in a sustainability context, the results enabled the formation of four hypotheses to stimulate pro-environmental behaviors. This research is part of a more general collaboration between our university and the city of Sierre to develop applied research in the field of eco-games. The specific research in this collaboration focused on elements of service design and UX design (for an application) to better consider the new notion of "phygital" experience. Without us being aware of it, our critical infrastructures have become "phygital", i.e., a combination of physical and digital, and therefore entirely dependent on electricity supply and computer/internet/telecommunication networks. The use of an eco-game app can be beneficial in terms of behavioral change, as we have seen in our literature review, but should not make us forget that it consumes energy. It is thus this type of research direction, based on this kind of paradox, that we want to focus on in the future.

References

1. Morganti, L., Pallavicini, F., Cadel, E., Candelieri, A., Archetti, F., Mantovani, F.: Gaming for earth: serious games and gamification to engage consumers in pro-environmental behaviors for energy efficiency. Energy Res. Soc. Sci. **29**, 95–102 (2017)
2. Douglas, B.D., Brauer, M.: Gamification to prevent climate change: a review of games and apps for sustainability. Curr. Opin. Psychol. **42**, 89–94 (2021)
3. Wu, X., Liu, S., Shukla, A.: Serious games as an engaging medium on building energy consumption: a review of trends, categories and approaches. Sustainability **12**(20), 1–16 (2020)
4. Ro, M., Brauer, M., Kuntz, K., Shukla, R., Bensch, I.: Making cool choices for sustainability: testing the effectiveness of a game-based approach to promoting pro-environmental behaviors. J. Environ. Psychol. **53**, 20–30 (2017)
5. Aguiar Castillo, L., Torres, J., De Saá-Pérez, P., Pérez-Jiménez, R.: How to encourage recycling behaviour? The case of WasteApp: a gamified mobile application. Sustainability **10**, 1544 (2018)
6. Neset, T.-S., Andersson, L., Uhrqvist, O., Navarra, C.: Serious gaming for climate adaptation-assessing the potential and challenges of a digital serious game for urban climate adaptation. Sustainability **12**(5), 1789 (2020)
7. Baca-Motes, K., Brown, A., Gneezy, E., Keenan, E., Nelson, L.: Commitment and behavior change: evidence from the field. J. Consum. Res. **39**, 1070–1084 (2013)

8. White, K., Habib, R., Hardisty, D.J.: How to SHIFT consumer behaviors to be more sustainable: a literature review and guiding framework. J. Mark. **83**(3), 22–49 (2019)
9. Negrua, A.L., Toader, V., Sofic, A., Tutunea, M.F., Rus, R.V.: Exploring gamification techniques and applications for sustainable tourism. Sustainability **7**(8), 11160–11189 (2015)
10. Polaine, A., Løvlie, L., Reason, B.: Service design: from insight to implementation. Rosenfeld Media (2013)

Challenges of Municipal Solid Waste Management in Jalandhar, Punjab (India): A Case Study

Davinder Singh and Sanjeev Kumar

Abstract Jalandhar is the largest city in the Doaba province of Punjab, India. Being the third most populous city in the state, the daily disposal of Municipal solid waste is not an easy task for the authorities involved. The study targets the type of waste produced, method of transportation, and final disposal technique used by the Municipal corporation of Jalandhar city. For this paper, official data has been collected, and the site has been visited in person to comment on the ground situation of the disposal grounds. The environmental, social, economic, and physio-chemical factors have been considered to evaluate the standards. The case study suggests an urgent need to change the method of disposal and scientifically design a waste management plan to achieve the norms laid by the national plans.

Keywords Landfill · MSW · Solid waste · Disposal · Sustainability

1 Introduction

Municipal solid waste management (MSWM) of a city directly relates to the rate of urbanization, economic growth, environmental quality, and community safety. Studies commented on the effect of poor MSWM policies causing severe impact on society, environment, and economy, suggesting the need to tackle the problem in a technical manner considering all the mentioned aspects [1–3]. Compared to the developed nations, the developing ones are performing very poorly due to a lack of capital and technology investments in the field of MSWM [4]. Although the per capita waste generation is considerably lesser than the developed nations, developing countries seek cost-efficient alternatives to deal with this challenge [5, 6]. Population growth and exponential urbanization are the two major contributors to this issue

D. Singh (✉) · S. Kumar
Dr. B. R. Ambedkar National Institute of Technology Jalandhar, Jalandhar, Punjab 144027, India
e-mail: singhdj@nitj.ac.in

S. Kumar
e-mail: sanjeevktl1993@gmail.com

E. Strauss (ed.), *Proceedings of the 7th International Conference on Civil Engineering*, Lecture Notes in Civil Engineering 371, https://doi.org/10.1007/978-981-99-4045-5_18

[7, 8]. Urban management bodies in developing countries face many challenges since they lack adequate funds for all waste management activities [9, 10]. Since waste management is cost-driven and cannot generate significant revenues, there is urgent necessity to address this issue to overcome the challenges and develop an integrated MSWM plan to avoid massive landfilling.

In India, land use is limited, but landfilling is considered the most popular method due to its easy implementation and low-costs. With options like incineration, pyrolysis, vermicomposting, and others, landfilling doesn't provide a sustainable solution [11–13]. The conditions are almost the same for different states and provinces. Addressing the need for an effective MSWM plan, the paper reports the real-time situation of MSWM in Jalandhar city. The study reveals that one-third of the waste generated by the various parties never reaches the waste treatment facility.

Furthermore, municipalities mainly concentrate on collecting but don't provide advanced treatment. By integrating the informal sector into the main streamlining of MSWM, resourceful material recovery could be achieved. Municipalities and government bodies in cities like Alleppey, Surat, Pune, and Panjim have worked for better collection and management systems [11]. These cities are an example for the authorities to focus on advanced treatment to achieve the title of sustainable cities. Therefore, the present study presents the possible challenges, opportunities, and strategies for a better MSWM in Jalandhar. Thus, the problem of MSWM is becoming obstinate and affecting the various stakeholders and their efforts.

1.1 Significance of the Study

Several studies have already commented on the conditions and policies of SWM in different cities of Punjab. The study targets Jalandhar as a modal area for the state due to its prime location 90 km from Amritsar and almost 375 km from Delhi, the capital of India. Jalandhar is well known for its sports industry and consists of significant districts like Hoshiarpur, Kapurthala, and Jalandhar. The study targets the waste management of the city area, which consists of one million people.

1.2 Various Contributors of MSW from the City

The primary sources of MSW are residential, commercial, institutional, C&D, municipal operations, industrial, and agriculture activities. Figure 1 represents the daily classification of generated waste by weight (%) through various means.

Fig. 1 Classification of generated waste by weight (%)

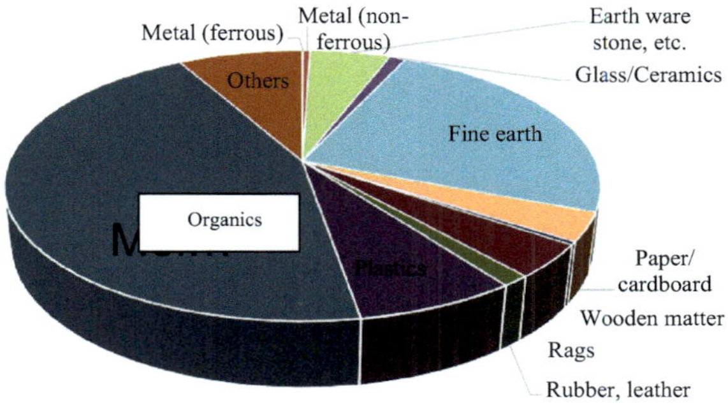

2 Comparison to Model Cities

Under the "Swachh Bharat Abhiyan" scheme, the Central Pollution Control Board (CPCB) of India ranks the city based on its yearly cleanliness performance. The rank for the city has degraded by 42 notches as it ranked 161st in the year 2021, which was 119th in the year 2020. Since different states of the country try their best to perform their best on this index, some of them invest in new technologies and methods for effective MSW disposal. Table 1 gives the comparison of figures of the cities and Jalandhar city. Surat and Pune generate energy from the waste and also work on biogas generation and composting from MSW. Alleppey, Panjim, and Bobbili lack waste to energy generation facilities and are hence only limited to decentralized biogas and composting facilities. These operations allow job opportunities in the formal and informal sections related to MSWM and help in revenue generation to make better policies for future requirements.

Table 1 Data related to model cites with better MSWM policies and implementation

City	Population (lakhs)	Area (km^2)	Total waste generation (Mt/day)	Waste fraction	
				Recyclables (inorganic) (%)	Compostable (organic) (%)
Surat	44.6	326.5	2000	11	56.87
Panjim	0.4	36.0	29.7	17	61.75
Alleppey	1.9	46.2	58.0	16	75
Pune	35	244	2100	16	62.44
Bobbili	0.5	25.6	14.0	14	70
Jalandhar	10.9	110	47.0	18	68

Fig. 2 **a** Satellite image of Wariana dumping yard and **b** open dumping at Wariana dump yard

3 Status of Recycling and Recycling Process in Jalandhar City

Jalandhar Municipal Corporation has signed an MOU with Punjab Grow More Fertilizers Ltd. A factory was established in the village of Basti Bawa Khel. It was the MCJ junkyard. MSW organized in this process turns into manure during composting. The site has 14 acres of land with 600 t/day. About 100 t/day SW is transferred to the Wariana site. About 250 tons used to be dumped at the SuchiPind site (now closed). Two requirements must be met for all MSW produced in Jalandhar to be used for organic fertilizer production. First is the requirement of more landfills, with a minimum 30 year lifespan; the other is good marketing by the government to sell organic fertilizers [14]. Figure 2 represents the satellite image of Wariana Dumping Yard and Open dumping at the Wariana dump yard.

4 Recommended Measures

4.1 Integrated Solid Waste Management

Implementing integrated solid waste management (ISWM), a scheduling approach for improved waste management techniques is needed. ISWM consists of five significant steps, including the set objectives, identifying alternatives, comparing alternatives, selecting suitable options, and implementing. The concept has worked for both developing and developed nations to scientifically address the problem of waste management. The municipal council should support interested townships in ISWM planning and implementation.

4.2 Effective Waste Collection

The city spends enormous funds on pickup containers to collect garbage from home and the streets. The garbage collection system must be up to date to reduce the difficulties caused by system failure. This would be beneficial to improve the efficiency with which garbage is collected. Timely pickup using Trucks to collect rubbish from the construction site can avoid delays in the waste collection. The local council has been accused of using a marginalization strategy against the city's residents for many years. The state government needs to be more sensitive to garbage collection and can better manage the problem. To incentivize human contributions, better budget allocations are needed for lorries and containers and to cover current expenses such as drivers and machine operators.

4.3 Public Awareness

The survey and verbal interview revealed that most people lack basic solid waste management abilities, resulting in widespread public carelessness. The city council needs to start an education and outreach plan to educate citizens about the relevance of solid waste management to public health and the environment. Basic SWM skills and the benefits of a sound SWM system can be highlighted in such plans. Another way to educate the public about solid waste is through forums and billboards. Interactive billboards to educate the public about the detrimental effects of littering on the environment can create awareness.

4.4 Composting

Composting of organic waste generated by households and recreational facilities such as pubs and restaurants should be encouraged, with financial incentives. Jalandhar city has the potential to consume the compost since the major economic activity is agriculture happening in the nearby rural areas. Thus, the compost can fetch enough revenue for the stakeholders to manage the composting operations effectively. Aerobic composting requires land and manual efforts; thus, it can create employment opportunities for new eco-based startups.

4.5 Reuse, Recycling, and Waste Recovery

The concept of 3R needs to be implemented by involving waste generators and interested consumers on the same platform. There is an urgency to formalize and initiate a conversation between the SWM stakeholders and the community groups. Minimizing the waste to be treated can save man, money, and power; hence more stress is needed on the recycling and reuse of MSW. Credit-based performance can be one exciting way to initiate this conversation. Incentives to the groups, townships, and other organizations can support the implementation of reuse, recovery, and recycling among the public.

4.6 Solid Waste Segregation Plant

A centralized facility meant to segregate sufficient capacity should be constructed to handle the daily waste. The facility will ensure proper segregation of hazardous or non-hazardous, dry or wet waste, organic or inorganic, and combustible or incombustible waste. The same can be shared with the interested parties who can consume it as a raw material for further processing into recycled goods. The facility can generate revenue by trading the segregated waste to these stakeholders. This will save the burden on existing landfills and ensure proper treatment and dumping of MSW in Jalandhar city.

4.7 MSW Incineration Plant

The phenomenon of converting energy from waste is well known to humankind. The same concept can be practiced in a scientifically developed facility to maximize the energy output from MSW of high calorific value [4, 17]. The waste which needs to be the raw feed for such operations comes from the segregation facility so that the other components can be handled by other suitable means. Figure 3 gives the schematic diagram of the typical MSW handling/process. Incineration plants are pretty common in major cities in India, and Jalandhar can also plan for one. The residue obtained after burning can be used as raw feed for other industries, and interested stakeholders can invest in such plants for maximum efficiency [18].

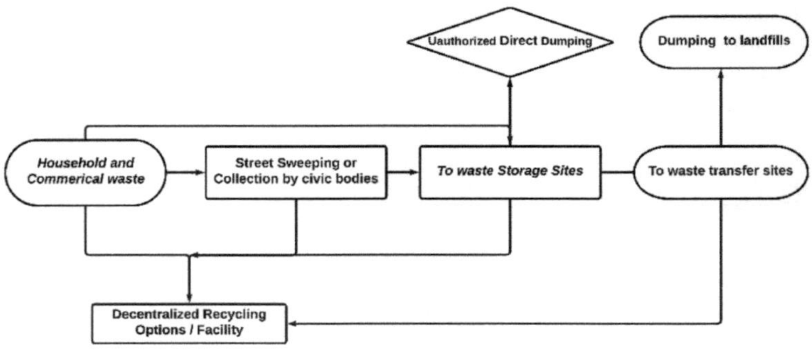

Fig. 3 Schematic diagram of typical MSW handling/process

4.8 *Legislation and Enforcement*

The proper implementation of rules and regulations was found missing on the ground level. There were provisions for hefty fines on offenders responsible for open dumping and burning, but only a fraction of cases are reported to authorities. People must deal with this situation actively, and the council must encourage people to report such cases. Municipal authorities can work with non-governmental groups and other non-profit organizations to educate people about these rules. This can also work for the timely address of MSW-related grievances and will lead to a clean and green Jalandhar.

5 Discussion

The present work reports that a lack of resources, budget allocations, planning, and poor administration is resulting in the failure of the present MSW collection and disposal model in the city. There is an urgent need to address the issues of sub-marginal communities associated with waste collection and recycling and to create general awareness among the city residents. Timely Environmental impact assessment of the adopted methods can assess and address the necessary changes in the planning and implementation. Although the government and non-governmental stakeholders are trying their best to work efficiently, an effective mechanism is missing. Compared to the nearby cities, MSW disposal conditions are not different. The report suggests involving a public–private partnership model, which can be initiated at a smaller scale as a trial. Human resources need to be trained and equipped with the latest safety equipment since they are directly exposed to MSW in pandemic and normal conditions. As suggested above in the recommendations, a credit-based incentive program, penalties on the offenders, and user charges should be implemented to promote MSWM in the city.

References

1. Bundhoo, Z.M.A.: Solid waste management in least developed countries: current status and challenges faced. J. Mater. Cycles Waste Manag. **20**, 1867–1877 (2018)
2. Mohee, R., Mauthoor, S., Bundhoo, Z.M.A., Somaroo, G., Soobhany, N., Gunasee, S.: Current status of solid waste management in small island developing states: a review. Waste Manag. **43**, 539–549 (2015)
3. Mir, I.S., Cheema, P.P.S., Singh, S.P.: Implementation analysis of solid waste management in Ludhiana city of Punjab. Environ. Challenges **2** (2021)
4. Alam, P., Singh, D., Kumar, S.: Incinerated municipal solid waste bottom ash bricks: a sustainable and cost-efficient building material. Mater. Today Proc. (2021)
5. Guerrero, L.A., Maas, G., Hogland, W.: Solid waste management challenges for cities in developing countries. Waste Manag. **33**, 220–232 (2013)

6. Karak, T., Bhagat, R.M., Bhattacharyya, P.: Municipal solid waste generation, composition, and management: the world scenario. Crit. Rev. Environ. Sci. Technol. **42**, 1509–1630 (2012)
7. Kumar, S., Deswal, S.: Comparative assessment of Kurukshetra city waste dumping sites using RIAM analysis: a case study. Lect. Notes Civ. Eng. **154**, 31–38 (2022)
8. Kumar, S., Singh, D.: Municipal solid waste incineration bottom ash: a competent raw material with new possibilities. Innov. Infrastruct. Solut. **6** (2021)
9. Mesjasz-Lech, A.: Municipal urban waste management—challenges for polish cities in an era of circular resource management. Resources. **10**, 55 (2021)
10. Ngullie, N., Maturi, K.C., Kalamdhad, A.S., Laishram, B.: Critical success factors for PPP MSW projects—perception of different stakeholder groups in India. Environ. Challenges **5**, 100379 (2021)
11. Kumar, A., Agrawal, A.: Recent trends in solid waste management status, challenges, and potential for the future Indian cities—a review. Curr. Res. Environ. Sustain. **2**, 100011 (2020)
12. Ghosh, S.K.: Sustainable SWM in developing countries focusing on faster growing economies, India and China. Procedia Environ. Sci. **35**, 176–184 (2016)
13. Das, A.K., Mukherjee, J., Chatterjee, U.: Importance-performance analysis to assess community role in solid waste management in the Hooghly district, West Bengal. Innov. Infrastruct. Solut. **7**, 1–20 (2022)
14. Puri, A., Kumar, M., Johal, E.: Solid-waste management in Jalandhar city and its impact on community health. Indian J. Occup. Environ. Med. **12**, 76 (2008)
15. Pastapure, V., Singh, D., Kumar, S.: Proceedings of Indian geotechnical and geoenvironmental engineering conference (IGGEC) 2021 Vol. 2 Effects of open dumping of municipal solid waste on surrounding soil characteristics: A review springer nature Singapore, Singapore 47–54 (2023)
16. Ranjan, S., Singh, D., Kumar, S.: Proceedings of Indian geotechnical and geoenvironmental engineering conference (IGGEC) 2021 Vol. 2 Analysis of landfill leachate and contaminated groundwater: A review springer nature Singapore, Singapore 55–62 (2023)
17. Kumar, S., Singh, D.: Transforming waste into sustainable solution: Physicochemical and geotechnical evaluation of cement stabilized municipal solid waste incinerator bottom ash for geoenvironmental applications. Process Saf. Environ. Prot. 176685–695. https://doi.org/10.1016/j.psep.2023.06.026. (2023)
18. Kumar, S., Singh, D.: From waste to resource: Evaluating the possibility of incinerator bottom ash composites for geotechnical applications. Int. J. Environ. Sci. Tech. https://doi.org/10.1007/s13762-023-04919-4

Architectural Design and Structural Mechanics

Evaluating Energy-Saving Potential of Passive Design Technologies Based on Residential Architectural Prototypes

Jiuwei Liu, Yuanli Ma, and Wu Deng

Abstract This research project explores the energy-saving potential of passive design technology under specific building prototypes. Five passive technologies were determined to be evaluated by the energy-saving effectiveness, which are building orientation, airtightness, external wall U-value, roof U-value and window U-value. Firstly, nine prototype models are built in IESVE to perform energy consumption. Based on the orthogonal table, a total of 144 groups of energy consumption data were obtained and conducted into SSPS to do a linear regression analysis to obtain the sensitivity ranking of the five passive technologies in each prototype. Generally, for the single passive technology, Roof U value and orientation were the most effective passive technical means to reduce energy consumption, followed by Windows U value. Air tightness and wall U value are less significant factors. When using the combination of passive technology to achieve the desired energy-saving goal of 65%, among the pairwise solutions, only the combination of roof and window optimization solutions was successful, and both techniques were verified as the most effective strategies during the individual test stage. The overall results of this research can provide architects with some reference when designing new residential buildings or renovating existing residential buildings. In future studies, the researcher can carry out more building prototypes for other regions and related sensitivity analysis of other design technologies, to provide an integrated framework for building retrofit and green building design.

Keywords Architectural prototypes · Passive design technology · Sensitivity analysis

---

J. Liu
Department of the Built Environment, College of Design and Engineering, Kent Ridge Campus, National University of Singapore, Singapore, Singapore

Y. Ma (✉)
Pan Tianshou College of Architecture, Art and Design, Ningbo University, No. 616 Fenghua Road, Jiangbei District, Ningbo, Zhejiang Province 315211, China
e-mail: Yuanli.Ma@nottingham.edu.cn

J. Liu · Y. Ma · W. Deng
Department of Architecture and Built Environment, University of Nottingham Ningbo China, Ningbo, China

E. Strauss (ed.), *Proceedings of the 7th International Conference on Civil Engineering*, Lecture Notes in Civil Engineering 371, https://doi.org/10.1007/978-981-99-4045-5_19

1 Introduction

In the past decade, environmental sustainability and energy conservation have become common problems faced by all countries and regions in the world, and China becomes the world's largest emitter of CO_2 [1]. In September 2020, President Jinping Xi announced "double carbon" targets at the United Nations General that China aims to achieve carbon peak by 2030 and achieve carbon neutral by 2060. Following this concern, reducing building energy consumption is essential. China's total building energy consumption exceeds other developed countries, which account for 28% of the country's total energy consumption by 2018 [2]. The high energy consumption ratios also contain great potential for energy conservation and consumption reduction in buildings, which has attracted extensive attention from researchers and engineers, mainly focusing on passive design strategies and energy efficiency improvement. However, in China, different building types involving the classification of common parameters in buildings, such as building size, style and construction materials, may affect the effectiveness of various passive technologies, then affect the designers' implementation of specific energy optimization schemes [3]. For different types of buildings, the performance of the same design technology will be different. Therefore, it is necessary to investigate the design technologies with architectural typologies organically to develop more effective energy optimization solutions.

Architectural typology classifies and integrates existing buildings to find "prototypes" with similar internal forms and geometric structures, which represents a series of buildings with similar characteristics [4]. The architects classify and prototype buildings according to the size, style and materials of the towns in the region, gradually shifting the research focus from simple architectural forms to architectural design and performance evaluation, such as thermal comfort, carbon emissions, sound and light [5]. While sensitivity analysis is an uncertainty analysis technique. From the perspective of quantitative analysis, it analyzes the changes of various measured quantities (parameters) when the model size or independent model parameters change within a specified range. Its essence is to change the value of relevant variables to explain the law of key indicators affected by these factors [6].

At present, in most cases of building energy optimization, sensitivity analysis on passive design strategies were not applicable to urban scale building stocks, most of these studies are based on an individual building, the result is not universal and representative. For example, Wang [7] explored the impact of using external shading strategies and adding fresh air systems and their efficient heat recovery equipment on energy consumption in the hot summer and cold winter (HSCW) zone where Ningbo is located. However, this study does not specify the building types, so it cannot prove that these two design strategies are representative of all types of buildings. Therefore, there is currently no reliable reference for the applicability of passive technology for mostly buildings in the HSCW zone. On the other hand, Zhang [8] and Li et al. [9] conducted sensitivity analysis of factors affecting household energy consumption based on univariate regression equation to the study of energy saving performance of passive houses in the HSCW zone. The limitation of their study was that the univariate regression equation could not consider the influence of other factors on the results. The accuracy of multiple regression equation is much higher than that, but at present, sensitivity analysis based on multiple regression model is rarely applied to building performance analysis [8].

This research focused on the HSCW zone, whose heating load is relatively low but fluctuates greatly compared with cold zone. According to the Chinese government policy, this area was excluded from the district heating list, resulting in poor indoor thermal environmental quality in winter. In this background, aiming at different types of buildings, it is of great practical significance to explore the passive design strategies applicability by sensitivity analysis under such climatic conditions. Specifically, this research mentioned the following objectives:

- Identify applicable passive design technologies/factors with corresponding building energy consumption based on 9 building prototypes.
- Obtain sensitivities ranking of each identified technology for each prototype.
- Propose applicable combinations of passive design technologies with 65% energy saving goal.

2 Research Methods

2.1 Flow Chart

The workflow (see Fig. 1) begins with inputting prototype information to build models and set related design parameters in IESVE. During the modeling process, energy simulation data can be obtained after the identification of tested design technologies. Orthogonal methods are then used to carry out sensitivity analysis. The final post-modelling process will demonstrate the effect of technology combination, and how it echoes the results of single technology sensitivity. The following sections will explain the methodology in detail.

2.2 Modeling

In the previous research, Ma et al. [5] used stratified sampling and the K-means clustering method to create 9 architectural geometric prototypes, which were used as basic models for this research's sensitivity analysis. Then the IESVE software is used to simulate energy consumption for nine architectural geometric prototypes (detailed information see Table 1). Some detailed design profiles (human activity, HVAC system, lighting, etc.) for model input are referenced from Code for Thermal Design of Civil Buildings [10], and these profiles will be kept constant when changing the value of the passive design factor.

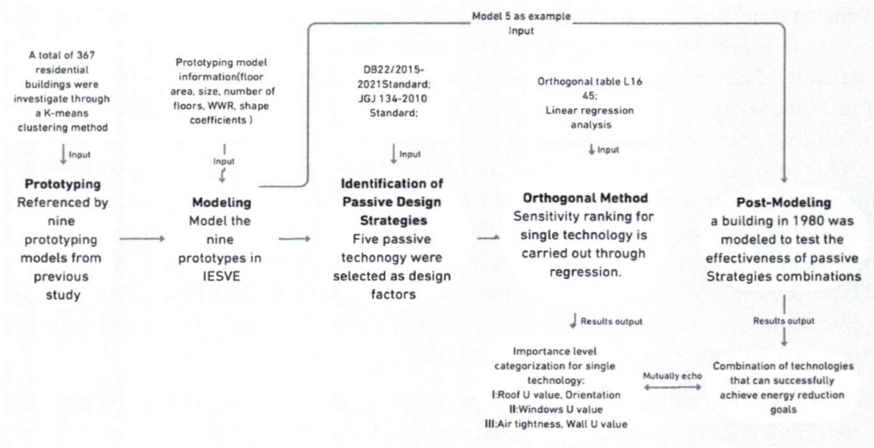

Fig. 1 Overview of the research flow chart

Table 1 Prototyping model information [5]

Model 1 (*Length; Width; Height(m):* 14.62, 9.09, 8.99)	**Model 2** (26.65, 8.23, 9.23)	**Model 3** (50.79, 11.27, 9.75)
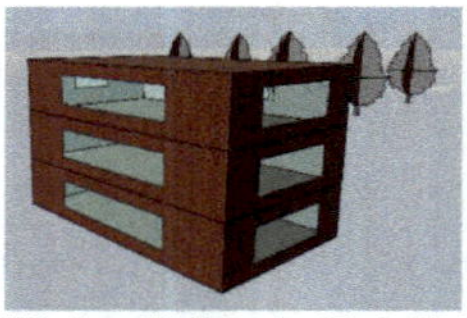 Window Wall Ratio (*E, S, W, N*): 0.08, 0.18, 0.31, 0.28	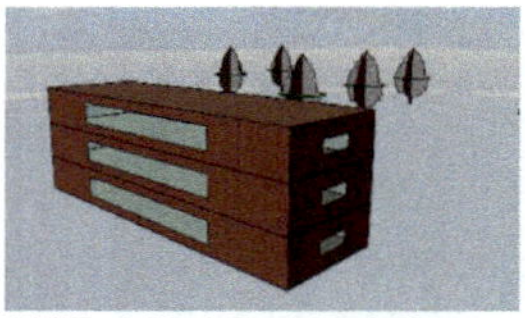 WWR: 0.09, 0.28, 0.08, 0.23	 WWR: 0.1, 0.27, 0.07, 0.22
Model 4 (29.31, 12.82, 14.91)	**Model 5** (52.85, 14.11, 16.83)	**Model 6** (43.54, 11.47, 21.62)
 WWR: 0.09, 0.39, 0.08, 0.2	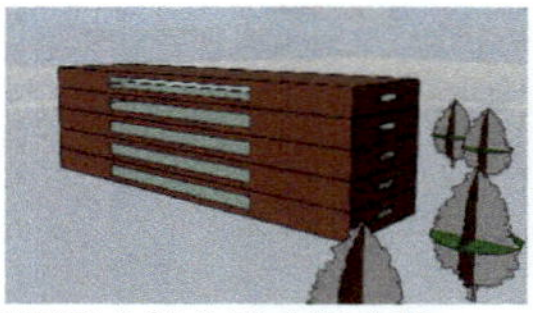 WWR: 0.05, 0.25, 0.05, 0.16	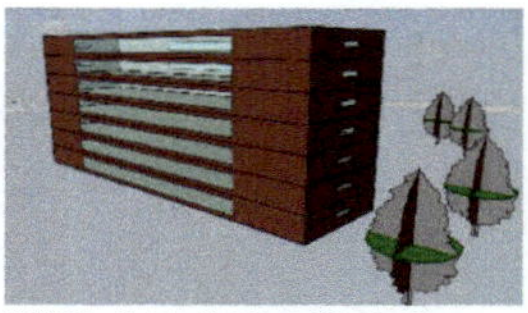 WWR: 0.04, 0.39, 0.03, 0.19
Model 7 (26.67, 15.53, 32.45)	**Model 8** (39.1, 16.14, 52.92)	**Model 9** (35.93, 18.34, 96.23)
 WWR: 0.09, 0.31, 0.07, 0.23	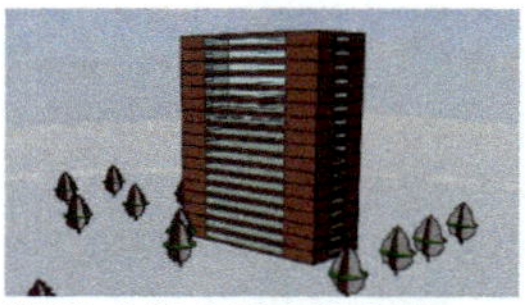 WWR: 0.09, 0.35, 0.09, 0.25	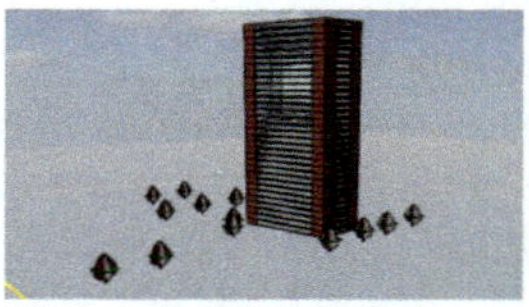 WWR: 0.04, 0.39, 0.03, 0.19

2.3 Identification of Passive Design Strategies

Combined with the performance parameters and climate analysis, six main points need to be considered when building a passive house in the HSCW zone: ventilation and cooling requirements in summer, thermal insulation, shading, dehumidification demand, anti-cold requirements in winter and solar energy utilization [9]. In response to these considerations, this research performed a multivariable sensitivity analysis by varying the values of five passive strategies factors used in the HSCW region as below.

Orientation

It has been agreed that it is appropriate for buildings in the HSCW zone to be oriented directly or close to north and south [11]. Generally, southern orientation is consistent with the passive strategy of the HSCW area. For the numerical setting of the orientation variable, it is assumed that there is a deviation around 15° for south orientation. For the deviation to the southwest, the sign is positive; for the southeast deviation, the sign is negative. The deviation to due south is 0. That is, the testing range for orientation varies from -15 to 15.

Airtightness

Chinese DB22/2015–2021 standard specifies that air permeability per unit seam length of energy saving residential buildings should be less than 2.5 m^2/mH [12]. At the same time, with the passive house development, the MoHURD issued "Guidelines" on October 2015 that the airtightness requirement should be less than 1 ACH in HSCW zone [10]. Based on the above standards, the air tightness range of the prototype models in this study is determined to be 0.25–1 ACH.

Thermal Insulation (Window)

According to the cluster model information of residential buildings in Table 1, the overall window/wall ratio of the model is between 0.16 and 0.39. Therefore, according to the provisions of JGJ 134 2010 on the window heat transfer coefficient between this area, the range of U value is 2.0–2.8 W/m^2 K [13]. To have the same amount of variation for each subsequent level, the window U-value range used in this research is determined to be 2.0 to 2.9 W/m^2 K.

Thermal Insulation (Wall)

Chinese JGJ 134-2010 standard regulates the heat transfer coefficient range that is limited between 0.8 and 1.5 W/m^2 K for external wall [13]. According to the Zhejiang residential building energy saving design standard (DB22/2015–2021), the U value range of the exterior wall is stipulated in detail as 0.7–1.0 W/m^2 K [12]. In this research, the external wall U-value range of the prototype models was set from 0.7–1.6 W/m^2 K.

Thermal Insulation (Roof)

Chinese JGJ 134-2010 standard regulates the range of U value for roof is limited between 0.5 and 1.0 W/m^2 K [13]. DB22/2015–2021 standard regulates that the same value should be less than 1 W/m^2 K [12]. Therefore, the roof U-value range of the prototype models in this study was set from 0.5 to 1.1 W/m^2 K.

2.4 Orthogonal Method

The above passive technologies were all divided into 4 levels, and the performance of corresponding building measures from Level 1 to Level 4 and gradually improves (see Table 2). To obtain the optimal system of building energy-saving measures suitable for each model, the orthogonal experimental design method was used to combine the five technologies and multiple regression analysis for the nine representative residential building models. The *orthogonal table L16 45* meets the test conditions and is applied to IESVE energy consumption simulation software to simulate building total energy consumption respectively.

Every prototyping model goes through 16 times simulation. As a result, a total 144 groups of energy consumption data were imported into SPSS for linear regression analysis to obtain single passive strategy sensitivity ranking.

3 Results and Discussion

3.1 Energy Consumption for Modelling Process

As a result, all the simulated building energy consumption data are basically between 10 and 15 kWh/m^2. And the majority are around 10 kWh/m^2, which is in line with the technical standards for near-zero energy buildings, which require residential buildings to consume less than 25 kWh/m^2 of electricity [7]. As the variation range of all passive technologies is set according to the various Chinese building standard

Table 2 Passive design factor level

	Orientation (degree)	Airtightness (ACH)	Thermal insulation		
			U value of window (W/m^2 K)	U value of wall (W/m^2 K)	U value of roof (W/m^2 K)
Level 1	−15	0.25	2.0	0.7	0.5
Level 2	−5	0.5	2.3	1.0	0.7
Level 3	+5	0.75	2.6	1.3	0.9
Level 4	+15	1	2.9	1.6	1.1

documents, and the profile of building heating and cooling documents is not in a long-term working state (it is not opened most of the time on weekdays), the total energy consumption value simulated has even reached the standard of passive housing. Given the initial setup conditions, this is not unusual. Therefore, IESVE simulation data are regarded as accurate and reliable under specific design profile conditions.

3.2 Single Passive Technology Sensitivity

The energy consumption data are imported into SPSS for linear regression analysis. Taking the model 1 (*Low-rise building; detached; point-style*) among the nine architectural prototypes as an example analysis to explain and present the correlation between sensitivity of individual factors and energy consumption. After the model 1 energy consumption data was imported into SPSS for linear regression analysis, and the results are shown in Fig. 2.

The determination coefficient R^2 is used to represent the percentage of the model obtained by fitting that can explain the change of dependent variable. The closer R^2 is to 1, the better the fitting effect of regression model is. $R^2 = 0.995$ in model 1, indicates that the fitting degree of the model effect is very high. The standardized regression coefficient of each variable indicates their corresponding sensitivity for building energy consumption. If the coefficient is positive, there is a positive correlation between variables, and vice versa. The absolute value of the coefficient can then be used as a direct comparison of sensitivity.

Model Summary[b]

Model	R	R Square	Adjusted R Square	Std. Error of the Estimate	Durbin-Watson
1	.997[a]	.995	.992	.43841	2.426

a. Predictors: (Constant), Uroof, Uwall, Uwindow, Airtightness, Orientation
b. Dependent Variable: Energy

Coefficients[a]

Model		Unstandardized Coefficients		Standardized Coefficients	t	Sig.	Collinearity Statistics	
		B	Std. Error	Beta			Tolerance	VIF
1	(Constant)	26.075	1.033		25.231	.000		
	Orientation	.218	.098	.051	2.224	.050	1.000	1.000
	Airtightness	12.132	.392	.712	30.939	.000	1.000	1.000
	Uwindow	3.842	.327	.270	11.757	.000	1.000	1.000
	Uwall	8.865	.327	.624	27.129	.000	1.000	1.000
	Uroof	3.215	.490	.151	6.559	.000	1.000	1.000

a. Dependent Variable: Energy

Fig. 2 Example regression results of model 1

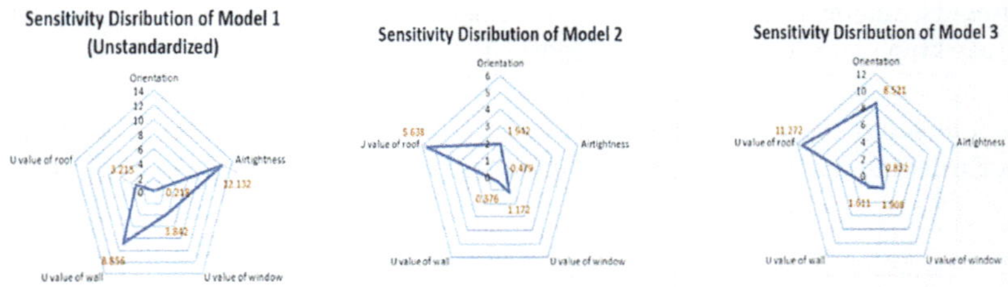

Fig. 3 Sensitivity distribution of model 1, 2 and 3 respectively

Low-rise Building (model 1, 2 and 3) (see Fig. 3)
For model 1, the sensitivities of passive design technologies can be ranked as: $Air\ tightness > Wall\ U_{value} > Window\ U_{value} > Roof\ U_{value} > Building\ orientation$. For model 2, it ranks as: $Roof\ U_{value} > Building\ orientation > Window\ U_{value} > Air\ tightness > Wall\ U_{value}$. For model 3, it ranks as: $Roof\ U_{value} > Building\ orientation > Window\ U_{value} > Wall\ U_{value} > Air\ tightness$

Multi-story and Mid-rise Building (model 4, 5 and 6) (see Fig. 4)
For model 4, it ranks as: $Roof\ U_{value} > Building\ orientation > Window\ U_{value} > Air\ tightness > Wall\ U_{value}$. For model 5, it ranks as: $Building\ orientation > Roof\ U_{value} > Window\ U_{value} > Wall\ U_{value} > Air\ tightness$. For model 6, it ranks as: $Roof\ U_{value} > Building\ orientation > Window\ U_{value} > Air\ tightness > Wall\ U_{value}$

High-rise Building (model 7, 8 and 9) (see Fig. 5)
For model 7, it ranks as: $Roof\ U_{value} > Building\ orientation > Window\ U_{value} > Air\ tightness > Wall\ U_{value}$. For model 8, it ranks as: $Roof\ U_{value} > Building\ orientation > Window\ U_{value} > Air\ tightness > Wall\ U_{value}$. For model 9, it ranks as: $Building\ orientation > Roof\ U_{value} > Window\ U_{value} > Air\ tightness > Wall\ U_{value}$

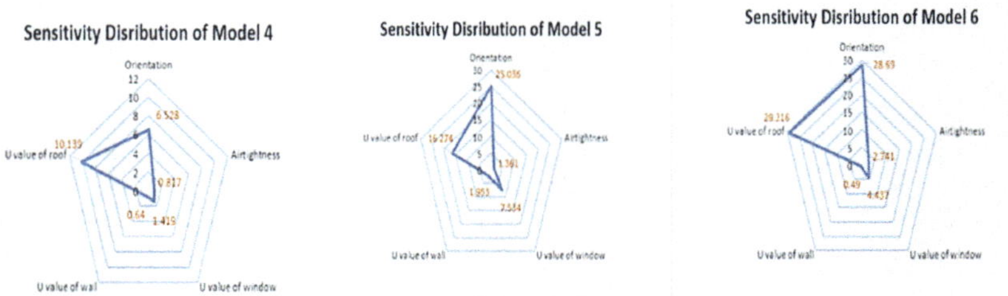

Fig. 4 Sensitivity distribution of model 4, 5 and 6 respectively

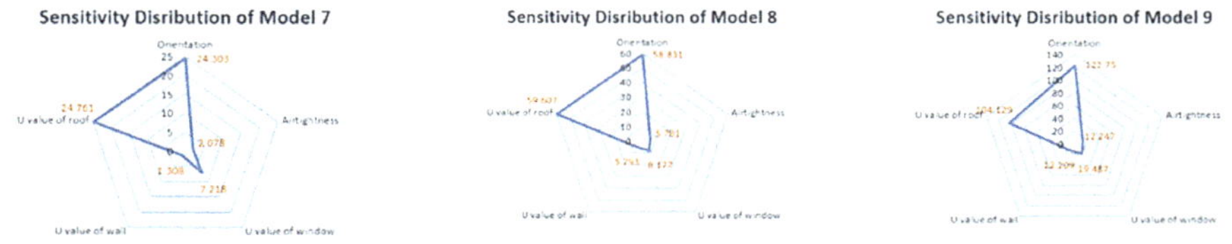

Fig. 5 Sensitivity distribution of model 7, 8 and 9 respectively

Table 3 Importance level categorization of the effect of reducing building energy consumption

Importance level (from high to low)	Passive technologies
I	U-value of roof, orientation
II	U-value of window
III	Air tightness, U-value of wall

Overall Analysis for All Building Prototype Results

In general, the sensitivity ranking of the five passive technologies in the nine architectural prototypes tends to be similar. Basically, the five passive technologies can be grouped into below three importance levels (see Table 3).

According to the above summarization of the importance levels of classification, architectural designers can have a general basis for the architectural design planning of the HSCW region just like Ningbo. Under the limited cost control, the corresponding passive technology design scheme can be reasonably applied to reduce energy consumption and operating cost of the building.

3.3 Passive Technologies Combinations to Reduce Building Energy Consumption to Set Goals

In the following post-modelling, a building in 1980 will be renovated according to the passive house standard in Shanghai, and the passive technology involved in the passive house standard will be used to realize the goal of achieving the energy saving rate of 65%, since the updated edition of '*Design Standard for Energy Efficiency of Public Buildings* GB 50189-2015' released in 2016 anticipates that new buildings should reach 65% energy reduction of 1980's baseline [14]. Here, the building prototype model 5 mentioned above is taken as an example, the building Envelope thermal property will be set according to '*Design Standard for Energy Efficiency of Public Buildings* GB 50189-1980 (the earliest version issued in 1980)' to simulate the initial energy consumption, in order to represent a building built in 1980 [15].

According to Schulz et al. [16], the properties of passive houses in Shanghai are summarized based on their research on Chinese passive housing development. These characteristics of passive houses in Shanghai will be used as an optimized standard to the corresponding features (i.e., each passive design is applied). According to the GB50189-1980 standard, the original settings of the 1980s building and the optimization scheme of each index are shown in the following Table 4 and marked.

In addition to the passive design methods mentioned above, some other aspects are also upgraded according to the properties of passive houses in Shanghai [16]. These optimizations are assumed to be implemented when studying the energy-saving effect of the above four measures. Then, using a progressing approach to identify the combinations of passive strategies that can be achieved. Firstly, by using a single strategy, then a combination of two, then a combination of three, and so on, applying either a single passive technology or a combination of passive technologies to the original building performance, to see which combinations of optimization can achieve the goal of the aforementioned 65% reduction in energy consumption. To check if a particular technology is still as effective in reducing energy consumption when it combined with other technology as when it is tested alone.

The initial energy consumption of building in 1980's is simulated as 99.86 kWh/m^2. After quantitative optimization measures are being determined, the total energy consumption after these default optimizations reduced to 54.67 kWh/m^2. Then a single passive technology or a combination of technologies is applied, and energy consumption simulation results are shown in the Table 5.

From the results of single optimization, the most significant strategy in energy consumption reduction is ① (optimize roof U-value), followed by ③ (optimize window U-value) and ④ (optimize air change rate), and the least one is ② (optimize wall U-value). This result is the same as the sensitivity ranking result of model 5 previously analyzed. However, the target of 65% energy efficiency cannot be achieved.

In the subsequent application of strategies combination, the building energy consumption has been further reduced. In the combination of the two strategies, only the combination of ① and ③, can the building achieve the target of reducing energy consumption by 65%, which are the most significant strategies for reducing energy

Table 4 Optimization scheme of each index referenced by passive houses in Shanghai [16]

Optimization label	Properties	Optimized values referenced by shanghai passive houses	Original value referenced by GB 50189-1980 [15]	Optimization of change (from original to optimized)
①	Roof: U value	0.17 (W/m^2 K)	1.5 (W/m^2 K)	1.5 → 0.17 (W/m^2 K)
②	Wall: U value	0.24 (W/m^2 K)	2 (W/m^2 K)	2 → 0.24 (W/m^2 K)
③	Window: U value	0.7 (W/m^2 K)	6.4 (W/m^2 K)	6.4 → 0.7 (W/m^2 K)
④	Air change rate	0.6 ACH	1 ACH	1 → 0.6 ACH

Table 5 Energy consumption simulation results for each implementation schedule

	Implementation schedule	Energy saving (compared to initial 1980's one) (kWh/m^2)	Energy saving (%)	65% reduction goal met or not
Single optimization	①	60.58	61	
	②	48.59	49	
	③	56.81	57	
	④	50.31	50	
Combine two optimizations	①②	64.22	64	
	①③	70.21	70	✓
	①④	62.39	62	
	②③	57.81	58	
	②④	54.32	54	
	③④	62.23	62	
Combine three optimizations	①②③	76.13	76	✓
	①②④	68.23	68	✓
	①③④	80.23	80	✓
	②③④	65.32	65	✓
Combine four optimizations	①②③④	81.52	82	✓

consumption in model 5 when tested individually. This verifies that when passive technologies used together, they can still be as effective as they were when tested separately. And the following three strategy combinations and four strategy combinations all successfully achieved this goal. It can be found that when the combination of passive strategies is applied, the energy reduction effect obtained in the numerical value is similar to the superposition of the single strategy in its energy reduction value, and there will be a certain fluctuation, but there will not be a reversal effect. Generally, when the passive technologies are combined, the effectiveness in energy consumption reductions for specific technology is similar as when it is used alone.

4 Conclusion

In conclusion, this research firstly identifies a theme to study the sensitivity of passive technology based on architectural prototypes to better save energy in buildings and contribute to China's strategic goal of achieving carbon neutral by 2060. Through IESVE modelling of 9 building prototypes in the Ningbo area, 16 groups of data corresponding to each building, and 144 groups of energy consumption data in total were obtained according to the orthogonal method. After that, linear regression analysis was conducted by SPSS software to obtain the ranking of sensitivity of 5 passive strategies to reduce building energy consumption under 9 building prototypes. In addition, using passive technologies combination to achieve the desired energy-saving goal is carried out in the subsequent part, and the result also echoes the previous conclusion of single technology sensitivity.

The concluding remarks towards the results of sensitivity analysis and the value of this study are:

- The sensitivity ranking of the five passive technologies in each of the nine building prototypes is relatively similar, with slight variations, but in general: Roof U value and orientation are considered to be the most effective passive technical means to reduce energy consumption, followed by Windows U value. Air tightness and wall U value are relatively low factors towards building energy saving.
- The combined use of passive technologies will further reduce building energy consumption, the energy-saving potential that single passive technologies showed is also applicative when combined with other strategies. To achieve higher energy-saving goals, technology integration is essential.
- The conclusions of passive technology sensitivity involved in this study provides an effective method to establish a comprehensive framework for specific renovation strategies of residential buildings, which is embodied in the following aspects:
  - Architects use the results of these prototypes and the corresponding sensitivity analysis to design new residential buildings and renovate existing residential buildings in the HSCW region.
  - Researchers can carry out more building prototypes for other regions and related sensitivity analysis of other passive design technologies.
  - Designers can propose more cost-effective design solutions and design buildings with more economic benefits.
  - The sensitivity analysis results can provide a reference standard for the implementation of the government building planning policy.

In general, the implementation strategy of passive technology in China needs more relevant research and a large amount of simulation data for theoretical support, to test the passive technology suitable for each region. With the continuous exploration of passive technology, there will be more and more buildings that satisfy passive house standards appear in China. On the one hand, the development of efficient buildings will bring huge economic benefits to China's construction market. On the other hand, it will also greatly promote the early realization of China's goal of carbon neutrality.

References

1. Cao, X., Dai, X., Liu, J.: Building energy-consumption status worldwide and the state-of-the-art technologies for zero-energy buildings during the past decade. Energy Build. **128**(1), 198–213 (2016)
2. Dun, M., Wu, J.: Forecasting the building energy consumption in China using grey model. Environ. Process. **7**(1) (2020)
3. Dascalaki, E.G., Droutsa, K.G., Balaras, C.A., Kontoyiannidis, S.: Building typologies as a tool for assessing the energy performance of residential buildings—a case study for the Hellenic building stock. Energy Build. **43**(12), 3400–3409 (2011)
4. Briggs, R.S., Crawley, D.B., Schliesing, J.S.: Energy Requirements for Office Buildings, vol. 1, Existing Buildings. GRI-90/0236.1 by Battelle, Pacific Northwest Laboratory, for Gas Research Institute (1992)
5. Ma, Y., Deng, W., Xie, J., Heath, T., Xiang, Y., Hong, Y.: Generating prototypical residential building geometry models using a new hybrid approach. Build. Simul. **15**(1), 17–28 (2021)
6. Saltelli, A., Annoni, P., Azzini, I., Campolongo, F., Ratto, M., Tarantola, S.: Variance based sensitivity analysis of model output. Design and estimator for the total sensitivity index. Comput. Phys. Commun. **181**(2), 259–270 (2010)
7. Wang, L.: Exploration and practice of passive building technology system in hot summer and cold winter area. China Acad. J. Electron. Publ. House (in Chinese) (2018)
8. Zhang, X., Liu, J., Chen, X., Zhang, J.: Sensitivity analysis of influencing factors and prediction methods study of cooling (heating) load in regional buildings. Build. Sci. **29**(8) (2013)
9. Li, Y., Tang, P.: Application strategy of passive house in hot summer cold winter zone (in Chinese). Constr. Sci. Technol. **12**, 62–65 (2016)
10. Ministry of Housing and Urban-Rural Development of the People's Republic of China: GB 50176–2016: code for thermal design of civil buildings (2016)
11. Haase, M., Amato, A.: An investigation of the potential for natural ventilation and building orientation to achieve thermal comfort in warm and humid climates. Sol. Energy **83**(3), 389–399 (2009)
12. Ministry of Housing and Urban-Rural Development of the People's Republic of China: DB22/ 2015–2021: design standard for energy efficiency of residential buildings (2021)
13. Ministry of Housing and Urban-Rural Development of the People's Republic of China: JGJ 134-2010: energy saving design standard for residential buildings in areas with hot summer and cold winter (2010)
14. Ministry of Housing and Urban-Rural Development of the People's Republic of China: GB 50189-2015: design standard for energy efficiency of public buildings (2016)
15. Ministry of Housing and Urban-Rural Development of the People's Republic of China: GB 50189-1980: design standard for energy efficiency of public buildings (1980)
16. Schulz, T., Feist, W., Schnieders, J.: Passive Houses in Chinese Climates (2016). https://phichina.com/sites/phichina.com/files/documents/Bericht_ChinaKlima_EN_Kurzversion_161102_mit_cover(CN+EN).pdf. Accessed 18 April 2022

3D Modeling of Folded Footings with Ring Beam on Sand Using Various Folding Angles

Ahmed E. Gomaa, Ahmed M. M. Hasan, Yasser M. Mater, and Sherif S. AbdelSalam

Abstract Folded isolated footings represent an alternative to traditional isolated footings to support structures on weak soils. The reinforced concrete used in folded footings can be optimized by minimizing the tensile stresses developing in the concrete section, reducing the resulting settlement and the redistribution of stresses on the supporting soil. This study presents a comprehensive numerical investigation of the performance of folded footings placed on cohesion-less soil. Six quarter-scaled footings supported on medium dense sand were modeled using finite element tools to analyze stress changes induced in concrete and soil. One flat footing was used as a control model and five folded footings with folding angles of 15°, 30°, 45°, 51.5°, and 60° with the horizontal were investigated. Results showed that the use of folded footings decreased the internal stresses by up to about 90%. In addition to increasing the depth of stress influence in the soil and enhancing the bearing capacity. Moreover, the total settlement occurring in the supporting soil decreased by about 25%. Finally, design charts were provided to enhance the structural and geotechnical performance of folded footings.

Keywords Folded footing · Shell footing · Numerical modeling · Soil-structure interaction

A. E. Gomaa · A. M. M. Hasan · Y. M. Mater (✉) · S. S. AbdelSalam
Civil, Infrastructure Engineering and Management Program, School of Engineering and Applied Science, Nile University, Giza 12588, Egypt
e-mail: Ymater@nu.edu.eg

A. E. Gomaa
e-mail: Ah.Emad@nu.edu.eg

A. M. M. Hasan
e-mail: Amahmoud@nu.edu.eg

S. S. AbdelSalam
e-mail: Sabdelsalam@nu.edu.eg

E. Strauss (ed.), *Proceedings of the 7th International Conference on Civil Engineering*, Lecture Notes in Civil Engineering 371, https://doi.org/10.1007/978-981-99-4045-5_20

1 Introduction

Shell structures provide maximized structural benefits and utilize the least amount of material. Shells can be used as structural foundations in four main categories, folded, conical, HYPAR, and inverted foundations [1]. Folded footings are more effective in terms of ease of construction and labor cost, but the material consumption is relatively higher compared to conical and inverted footings. The use of folded footings enables different distributions of contact pressure on the soil [2, 3]. Therefore, they are considered as an optimum option for transferring large loads to soils with weak properties [4].

In opposite to flat footings, folded footings create deeper failure surfaces with a higher depth of stress influence in the soil beneath the footing [5]. This means that using folded footings yields a higher soil bearing capacity when compared to regular flat footings. This variation is due to the stiffness and geometry of the shell elements. In the case of folded footings, increasing the thickness of the folded plates results in increasing the contact pressure on soil [5–7], and it was found that the use of folded footings yields a smaller settlement compared to flat footings [7–9].

In this study, the main goal is to model the interaction between folded footings and soil based on Finite Element Method (FEM) to assess the structural and geotechnical benefits and limitations. The numerical software SAP2000 and PLAXIS 3D were utilized to simulate various folded footings supported on dense sand using folding angles equal to 0, 15, 30, 45, 51.5, and 60° with the horizontal. Outcomes focused on changes in the internal stresses within the concrete section of the footing as well as stress influence depth and soil settlement.

1.1 Material Properties

This section presents the material properties used for the constitutive models of the finite element analysis (FEA). Table 1 shows three main materials that were adopted during the simulation. The following values are used on both software tools, SAP2000 and PLAXIS 3D.

Loose sand was used for modeling the soil inside the cavity of the folded footing to simulate the effect of lightly compacted sand due to in situ construction limitations. While the medium dense sand was assigned to the remaining soil under the footings in all models. Properties of both materials are presented in Table 1. Sand was modeled as Mohr–Coulomb model with a cohesion (c) equal to zero. In the SAP2000 model, a reinforced concrete folded shell footing was simulated using the linear elastic model and given a compressive strength of 25 MPa in all models. The models herein adopted changing the height of the footing to satisfy the required folding angle.

Table 1 Material properties used for the FEA

Property[a]	Loose sand	Medium dense sand	Reinforced concrete
Constitutive model	Mohr Coulomb	Mohr Coulomb	Linear elastic
E_s (kN/m^2)	3500	5000	22,000,000
ϕ (deg.)	30	33	–
c (kN/m^2)	0	0	–
e	0.5	0.5	–
γ_d (kN/m^3)	17	18	25
v	0.2	0.2	0.3

[a] Es, young's modulus of dry sand (elastic modulus); ϕ, internal friction angle of sand; c, sand cohesion; γd, dry unit weight; e, void ratio; and v, Poisson's ratio

1.2 Finite Element Modeling

The meshing in the structural model was created using the divide tool of SAP2000 with a mesh of 500 shells per model to assure the same degree of mesh refinement in all models as shown in Fig. 1. The in-plan dimension of all footings was 0.5 × 0.5 m. Each model was surrounded by an edge beam (or a ring beam) with dimensions of 0.1 × 0.1 m as shown in Fig. 1. The six footing models are presented in the figure, where the flat footing acted as a control model to be compared with the 15, 30, 45, and 60° models. Historically [10], the Great Pyramid of Giza, Khufu Pyramid, has an angle of 51.5° from horizontal. Accordingly, the sixth model was taken into consideration in the analysis. Two values of modulus of sub-grade reaction of Winkler model (k_{sub})—one for Dense Sand and one for Loose Sand—were calculated and applied in SAP2000 to simulate the soil interaction.

$$q_{\text{ult}} = cN_c + \gamma_1 DN_q + 0.5B\gamma_2 N_\gamma \tag{1}$$

Figures 2a, b represent the mesh of the flat model and the 30 degrees folded footing model on PLAXIS 3D. The mesh was generated by the software and set to be "Medium". All continuums were modeled using an unstructured mesh of 10-node tetrahedral elements. The maximum prescribed load for modeling in both SAP2000 and PLAXIS 3D was selected to be 20 kN. This load was considered after estimating the ultimate bearing capacity of soil under the flat footing model and according to Eq. 1 by Terzaghi. In this equation, (q_{ult}) is the ultimate bearing capacity, (N_c), (N_q), and (N_g) are bearing capacity factors related to the angle of friction (f′). The parameters (D) and (B) represent founding depth and breadth of foundation, respectively. Finally, the constant (c) reflects the apparent cohesion intercept of the soil.

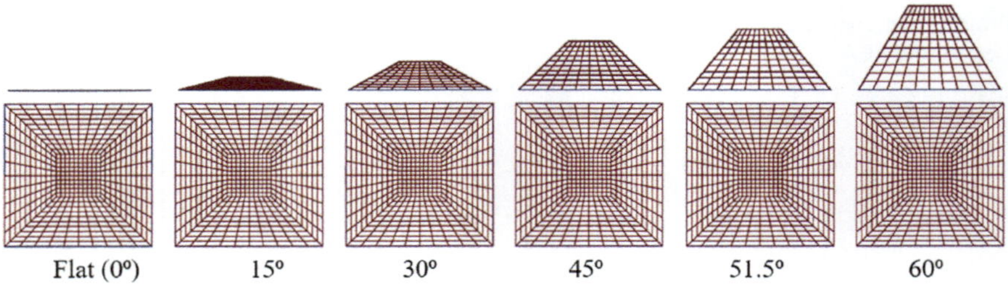

Fig. 1 Structural models for the footings using SAP2000

Fig. 2 Deformed mesh from PLAXIS 3D

(a) Flat footing

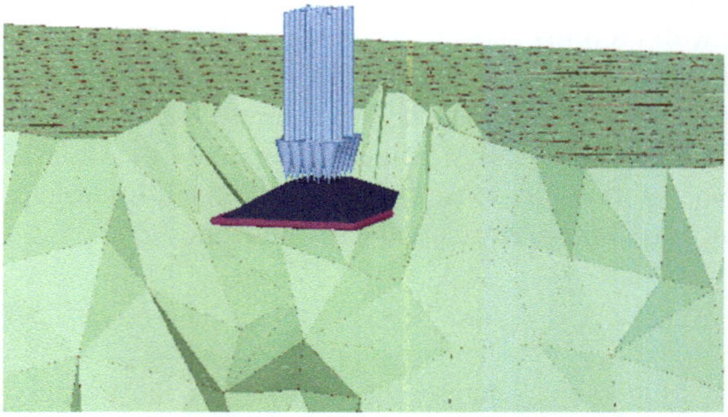

(b) 30° folded footing

2 Results and Discussion

2.1 Effect of Folding Angle on Internal Straining Actions

This section includes the structural analysis results obtained from SAP2000, to investigate the effect of using various folding angles on the straining actions formed inside the folded isolated footings. Figure 3a shows the top view of the six models and displays the moment distribution of each shell in kN.m/m. It can be noticed

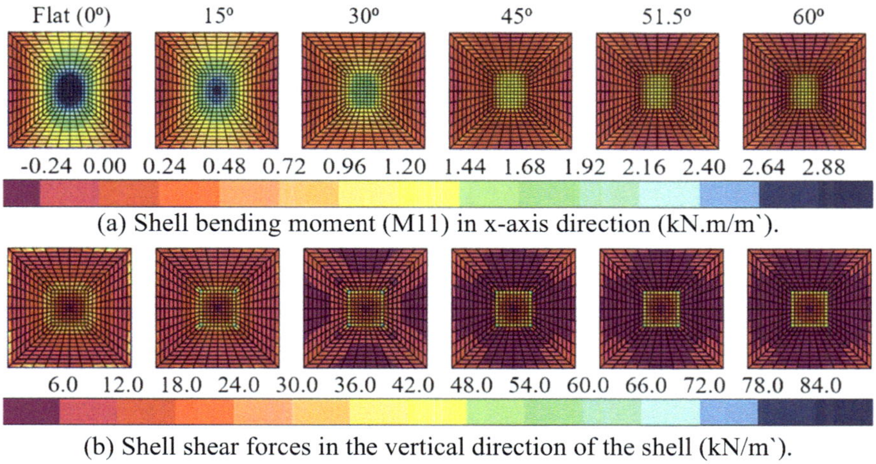

Fig. 3 Analysis results for the footing models using SAP2000

that the minimum moment in the x-direction was achieved in the footing with the maximum folding angle. Further, the absolute shear force developed in the shell elements followed the same decreasing trend as shown in Fig. 3b.

Figure 4a shows both the normal and shear forces developed within all the folded footing models. As presented, the shear force dropped with increasing the folding angle of the footings. The maximum shear on the shell was obtained in the flat footing model with a shear force equal to 150 kN/m and dropped to about 10 kN/m using a folding angle of 60°. On the other hand, the normal force firstly started at zero in the flat footing model to reach its maximum at a folding angle of 30°. Then the normal force started to decrease again with increasing the folding angle. This normal force decay is because of horizontal confinement force induced by the ring beam, which acts as statically indeterminate horizontal force components opposing the vertical load. Upon that, the 30° model that provided the highest normal force inside the shell was considered optimum.

Figure 4b shows the relation between the developed stresses in the shell and ring beam elements at various folding angles using four profiles. The first represents the ring beam stresses, the second and third profiles represent the top and bottom stresses in the shells (folding plates), and the fourth represents the absolute summation of the top and bottom shell stresses. The figure shows that the normal stress in the ring beam is following a quadratic equation profile with a maximum value of 0.41 MPa at the 30° model.

The stress values on the top and bottom faces of the shell follow the same decreasing profile. The stress at the top face of the shell in flat model scored a maximum value of −2.3 MPa, then decreased with increasing the folding angle to reach its minimum of −0.22 MPa at 60° folding angle. Likewise, tensile stresses at the bottom face started at a maximum value of 4.70 MPa, then decayed by increasing

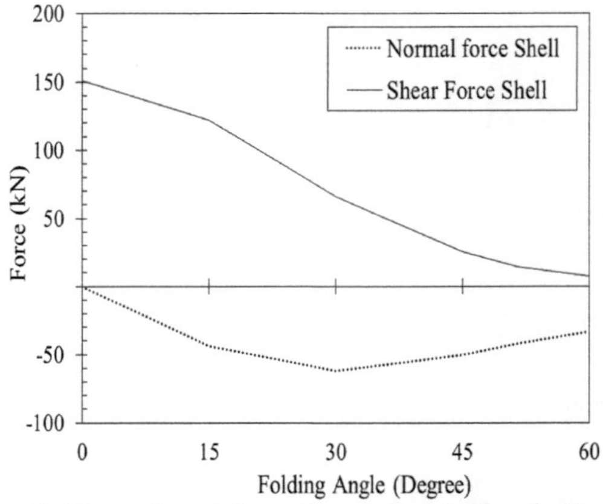

(a) Normal and shear forces inside the shell.

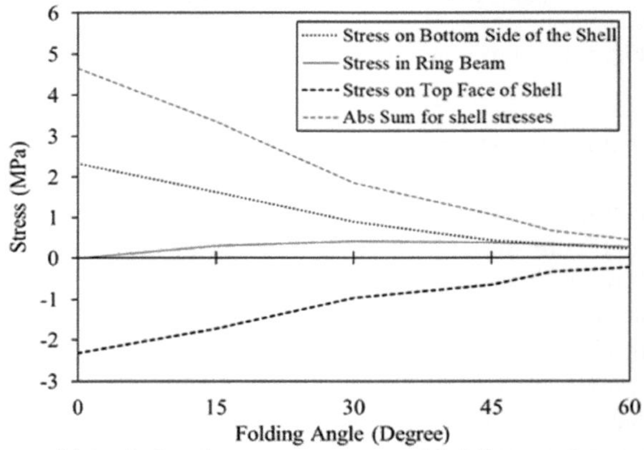

(b) Relation between stress and folding angle.

Fig. 4 Straining action inside the footing shell and ring beam elements

the folding angle to reach a minimum value of 0.43 MPa at the 60° folding angle. The absolute curve in Fig. 4b was calculated by the summation of the absolute values of stress on the top and bottom faces to indicate the decay of the total stress in the shell elements versus the increase in the folding angle.

2.2 *Effect of Folding Angle on Supporting Soil*

This section includes modeling of soil-structure interaction using PLAXIS 3D. Results presented in Fig. 5a, b are taken from the 30° folding model, which represented the maximum normal force from SAP2000 results. This model was selected to demonstrate settlement and stresses within the soil, results of other models were summarized in the next section. The settlement profile of the 30° model in Fig. 5a

Fig. 5 Settlement and total stress under the 30-degree footing

(a) Settlement under the 30° footing.

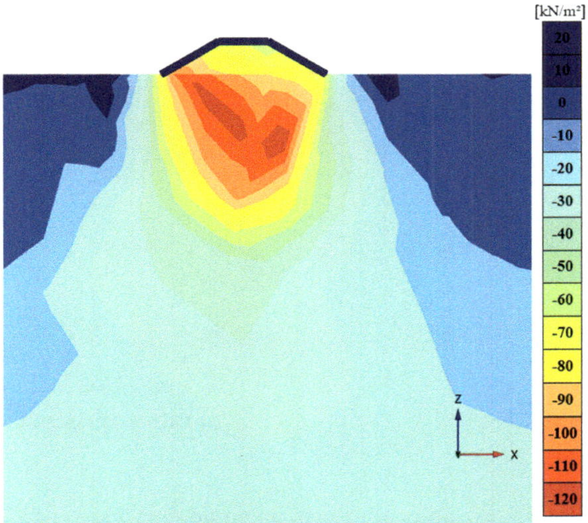

(b) Total stress z-dir under 30° footing

shows an initial settlement of 19 mm at the first influence zone. After that, settlement decays to reach 4 mm at depth 0.6 m from the ground level.

The total Cartesian stresses in the vertical direction (z-direction) displayed in Fig. 5b shows a high initial stress of -120 kN/m^2 at the contact surface between the soil and the folded footing. After that, the effective stress gradually decreases and then diminishes at 1.10 m below footing level. Figure 5b shows a stress confinement effect in the soil around the footing due to the 30° inclination that generated a horizontal pressure component which caused greater stresses concentration beneath the footing. Figure 6a plots relation between the soil settlement for different folding angle models.

Fig. 6 Settlement and influence depth for all footings predicted from PLAXIS 3D

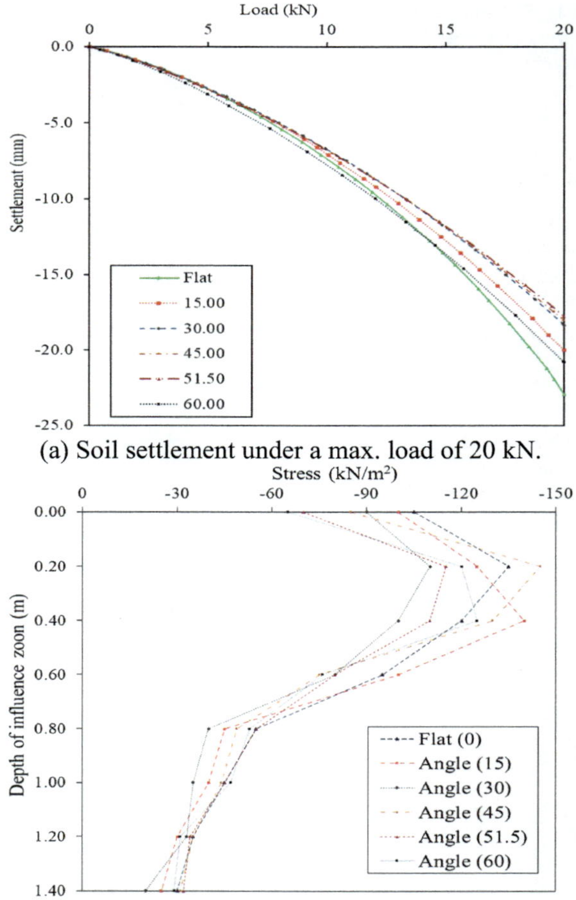

(a) Soil settlement under a max. load of 20 kN.

(b) Influence depth under various stress values.

The figure shows a maximum settlement value of 22.5 mm in the case of the flat footing model. While the minimum settlement was 17.6 mm under the 51.5° model. The settlement increases again in the 60° model to reach 20.8 mm. The effect of changing the folding angle on the soil stresses is presented in Fig. 6b, and it shows the stress distribution in z-direction under each footing. The curves in the figure have maximum stress of -147 kN/m^2 at 0.2 m from the ground surface under the 51.5° folded footing. This increase in stress distribution can be explained by the concept of confinement, as increasing the folding angle tend to increase lateral pressure on the loose sand within the cavity of the folded footing, which transfers higher stress through a larger influence depth. It can be observed that all the models achieved almost the same stress at a depth of 1.4 m with an average value of -30 kN/m^2. Hence, the maximum stress influence depth was about 2.8 $\times$ footing width (B).

2.3 Design Charts for Folded Footings

Figure 7a presents a correlation between the characteristic compressive strength of concrete (fcu), and normal stress normalized by the concrete modulus of rupture (fr), presenting a guide for the design of sections subjected to normal stresses acting on RC folded footings with various fcu and folding angles. Figure 7a shows that at any value of fcu the stress decreases by increasing the folding angle. Figure 7b shows the effect of folding angles on the soil stresses (i.e., influence depth). The ultimate bearing capacity (Qu) is divided by stresses at various depths, and the influence depth is divided by the footing width (Z/B). This curve aims to provide a design chart that links the footing geometry and total stresses with the stress influence depth and bearing capacity.

Fig. 7 Design charts for selection of proper folded footing (structural and geotechnical)

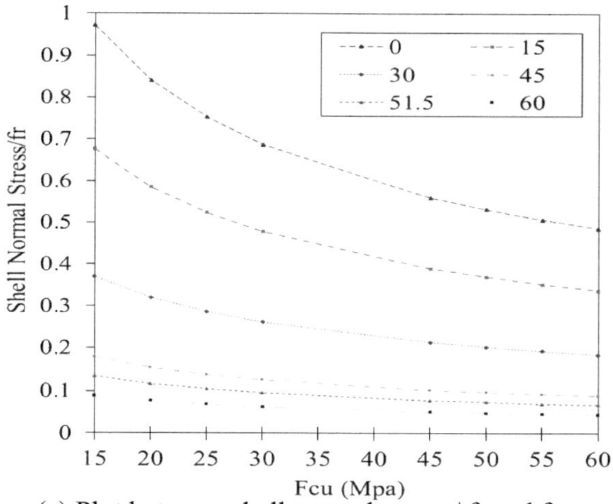

(a) Plot between shell normal stress / f_r and f_{cu}.

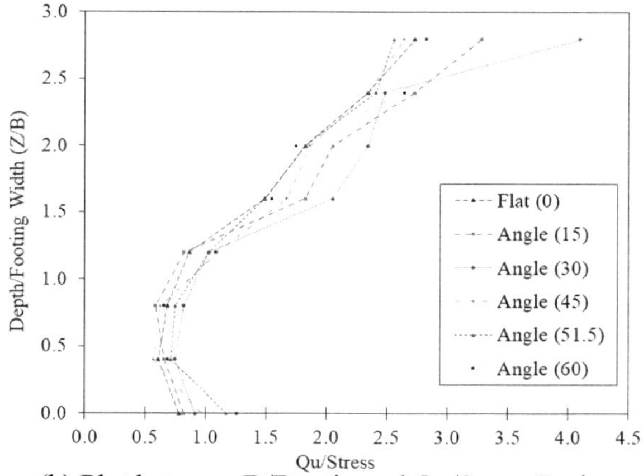

(b) Plot between D/B ratio and Qu/Stress Ratio.

3 Conclusion

This study investigated the interaction between cohesion-less soil and folded isolated footings using various folding angles. From the structural outcomes of the analysis, it was found that the normal force within the shell in the direction of the fold started with no normal force at the flat footing model, then increased to reach a maximum value in the 30° model, finally the values of the normal force gradually decreased to reach a balance point at the 60° model. This effect was due to the structural system redundancies, one of which is the ring beam. The normal stresses in the ring beam complement the change of normal forces within the shell. Overall, this promotes a lower cross-section thickness for steeper footings. From the geotechnical outcomes of the analysis is that the settlement decreases with increasing the folding angles. This led to a variation in the stresses under each model, more concentration at the edge of the folded footings, and an increased stress influence depth below the foundation level. The minimum stress under the footing was found in the 30° model.

To summarize, the folding action significantly reduces the settlements, straining actions in the footing shell and beam elements, and provides a more cost-effective structural design. Finally, the 30° folding model provides minimum stress under the footing, a smaller settlement compared to flat footing, and minimum flexural reinforcement for RC shell element design.

References

1. El-kady, M.S., Badrawi, E.F.: Performance of isolated and folded footings. J. Comput. Des. Eng. **4**(2), 150–157 (2017)
2. R. Rinaldi, M. Abdel-Rahman, A. Hanna: Experimental investigation on shell footing models employing high-performance concrete. In: International Congress and Exhibition Sustainable Civil Infrastructures: Innovative Infrastructure Geotechnology, Springer, pp. 373–390 (2017)
3. Shoukath, S., Rajesh, A.: Seismic performance of hyperbolic paraboloid and inverted spherical shell foundation. Res. J. Adv. Eng. Sci. **2**(2), 148–151 (2017)
4. Hassan, S.A., Al-Soud, M.S., Mohammed, S.A.: Behavior of pyramidal shell foundations on reinforced sandy soil. Geotech. Geol. Eng. **37**(4), 2437–2452 (2019)
5. D. Esmaili, N. Hataf: Determination of ultimate load capacity of conical and pyramidal shell foundations using dimensional analysis. Iran. J. Sci. Technology. Trans. Civ. Eng. **37**(C), 423 (2013)
6. Y. Idris, R. Dewi, Y. Sutejo, M.S. Agil Al: Bearing capacity of folded plate foundations in clay soil. J. Appl. Eng. Sci. **19**(3), 681–687 (2021)
7. R. Sidqi, M.N. Mahmood: Investigating the nonlinear performance of reinforced concrete shell foundations. In: IOP Conference Series: Materials Science and Engineering, vol. 978, no. 1, p. 012050. IOP Publishing (2020)
8. Abdel-Rahman, M.: Lateral loading resistance for shell foundations. HBRC J. **16**(1), 227–241 (2020)
9. T. Iyer, N. Rao:Model studies on funicular shells as rafts on sands. In: Proceedings, Symposium on Shallow Foundations, Bombay, India, vol. 1, pp. 149–156 (1970)
10. Grigoriev, S.A.: Inclinations of Egyptian pyramids and finding of the divine essence. Archaeoastron. Anct. Technol. **3**(1), 1–27 (2015)

Approximate Estimation for Global Buckling Load of Cylindrical Single-Layer Grid Shells: Fitting of Envelope Equations Based on Regression Analysis

Baoxin Liu, Pei-Shan Chen, Jialiang Jin, and Xiangdong Yan

Abstract This study defines an imaginary stiffness G to represent the overall stiffness for the grid shells. Through the linear buckling analysis, four kinds of grid patterns with different geometric parameters, in total 9100 cases, are analyzed to get their buckling load factors λ. Based on the regression analysis, four envelope equations are fitted with G, λ and geometric parameters. The applicability of the equations is illustrated by the distribution of the difference $r_{\Delta\lambda}$.

Keywords Single-layer grid shells · Imaginary stiffness · Buckling load · Regression analysis

1 Introduction

As a principal stability problem for the single-layer grid shells, global buckling has been a hot issue for a long time [1]. Currently, two main analytical methods are used to evaluate the global buckling capacity for the single-layer grid shells: continuum

B. Liu
Department of Civil Engineering and Architecture, Kyushu Institute of Technology, Kitakyushu 804-8550, Japan
e-mail: wuyezi894427800@gmail.com

P.-S. Chen (✉)
Graduate School of Engineering, Kyushu Institute of Technology, Kitakyushu 804-8550, Japan
e-mail: chen@civil.kyutech.ac.jp

J. Jin
Graduate School of Environmental Studies, Nagoya University, Nagoya 464-8601, Japan
e-mail: j1822238004@gmail.com

X. Yan
Department of Civil and Environmental Engineering, University of Pittsburgh, Pittsburgh 15216, USA
e-mail: yan.xiangdong760@mail.kyutech.jp

E. Strauss (ed.), *Proceedings of the 7th International Conference on Civil Engineering*, Lecture Notes in Civil Engineering 371, https://doi.org/10.1007/978-981-99-4045-5_21

shell analogy [2] and discretely finite element analysis [3]. However, the continuum shell analogy has limits on calculation accuracy, while the finite element analysis may be difficult to use for common engineers. Hence, the authors aim to propose an evaluation equation, which can approximately calculate the global buckling load of a single-layer space frame. Unlike the conventional mechanical buckling analysis, this study utilizes regression analysis to fit the evaluation equation.

Recently, the regression analysis of statistics, related to large-scale data analysis, has been applied in various filed, such as regression analysis on the compressive strength of slag-metakaolin geopolymer pastes [4], but its application on the evaluation of buckling load for the grid shells is very few. In the past research [5, 6], the clear statistical relationship of grid shells can be found between the buckling load and geometric parameters. Accordingly, the authors assume that within the conventional structural design (conventional shape, span, height, and so on), the relationship between geometric parameters and buckling loads may be expressed into formulations and/or equations. Therefore, the authors attempt to combine the buckling analysis with regression analysis to fit one or a series of envelope equations to estimate the global buckling load with some geometric parameters.

For such a purpose, large amounts of linear buckling analysis are carried out with considering geometric parameters as much as possible. A buckling analysis program called CPS, developed by author Chen, is applied to perform linear buckling analysis with automatically displaying the buckling modes and simulating the deformation process. In addition, this study proposes an imaginary stiffness G to express the overall stiffness of the whole structures, which is related to the maximum displacement $d_{\max}$ gained by general elastic linear structural analysis. According to the regression analysis based on Tablecurve 3D, the relationship among the parameters, imaginary stiffness G and buckling load can be fitted as the form of envelope equations. The envelope equation is a new concept, though it is a kind of approximate estimation, its significance is to provide a limiting standard for the designers, where the analysis result cannot be less than the value calculated by the envelope equations.

As the first stage of the present study, the authors target regular geometric parameters such as mesh-density n, arc-angle θ, span L, and diameter of member section $\varnothing$ with 4 kinds of grid patterns, and show the process of envelope equations fitting in this paper. The applicability of the equations is illustrated by the distribution of the difference $r_{\Delta\lambda}$, the frequency f is almost concentrated in the range of $r_{\Delta\lambda} = \pm 20\%$, which means the envelope equations still have a high applicability. In addition, the influence of imperfection, geometric non-linear, material non-linear, etc., will be investigated in the future.

2 Basic Principle

This study aims to fit the envelope equations for approximately estimating the global buckling load λ with some geometric parameters and imaginary stiffness *G*, where the equations can be expressed in form as $\lambda = \mathrm{f}$ (geometric parameters, G). Thus, the definition of imaginary stiffness *G*, and the application of linear buckling analysis and regression analysis are introduced in the following sections. To make it easier to understand how the envelope equation is fitted, the process is demonstrated in the final.

2.1 Geometric Parameters and Imaginary Stiffness

In general, the main reasons for the instability of single-layer grid shells are that their structural behavior is highly affected by geometric parameters and stiffness [5, 6]. As shown in Fig. 1, the geometric parameters of a single-layer grid shell are introduced as span L_x and L_y, angle θ (related to Rise *h*), mesh-density n_x and n_y, the diameter of member sections $\varnothing$.

The former researches, based on continuum shell analogy, defined the stiffness of a single-layer grid shell with the member sections [7–9]. In the actual structural design, the member sections of a grid shell may not be uniformly distributed, especially for the form optimization, so the authors hope to decline the influence of the member section. In general, the maximum displacement of a grid shell (Fig. 2) can be used as a judgment of the overall stiffness. It is easy for common designers to get the maximum displacement through the linear elastic analysis. Consequently, according to the statistical analysis results based on general linear elastic analysis, an imaginary stiffness *G* related to the maximum displacement d_{max} is proposed as Eq. 1.

$$G = \frac{F}{\delta} = \frac{pL_xL_y}{d_{\max}/n_xn_y} \times 10^{-11}(\mathrm{N/mm}) \quad (1)$$

Fig. 1 Geometric parameter

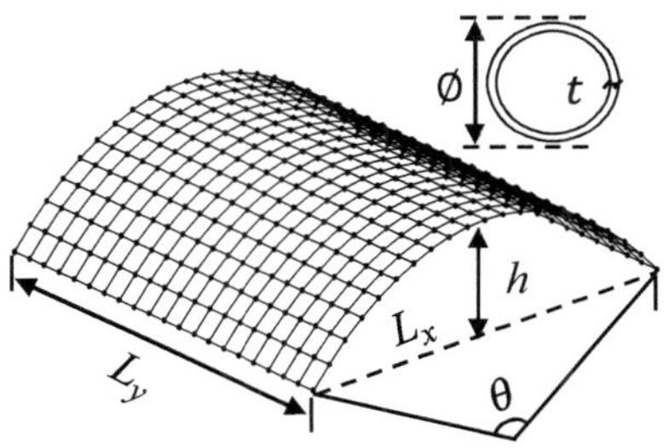

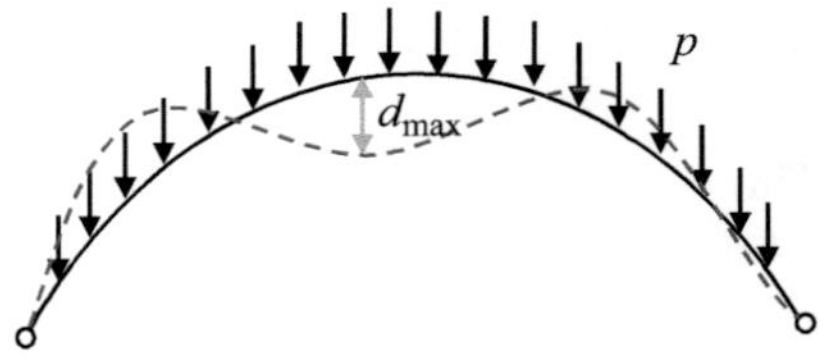

Fig. 2 Maximum displacement

where the value of pL_xL_y can be considered as F (total load), while the $d_{max}/(n_xn_y)$ can be assumed as δ (distribution of d_{max} on each grid).

2.2 Application of Linear Buckling Analysis and Regression Analysis

To determine the influence of geometric parameters on the buckling load factor λ and the imaginary stiffness *G*, large amounts of buckling analysis are necessary. The analysis models are settled as ideal linearly elastic structures, all initial imperfections are ignored, and steel member is assumed to work elastically [10], thus linear buckling analysis is applied in this study. In general, the basic formulation for linear buckling analysis is based on the generalized eigenvalue problem as follows [11]:

$$\left(K_L + \lambda^{(i)} K_G\right) U^{(i)} = 0 \tag{2}$$

where K_L and K_G are the matrices representing the linear and geometric stiffness, respectively, $\lambda^{(i)}$ is the linear buckling load parameter associated with the *i*th buckling mode $U^{(i)}$. In this study, the authors use the first eigenvalue value λ as buckling load factor.

To forecast the relationship between the geometric parameters, imaginary stiffness *G* and buckling load factors λ, regression analysis is applied. The authors try some mathematical regression models such as linear regression, logistic regression, and polynomial regression, and compare their coefficient of determination R^2 for knowing which is the most approximate to fit the envelope equations with high precision. The precision is judged by the coefficient of determination R^2 which is defined as follows:

$$R^2 = 1 - \sum_i (\lambda_i - f_i)^2 / \sum_i (\lambda_i - \bar{\lambda})^2 \tag{3}$$

As shown in Fig. 3a, λ_i is the point of analysis result, f_i is the result from the 3D curve of regression equations which are shown in Fig. 3b, $\bar{\lambda}$ is the average value of λ_i. In the best case, the modeled values exactly match the λ_i, which results in R^2 closing to 1 [12].

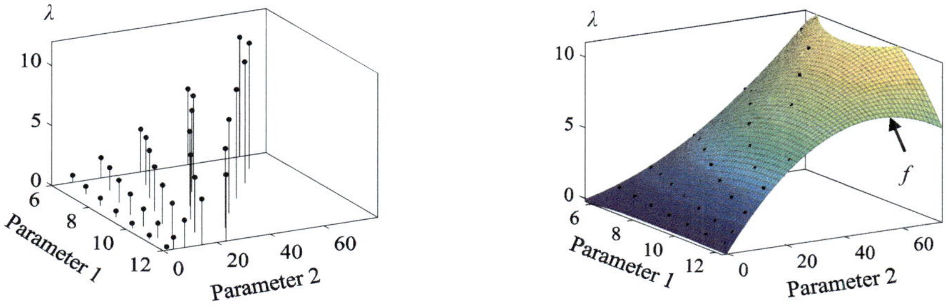

(a) Distribution of analysis result (b) Regression equations

Fig. 3 Sample of regression model

2.3 *Process for Fitting the Equations*

As mentioned before, the authors hope to decline the influence of the member section. Thus, the diameter of member section ∅ is not set as the geometric parameter to fit the equations, it is just used as the analysis case for the other geometric parameters to get *G*. In addition, as shown in Fig. 4, according to the influence of geometric parameters on λ and *G*, the process to fit the envelope equation is separated into two routes. Route-1 builds the foundational analysis model with the geometric parameters whose influence on λ and *G* are non-monotonic changed. Then regression analysis is utilized to find the best mathematical regression models according to the foundational analysis model, such that, the foundational equations λ_f can be derived with high

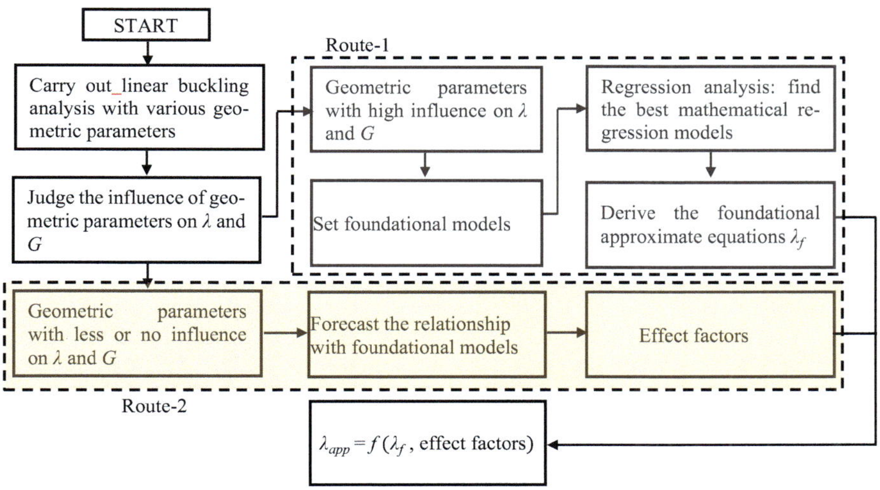

Fig. 4 The process of deriving equations

precision. In the route-2, effect factors are determined to adjust the foundational equations λ_f. The effect factors are statistical results according to the relationship between the foundational analysis models and geometric parameters (with monotonic changing influence on λ and G). As a result, the envelope equations can be expressed as $\lambda_{app} = f\left(\lambda_{f,\ \text{effect factors}}\right)$, so that the designers can check and estimate the buckling load easily through these equations without complex buckling analysis.

3 Numerical Analysis

3.1 Analytic Models

There are 4 kinds of grid patterns analyzed which are shown in Fig. 5, each grid pattern has the same number and process of analysis cases, as shown in Fig. 6 [13]. As the first stage of this study, the span L is set as $L_x = L_y$ and mesh-density n is set as $n_x = n_y$. Because the authors want to find the relationship between these geometric parameters and buckling load, the diameters of member section in each analysis case model are the same. The member material is SS400 with Young's modulus of $E = 2.05 \times 10^5$ N/mm^2.The vertical distribution load acting on nodes is 1 KN/m^2. The boundary conditions are 4-edge supporting at pinned-joint.

3.2 Numerical Analysis

This study considers 4 kinds of grid patterns, their influence on λ and G is shown in Fig. 7. It can be found that the trends of G for the pattern (i) are similar to the pattern

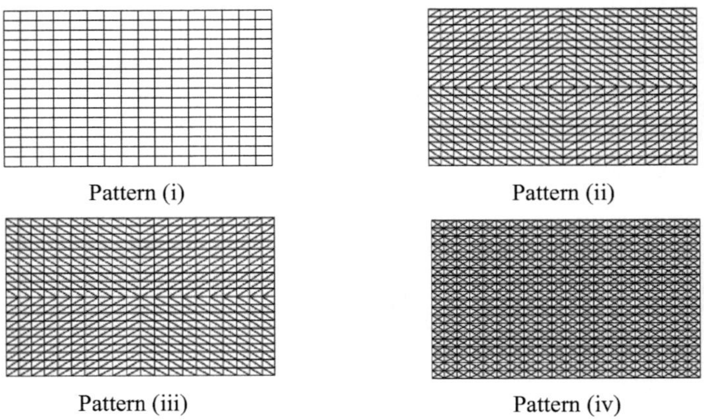

Fig. 5 4 kinds of grid patterns

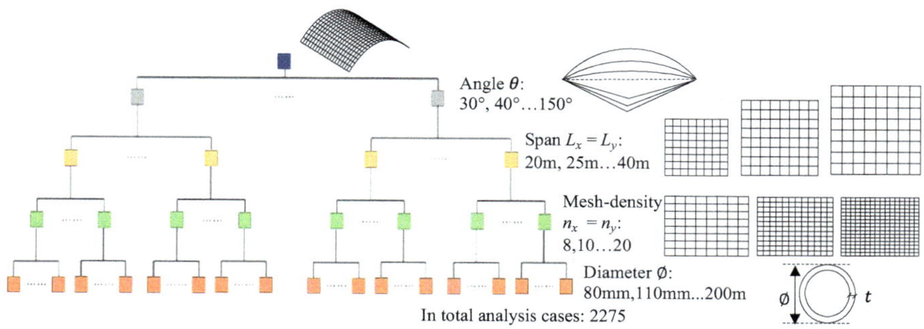

Fig. 6 A hierarchical tree illustrating the number and process of analysis cases for pattern (i)

(ii) from the angle $\theta = 30$–$60°$, while the pattern (iii) is similar to the pattern (iv) from angle $\theta = 30$–$90°$. Though both λ and G for all the grid patterns will firstly increase and then decrease with the increase of angle, their peak values are different, which means that grid patterns can't be built a certain relationship. Therefore, this study fits 4 kinds of envelope equations for each grid pattern. Due to the limitation of the paper space, the authors use the analysis result of grid pattern (i) to specifically demonstrate how the envelope equations are fitted.

3.3 Foundational Approximate Equations

Past research by authors [14] investigated the influence of geometric parameters, the results indicated that the changing trends of λ and G for different L and n are almost identical, only the influence of θ on λ and G are non-monotonic changes. In this paper, the authors also set the span $L = 20$ m with mesh-density $n = 8$ as the foundational analysis models and builds the mathematical regression models with angle θ, buckling load factor λ, and imaginary stiffness G to fit the foundational equations, while the influence of span L and mesh-density n are statisticed as effect factors in next section.

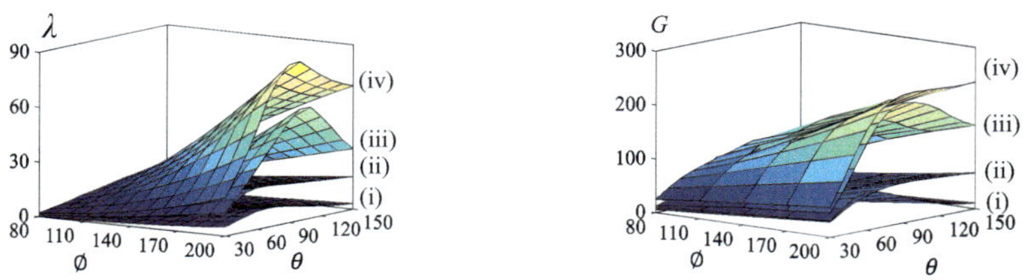

(a) Comparison on λ

(b) Comparison on G

Fig. 7 Influence of different grid patterns on λ and G

Because of considering the practical design, the authors reduced the range of angles to 60–120° during the process of fitting equations. In order to fit the equations with high precision, the authors analyze hundreds of regression models through the Tablecurve 3D, and show the 3 better regression models and their distribution of residuals in Fig. 8.

According to the coefficient of determination R^2, it could be found that R^2 of (c) is the most appropriate to fit the equations. Consequently, the foundational equations are derived as Eq. 4 based on the mathematical regression model (c), while the coefficients for each grid pattern are shown in Table 1. In addition, when applying this equation, vertical distribution load p in G should be transformed into 1 KN/m^2.

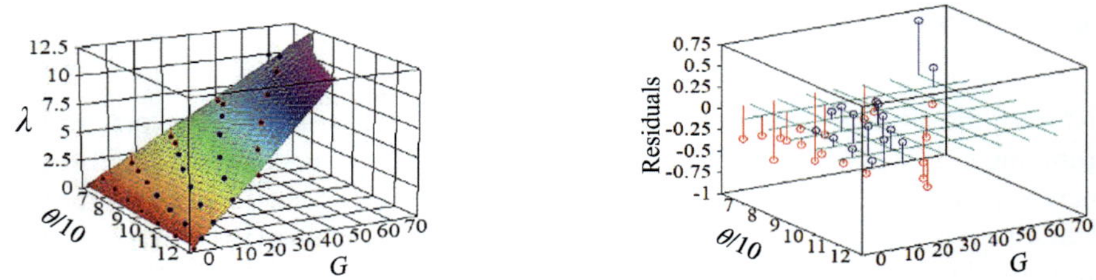

(a) $\ln z = a + b\ln x/x + c\ln y$ $R^2 = 0.989537$

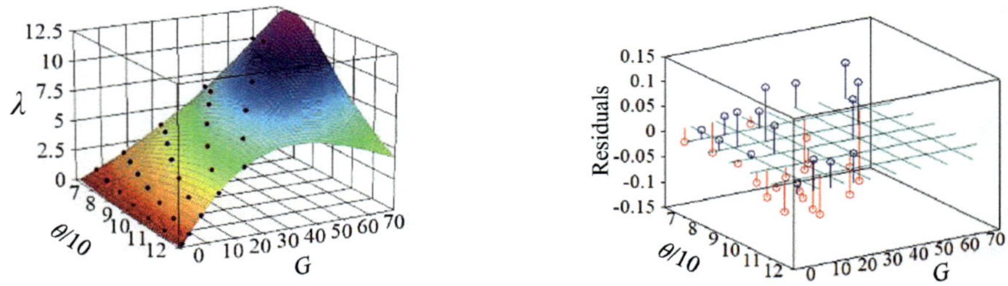

(b) $z=(a+cx+ey+gx^2+iy^2+kxy)/(1+bx+dy+fx^2+hy^2+jxy)$ $R^2 = 0.999259$

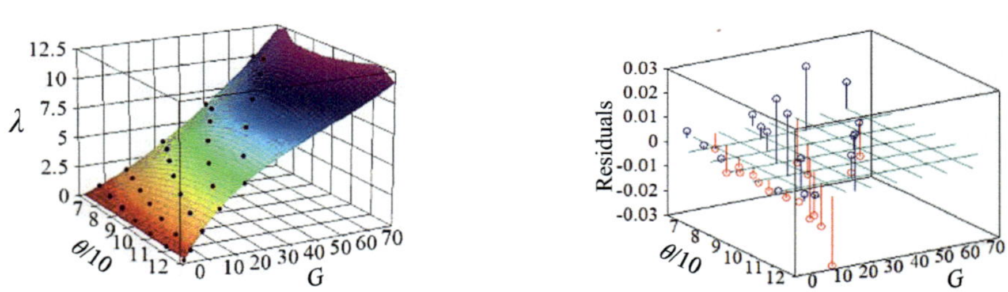

(c) $z=a+bx+cx^2+dx^3+ey+fy^2+gy^3+hxy+ix^2y+jxy^2$ $R^2 = 0.9999813$

Fig. 8 Regression model ($z = \lambda$, $y = G$, $x = \theta/10$)

Table 1 The coefficient for each grid patterns (i)–(iv)

Coefficients	(i)	(ii)	(iii)	(iv)
a	4.744	−4.153	219.3	157.4
b	−1.485	−1.04	−72.1	−43.66
c	0.142	0.222	7.595	4.18
d	−0.004283	−0.01357	−0.264	−0.1514
e	−0.5497	0.8544	−0.1028	−0.9798
f	0.005038	−0.003672	0.001663	0.005642
g	−1.92E-06	3.55E-05	−7.92E-07	−6.40E-06
h	0.1352	−0.055	0.006345	0.1202
i	−0.005352	0.004257	0.001324	−0.002061
j	−0.0006438	−0.0003992	−5.51E-05	−0.0002467

$$\lambda_f = a + b\left(\frac{\theta}{10}\right) + c\left(\frac{\theta}{10}\right)^2 + d\left(\frac{\theta}{10}\right)^3 + eG + fG^2 + gG^3 + h\left(\frac{\theta}{10}\right)G + i\left(\frac{\theta}{10}\right)^2 G + j\left(\frac{\theta}{10}\right)G^2 \tag{4}$$

3.4 Effect Factors

In Sect. 3.3, the foundational equations λ_f have been gained, but it can be only used for the foundational analysis models with span $L = 20$ m with mesh-density $n = 8$. Thus, to build the relationship between λ_f and spans, mesh-densities, the authors set the result of span $L = 20$ m and mesh-density $n = 8$ as the foundational models respectively, and then investigate the relationship between the result of foundational models with the other span and mesh-density on λ and G as shown in Fig. 9.

It's easy to find that the magnification of each span on $L = 20$ m and each mesh-density on $n = 8$ are almost constantly in a line, which means that the influence of span and mesh-density on λ and G is constant, and almost don't accept the influence from the other geometric parameters such as angle and diameters. Therefore, the authors take the average values of these magnifications out of simplifying the equations, and consequently propose and name four effect factors of span and mesh-density as follows:

$$\text{Attenuation factors of } \lambda \text{ for span } L \quad \alpha_{\lambda L} = \frac{\lambda_{\mathrm{Li}}}{\lambda_{20}} \tag{5}$$

$$\text{Attenuation factors of } G \text{ for span } L \quad \alpha_{\mathrm{GL}} = \frac{\mathrm{G_{Li}}}{\mathrm{G_{20}}} \tag{6}$$

$$\text{Enhancement factor of } \lambda \text{ for mesh-density } n \quad \beta_{\lambda n} = \frac{\lambda_{\mathrm{ni}}}{\lambda_8} \tag{7}$$

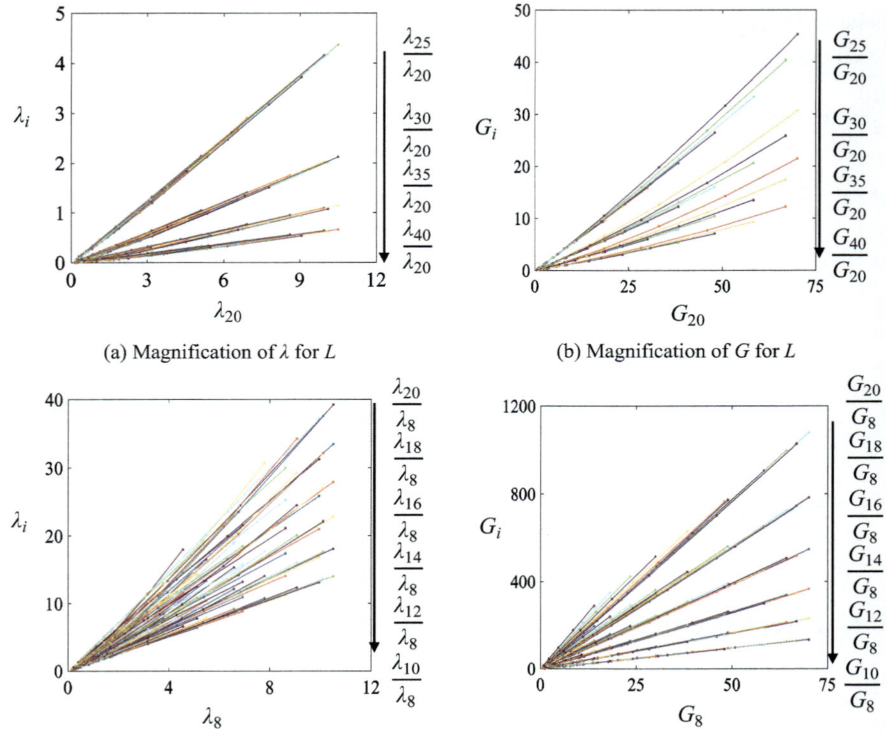

(a) Magnification of λ for L

(b) Magnification of G for L

(c) Magnification of λ for n

(d) Magnification of G for n

Fig. 9 The relationship of magnification for each span on 20 m and each mesh-density on 8

$$\text{Enhancement factor of } G \text{ for mesh-density } n \quad \beta_{\mathrm{Gn}} = \frac{\mathrm{G_{ni}}}{\mathrm{G_8}} \tag{8}$$

where λ_{20} and G_{20} are the analysis results of span $L = 20$ m while λ_8 and G_8 are the analysis results of mesh-density $n = 8$. As a result, the approximate equations can be expressed as Eq. 10. The values of effect factors $\alpha_{\lambda L}$, α_{GL}, $\beta_{\lambda n}$, β_{Gn} are shown in Tables 2, 3, 4 and 5. Here, G should be changed as $\alpha_{\mathrm{GL}}\beta_{\mathrm{Gn}}\mathrm{G}$.

$$\mathrm{G} \rightarrow \alpha_{\mathrm{GL}}\beta_{\mathrm{Gn}}\mathrm{G} \tag{9}$$

$$\lambda_{\mathrm{app}} = \alpha_{\lambda L}\beta_{\lambda n}\lambda_{\mathrm{f}}(\theta, \alpha_{\mathrm{GL}}\beta_{\mathrm{Gn}}\mathrm{G}) \tag{10}$$

Table 2 The attenuation factors $\alpha_{\lambda L}$ for each grid patterns (i)–(iv)

Span	(i)	(ii)	(iii)	(iv)
20	1	1	1	1
25	0.369	0.489	0.459	0.465
30	0.167	0.273	0.242	0.248
35	0.083	0.167	0.141	0.145
40	0.045	0.108	0.089	0.091
$L = 20$ m				

Table 3 The attenuation factors α_{GL} for each grid patterns (i)–(iv)

Span	(i)	(ii)	(iii)	(iv)
20	1	1	1	1
25	0.532	0.723	0.762	0.793
30	0.321	0.567	0.605	0.663
35	0.210	0.468	0.494	0.571
40	0.143	0.400	0.410	0.501
$L = 20$ m				

Table 4 The enhancement factor $\beta_{\lambda n}$ for each grid patterns (i)–(iv)

Mesh-density	(i)	(ii)	(iii)	(iv)
8	1	1	1	1
10	1.294	1.391	1.278	1.275
12	1.622	1.802	1.549	1.542
14	1.986	2.184	1.816	1.802
16	2.385	2.533	2.084	2.063
18	2.818	2.874	2.352	2.325
20	3.284	3.213	2.620	2.588
$n = 8$				

Table 5 The enhancement factor β_{Gn} for each grid patterns (i)–(iv)

Mesh-density	(i)	(ii)	(iii)	(iv)
8	1	1	1	1
10	1.891	1.954	1.915	1.915
12	3.245	3.389	3.323	3.279
14	5.188	5.411	5.250	5.177
16	7.866	8.131	7.859	7.707
18	11.446	11.669	11.254	10.970
20	16.119	16.147	15.482	15.046
$n = 8$				

4 Applicability Analysis

Though this study promises the high precision for the foundational equations based on regression analysis, it still exists little difference, because the coefficient of determination R^2 is just close to 1. In addition, out of considering briefness of the envelope equations, the authors take the average values for magnification of each span on $L =$ 20 m and each mesh-density on $n = 8$, which could result in a big difference. Therefore, in order to examine the applicability of the proposed equations, the authors calculate the approximate values λ_{app} through Eq. 10, and compare them with the results from linear buckling analysis λ_{an}. In the final, the distribution of difference is illustrated by distribution histograms of frequency f as shown in Fig. 10, where the $r_{\Delta\lambda}$ is the rate of difference and expressed as $r_{\Delta\lambda} = (\lambda_{app} - \lambda_{an})/\lambda_{an} = \Delta\lambda/\lambda_{an}$, while the f is the frequency of $r_{\Delta\lambda}$ and $\sum f = 1$.

From Fig. 10, it can be found that there exists some $r_{\Delta\lambda}$ even up to $\pm 100\%$, that is because the λ_{an} is very small, the difference $\Delta\lambda$ almost closes to the λ_{an}, moreover, frequency f is almost concentrated in the range of $r_{\Delta\lambda} = \pm 20\%$, which means the envelope equations still have high applicability.

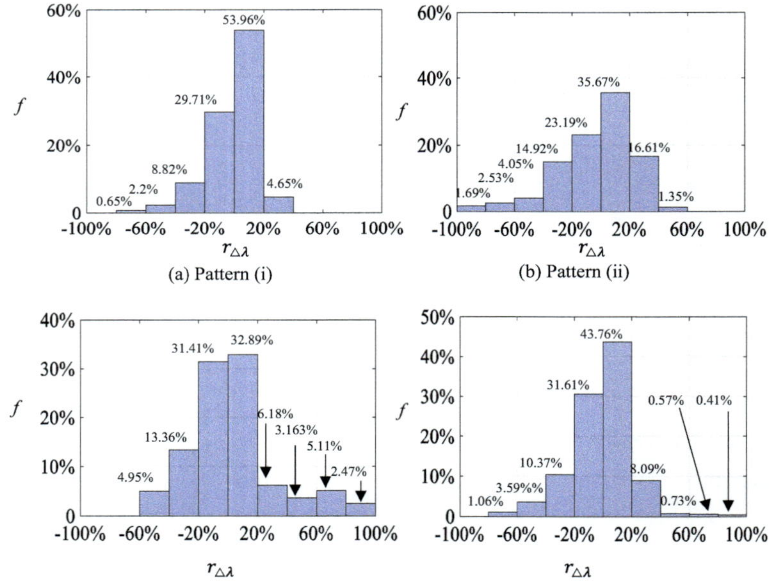

Fig. 10 Distribution of difference for each grid pattern

5 Conclusions

This study analyzes large amounts of single-layer grid shells with various geometric parameters by linear buckling analysis, and consequently proposes an approximate equation to evaluate the buckling load for a cylindrical single-layer grid shells with 4 kinds of grid patterns and pipe-shaped members. The significant findings obtained from this study are summarized as follows.

(1) For the influence of grid patterns, it is impossible to build a certain relationship for different grid patterns on λ and G due to their high irrelevance. Thus, whether in the study of the properties or the derivation of the equation on single-layer grid shells, the influence of the grid shape should be essentially considered.
(2) The envelope equations proposed in this paper consist of two parts which are foundational equations and effect factors. The authors only prove the high precision for the foundational equations based on the regression analysis, while the effect factors just are taken by average values which could lead to some big difference, but from the rate of difference $r_{\Delta\lambda}$, it can be found that the frequency f is almost concentrated in the range of $r_{\Delta\lambda} = \pm 20\%$ which means that the envelope equations are valid enough to evaluate the buckling load for the single-layer grid shells studied in this paper.

However, though the influence of span L and mesh-density n are demonstrated in this paper, the situation of span $L_x \neq L_y$ and mesh-density $n_x \neq n_y$ are not studied. Moreover, the influence of initial imperfection and structural design is not considered yet. In addition, the authors will not only extend this research to the other shapes and boundary conditions of grid shells, but also make an effort to apply the proposed methodology to the other filed in future.

Acknowledgements The author LIU would like to show his sincere gratitude to ASSURAN International Scholarship Foundation for supporting his life of studying abroad.

References

1. Chilton, J.: Space Grid Structures. Architectural Press, Great Britain (2000)
2. Wright, D.T.: Membrane forces and buckling in reticulated shells. J. Struct. Div. **21**, 175–201 (1965)
3. Karadeniz, H.: Stochastic Analysis of Offshore Steel Structures—An Analytical Appraisal. Springer-Verlag, London (2013)
4. Su, M., Zhong, Q., Peng, H.: Regularized multivariate polynomial regression analysis of the compressive strength of slag-metakaolin geopolymer pastes based on experimental data. Constr. Build. Mater. **303**, 124529 (2021)
5. Altuna-Zugasti, A.M., Lopez-Arancibia, A., Puente, I.: Influence of geometrical and structural parameters on the behaviour of squared plan-form single-layer structures. J. Constr. Steel Res. **72**, 219–226 (2012)

6. Chen, P.S., Mamoru, K.: Optimization for maximum buckling load of a lattice space frame with nonlinear sensitivity analysis. Int. J. Space Struct. **21**(2), 111–118 (2006)
7. Hangai, Y.: Structural behaviors of single-layer space frames (Part 1)—formulae for buckling loads of single-layer domes. Seisan Kenkyu **39**(12), 17–20, (1987) (In Japanese)
8. Chen, X., Wang, N., Shen, S.Z.: Ultimate loading capacity of braced domes. IASS-ASCE, Symposium, pp.539–546 (1994)
9. Kato, S., Yamashita, T.: Evaluation of elasto-plastic buckling strength of two-way grid shells using continuum analogy. Int. J. Space Struct. **17**(4), 249–261 (2002)
10. Yan, G., Fang, C., Feng, R., Hua, X., Zhao, Y.: Detection of member overall buckling in civil space grid structures based on deviation in normal strain along the member. Eng. Struct. **131**, 599–613 (2017)
11. Chair of Working Group 8.: Draft Guide to Buckling Load Evaluation of Metal Reticulated Roof Structures. IASS WG 8 for Metal Spatial Structures, Japan (2014)
12. Zhang, D.: A coefficient of determination for generalized linear models. Am. Stat. **71**(4), (2016)
13. Malek, S., Wierzbicki, T., Ochsendorf, J.: Buckling of spherical cap gridshells: a numerical and analytical study revisiting the concept of the equivalent continuum. Eng. Struct. **75**, 288–298 (2014)
14. Liu, B., Chen, P.S., Jin, J., Yan, X.: Approximate equation for estimating global buckling load of single-layer cylindrical space frames. In: Proceedings of AWAM International Conference on Civil Engineering, AICCE (2022)

Reinforced Concrete Structural Engineering and Durability of Concrete Structures

Predicting the Performance of Shear Wall Structures Using the Confidence Nets Model

Nouraldaim F. A. Yagoub and Wang Xuxin

Abstract The recent increase in earthquake engineering employs machine learning technologies to develop prediction models for structural behavior. A reinforced concrete shear wall is one of the most important structural parts of a structure for resisting lateral loads. However, predicting the behaviors of shear wall members and their influence on the structure has always been difficult. Recent studies suggest the use of artificial intelligence (AI) models in this field and considerable amount of attention in the earthquake engineering community, as they can map the complicated relationship between the anticipated damage and the input variables and have shown promising results. In this paper, we aim to push the accomplishments of AI models further by providing more reliable predictions supported by an estimate of a confidence score. Moreover, the proposed model is 185 times faster than the standard finite element analysis method. The model's predictive performance is also compared with the FEM method.

Keywords Artificial intelligence · Reinforced concrete shear walls · Deformation capacity · Predictive modeling

N. F. A. Yagoub · W. Xuxin (✉)
School of Civil Engineering, Southeast University, Nanjing 210096, China
e-mail: wangdisks@163.com

N. F. A. Yagoub
Department of Civil Engineering, Faculty of Engineering Science, University of Nyala, Nyala, Sudan

W. Xuxin
Full Key Laboratory of Concrete and Prestressed Concrete Structures of Ministry of Education, Southeast University, Nanjing 210096, China

E. Strauss (ed.), *Proceedings of the 7th International Conference on Civil Engineering*, Lecture Notes in Civil Engineering 371, https://doi.org/10.1007/978-981-99-4045-5_22

1 Introduction

Reinforcement concrete shear walls are the main lateral load-bearing components in high-rise buildings due to their strength, stiffness, and deformation capacity (ductility). Recent earthquakes have underlined the necessity of shear wall performance [1]. Concrete shear walls are a popular design choice for buildings subject to lateral loads like wind and earthquakes. Seismic behavior and failure mechanisms of vertical structural elements are influenced by section geometry and support conditions, structural material mechanical properties (concrete and steel), transverse and longitudinal reinforcement amount and detailing, and loading pattern. Similar to shear walls, flexure-controlled shear walls fail by yielding in flexure before achieving shear strength. Chinese standards specify semi-empirical formulae for estimating RC wall shear strength (JGJ 3–2010) [2], U.S. code (ACI 318-19) [3], and Europe code (EC-8) [4] Based on superposition and experiment data. These simple design formulas have a specified safety reserve, however, they don't effectively predict shear strength, making it impossible to predict RC wall failure scenarios. RC wall deformation has been studied less than failure modes and strength. Paulay and Priestly estimated the ultimate drift ratio of a flexure-controlled RC wall [5, 6]. Several scholars also proposed deformation capacity equations [7–9]. Physical, experimental, numerical, and statistical models of ultimate strength for squat RC walls have been studied. Building codes, engineering practice guides, and construction standards and guidelines adopted these research results [9]. Researchers developed models to predict such behavior. For instance, Chandra et al. [10] created a truss-based methodology for determining RCSW shear capacity. Lu and Henry [10] developed a finite element model to study how lightly reinforced concrete walls respond to earthquakes. They tested their model on RC walls with minimal vertical reinforcement to make sure it was proper. They said that the model did a good job of capturing both the global and local response parameters.

Research questions rooted in discovering new patterns and understanding complex relationships have led to studies in artificial intelligence. In recent years, artificial intelligence (AI), a data-driven approach, has received attention and has been used to predict how well RC walls can perform during earthquakes. Based on the prior test findings, we implement a technique called confidence net developed by Altayeb et al. [11, 12] that could accurately predict the ultimate load of the squat RC walls and produce a measure of the prediction interval. In order to solve difficult issues that are depicted as a set of inputs and a desired output, ANN is a computational approach that makes use of a large number of neural units to model the structures and activities of the human brain. In recent years, ANNs have been applied in a variety of fields, including meteorology, hydrology, and engineering, due to their high accuracy and ease of use [13–15]. In addition, Chithra et al. [16] and Khademi et al. [14] have shown that ANN performs better at making predictions than conventional techniques like regression techniques. Implementing an ANN, however, has significant drawbacks, such as the slow learning rate and the solution trap problem while attempting to identify the local minimum [15].

In this study, the technique proposed by Altayeb et al. is implemented on the ACI-Database to predict the maximum shear strength. The authors show that the proposed model can provide the much-needed speed of prediction for Abaqus simulations. There is a significant reduction in the time taken to obtain a result using the method, and the model's accuracy is assessed. Then several simulations using Abaqus are performed, and the model is evaluated against them.

In the first section, we introduce the dataset, model, training procedure, and the Abaqus simulation. The second section of the paper presents the results of using this method over ordinary neural network-based methods. The approach's advantages are also discussed. Finally, a conclusion is provided.

2 Methodology

2.1 Finite Element Model

Concrete materials were explained using the ABAQUS CDP model. This applies to monotonic or cyclic concrete components. All concrete components, including the cap beams, wall panel, and foundation, were created using an eight-node 3D brick element (C3D8R). In this study, Feng et al. [17] a softened concrete damage plasticity model used to specify CDP input data (stress–strain). This model examines compression softening's impact on stress–strain data and concrete degradation. Table 1 shows the compressive strength of each specimen. Assume the maximum concrete member strain is 0.003. ABAQUS' concrete damaged plasticity model requires five parameters. Dilation angle ψ, flow potential eccentricity $\in$, , biaxial to uniaxial $\sigma b0/\sigma c0$ strength, second stress to yield function Kc, and viscosity parameter μ. The parameters were 30, 0.1, 1.16, 0.667, and 0.005. Program description (SIMULIA 2008) and past research [18] corroborate this assumption. T3D2 was used to model the wall panels' steel reinforcing. Bars were set in concrete. According to Table 1, all truss elements were modeled using a bilinear elastic–plastic relationship. To emulate fixing, the footing's bottom surface was fully confined. Two-part model analysis was utilized to simulate the experimental loading sequence. Initiate axial load. Apply predefined lateral drifts using the testing procedures (pushovers).

2.2 Models' Description

In the first selected models, two types of walls were tested in the program: Type I (h/1 = 1), which had dimensions of 750 mm wide by 750 mm high by 70 mm thick, and Type II (hi 1 = 2), which had dimensions of 650 mm wide by 1300 mm high by 65 mm thick. In all instances, an upper and a lower beam were monolithically joined to the walls. the lower beam and the upper beam (1150 mm long, 150 mm

Table 1 Concrete, longitudinal, and transverse reinforcement characteristics

Type	Items	Model two (w1–w9)		Models (SW11–SW13, SW21–SW23)
Concrete	Strength (MPa)	27.4		50
	Module Es (Gpa)	32,700		33,446
	Poisson's ratio	0.2		0.2
Transverse bars longitudinal	Diameter	Ø5	Ø8	Ǿ4
	Es (Gpa)	210	210	210
	Yielding stress of steel (Mpa)	608.9	445.6	420
	Tensile strength of steel (MPa)	667.7	598.9	490
	Poisson's ratio	0.3	0.3	0.3
Longitudinal bars	Diameter	Ø10		Ǿ8, Ǿ6.25
	Module Es (Gpa)	210		210
	Yielding stress of steel (Mpa)	469.2		470
	Tensile strength of steel (MPa)	675.7		565
	Poisson's ratio	0.3		0.3

deep, and 200 mm thick) (1150 mm long, 300 mm deep, and 200 mm thick) [19]. In the second category, several models of slender RC walls (W1–W9) were created and tested to determine how an axial load can affect them. Wall w1 is referred to as a referencing wall. The walls dimensions were 700 mm by 100 mm, except for wall w4, which had a thickness of 75 mm. The concrete cover was 10 mm. The height of the walls was 1600 mm, except for wall w5, which had a height of 1180 mm. The dimensions of foundation and top beam were 425 mm * 400 mm * 1400 mm and 300 mm * 300 mm * 700 mm, respectively. The reference wall thickness w1 was 100 mm [20]. The vertical and horizontal reinforcements are 4, 6.25, 8, and 10 mm in diameter. Stirrups constrained the wall edges.

2.3 *Confidence Nets Model*

The confidence nets technique on the ACI 445B Shear Wall Database is used to predict the maximum shear strength. The ACI 445B Shear Wall Database is a large dataset that is compiled and vetted several times per year by the American Concrete Institute [21]. After obtaining the dataset, the authors first use the min–max normalization technique to map all the data to values between [0, 1]. Then the dataset is split into training and testing at a ratio of 3:1, respectively. The confidence nets model is composed of two main components: the predictor model (A Neural Network) and an error estimation model (XGBoost). The predictor model is responsible for predicting the outputs; however, the error estimation model is responsible for predicting the error produced by the predictor model on the inputs that will be fed. This means that the error estimation model must learn the relationship between a set of inputs and the error produced, subsequently understanding the weakness of the estimation model. The architecture of the confidence nets model is explained in detail in the works of Altayeb et al. [11], and Fig. 1 illustrates the model's training procedure.

The confidence nets model was trained using the Adam optimizer with a learning rate initialized at 0.003 for 400 epochs. The cost function used for training is the mean-squared error loss. The structure used for the confidence nets is a two-layer hidden layer with 100 neurons and a single output neuron. The activation function ELU was used between the hidden layers and after the final output layer.

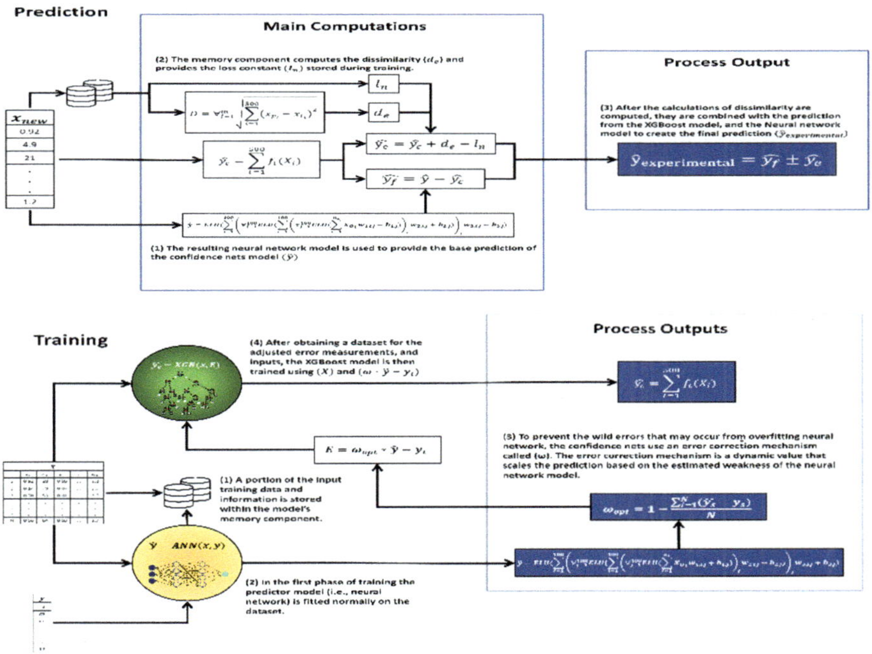

Fig. 1 Illustration of the training procedure and prediction process of the confidence nets model that was provided by the original Authors of the confidence nets [11]

3 Results

3.1 FEM Model

To verify the finite element model, shear walls are modeled in order to replicate the test wall specimens of [19, 20]. Results from the verification model and experimental findings are assessed in terms of the base shear-lateral drift action, as shown in Fig. 2a, b, and Table 1.

Table 2. FEM and Experimental comparison of tmax. The outcomes of the current numerical model are discovered to be in excellent agreement with the outcomes of the experiments. The experimental backbone curve's initial stiffness is found to be somewhat less than the stiffness of the computational models. However, the maximum difference in the base shear of the numerical models is only 3%. As a result, the numerical model created is thought to be appropriate for further research.

3.2 Confidence Nets Model

Figure 3 Predictions showing the performance of the confidence nets approach in comparison with a normal prediction provided by a deep neural network model of the same size shows 39 predictions on the test dataset comparing the confidence nets model and a deep neural network model of the same size. The confidence nets model is much more superior to the deep neural network model. In general, a deep neural network tries to follow the mean of the data. Since not much data is available, a deep neural network alone is not well suited for such tasks. The confidence net model, however, shows that it is more sensitive than a deep neural network model and can detect small changes in inputs, so the predictions are closer to the ground truth values. Table 3. Sampled predictions showing the properties of the new confidence nets approach shows sampled predictions from the training dataset. The table shows

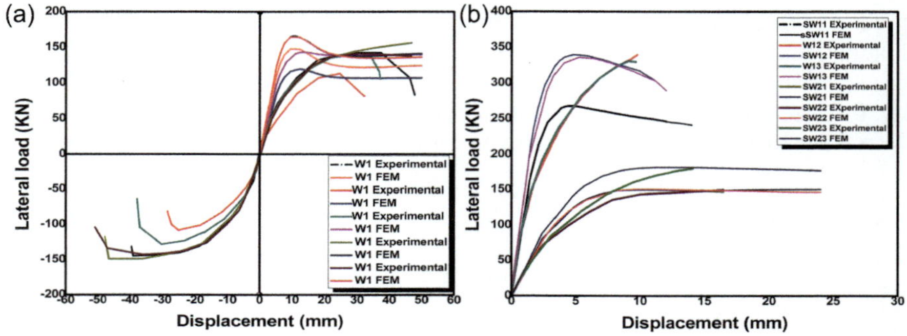

Fig. 2 **a**, **b** Load–displacement curves for Type I and Type II. walls

Table 2 FEM and experimental comparison of tmax

No.		Lateral loading		tmax	
		Experimental	FEM	Experimental	FEM
1	W1	142.9	147.9	2.04	2.11
2	W4	113.4	119.8	1.62	1.71
3	W6	139.8	140.5	1.9	2.01
4	W8	156.1	165.7	2.23	2.36
5	W9	142.4	164.6	2.03	2.35
6	SW11	260	268	4.9	5.34
7	SW12	340	340	6.47	7.04
8	SW13	330	336	6.28	6.95
9	SW21	148.9	150	3.01	3.55
10	SW22	150	188	3.55	3.44
11	SW23	180	181	4.20	4.18

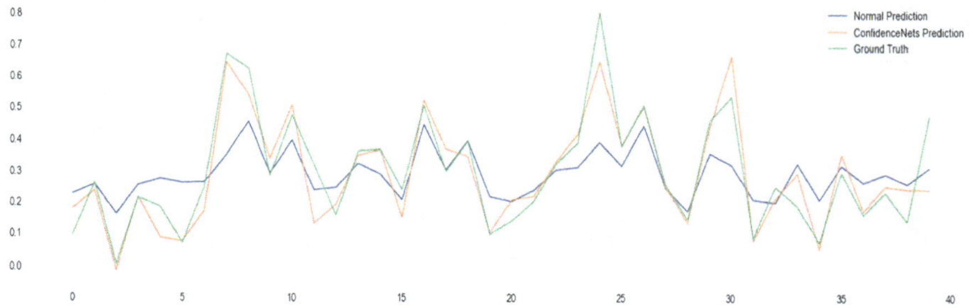

Fig. 3 Predictions showing the performance of the confidence nets approach in comparison with a normal prediction provided by a deep neural network model of the same size

much more detailed presentations of the predictions and how they compare. The main advantage provided by the confidence net model is the prediction interval. This prediction interval becomes increasingly useful in experimental cases. It gives insight into the model's confidence of the prediction of achieving the result stated.

Table 3 Sampled predictions showing the properties of the new confidence nets approach

	Normal prediction	Confidence nets prediction	Ground truth
92	0.267	0.076 ± 0.191	0.048
33	0.327	0.295 ± 0.033	0.191
19	0.22	0.105 ± 0.115	0.1
34	0.21	0.054 ± 0.156	0.074

(continued)

Table 3 (continued)

	Normal prediction	Confidence nets prediction	Ground truth
153	0.251	0.318 ± 0.067	0.158
3	0.257	0.218 ± 0.039	0.216
177	0.248	0.248 ± 0.0	0.13
192	0.305	0.439 ± 0.133	0.499
6	0.266	0.174 ± 0.092	0.249
95	0.315	0.489 ± 0.174	0.473
165	0.301	0.285 ± 0.016	0.207
137	0.297	0.352 ± 0.055	0.206
93	0.448	0.509 ± 0.062	0.608
96	0.201	0.153 ± 0.047	0.128
158	0.298	0.216 ± 0.081	0.273
174	0.373	0.646 ± 0.273	0.766
154	0.307	0.296 ± 0.011	0.197
8	0.457	0.542 ± 0.085	0.627
71	0.265	0.32 ± 0.055	0.291
159	0.321	0.548 ± 0.227	0.469

Table 4 A comparison between the FEM approach and the AI approach in terms of computational speed and performance

Parameter	FEM model	AI model (confidence nets)
Total no. of predictions/simulations	11	39
Average speed per simulation (s)	148	0.80
Fastest simulation (s)	74	0.63
Slowest simulation (s)	148	1.01
Accuracy (%)	97	91.6

3.3 Comparison

Table 4 shows a comparison between the two approaches: the FEM approach and the AI approach. In summary, the results of 11 simulations and 39 predictions are obtained from the two approaches, and the computational time taken is recorded. The FEM approach takes an average speed of 148 s to complete a prediction, while the confidence nets model takes an average speed of 0.8 s to provide a prediction on the same PC. The improvement in speed is almost 185× that of using the FEM approach. In terms of speed this is a significant improvement, meaning that using the AI approach, it takes only 0.5% of the amount of time taken by the FEM method to provide a result for the maximum shear strength. Moreover, there is a small variation between the fastest prediction and the smallest prediction time. In comparison, there is a large variation in the FEM model. In terms of accuracy, however, we can see that to obtain this significant improvement in speed there is a slight reduction in accuracy. While the FEM approach achieves about 97% accuracy, the AI approach achieves only 91.6%.

4 Conclusion

The confidence nets technique is used in this paper to predict the maximum shear strength using the ACI-Database. The authors show that the suggested model might provide the much-needed speed of prediction for Abaqus simulations. A significant reduction in the time taken to obtain a result using the method is achieved. Then several simulations using Abaqus are performed, and the model is evaluated against them.

1. The results of the analysis have been compared to those obtained from the experimental results, which have been demonstrated to have been successful.
2. The FEM method used in this study not only gave an appreciable estimation of reinforced concrete shear walls' strength properties but also possessed high accuracy. The FEM model can be a promising alternative for further analysis.
3. The confidence nets model implemented in this work provided a reliable level of accuracy of more than 91%.
4. A remarkable increase in speed is achieved using the confidence net model. The AI approach is 185 times faster than the FEM approach.

From this work, it seems that the AI approach can become an alternative for FEM methods in the near future. However, further investigation into the limitations of the AI approach must be studied.

References

1. Paulay, T: Design aspects of shear walls for seismic areas. Can. J. Civ. Eng. **2**(3), 321–344 (1975)
2. Ministry of Housing: Urban-Rural Development of the People's Republic of China, Technical Specification for Concrete Structures of Tall Building (JGJ 3–2010) (2010)
3. ACI Committee, others: ACI 318–19: Building Code Requirements for Structural Concrete and Commentary. Am. Concr. Inst. Farmingt. Hills, MI, USA (2019)
4. Code, P.: Eurocode 8: Design of structures for earthquake resistance-part 1: general rules, seismic actions and rules for buildings. Brussels Eur. Comm. Stand. (2005)
5. Paulay, A., Priestley, T., Synge M.J.N.: Ductility in earthquake resisting squat shearwalls. J. Proc. 257–269 (1982)
6. Paulay, T.: The displacement capacity of reinforced concrete coupled walls. Eng. Struct. **24**(9), 1165–1175 (2002)

7. Deger, Z.T., Basdogan, C.: Empirical expressions for deformation capacity of reinforced concrete structural walls. ACI Struct. J. **116**(6), 53–61 (2019)
8. Grammatikou, S., Biskinis, D., Fardis, M.N.: Strength, deformation capacity and failure modes of RC walls under cyclic loading. Bull. Earthq. Eng. **13**(11), 3277–3300 (2015)
9. Design of concrete structures: Seismic Design of Buildings to Eurocode, vol. 8, pp. 122–190. (2021)
10. Chandra, J., Chanthabouala, K., Teng, S.: Truss model for shear strength of structural concrete walls. ACI Struct. J. **115**(2), 323–335 (2018)
11. Altayeb, M., et al.: Confidence-nets: a step towards better prediction intervals for regression neural networks on small datasets (2022)
12. Altayeb, M., Wang, X., Musa, T.H.: An ensemble method for predicting the mechanical properties of strain hardening cementitious composites. Constr. Build. Mater. **286**, 122807 (2021)
13. Gao, W., Chen, D., Dai, S., Wang, X.: Back analysis for mechanical parameters of surrounding rock for underground roadways based on new neural network. Eng. Comput. **34**(1), 25–36 (2018)
14. Khademi, F., Jamal, S.M., Deshpande, N., Londhe, S.: Predicting strength of recycled aggregate concrete using artificial neural network, adaptive neuro-fuzzy inference system and multiple linear regression. Int. J. Sustain. Built Environ. **5**(2), 355–369 (2016)
15. Gordan, B., Jahed Armaghani, D., Hajihassani, M. , Monjezi, M.: Prediction of seismic slope stability through combination of particle swarm optimization and neural network. Eng. Comput. **32**(1), 85–97 (2016)
16. Chithra, S., Kumar, S.R.R.S., Chinnaraju, K., Ashmita, F.A.: A comparative study on the compressive strength prediction models for High Performance Concrete containing nano silica and copper slag using regression analysis and artificial neural networks. Constr. Build. Mater. **114**, 528–535 (2016)
17. Feng, D.C., Ren, X.D., Li, J.: Cyclic behavior modeling of reinforced concrete shear walls based on softened damage-plasticity model. Eng. Struct. **166**, 363–375 (2018)
18. Chao, L., Hao, H., Bi, K.: Numerical study on the seismic performance of precast segmental concrete columns under cyclic loading. Eng. Struct. **148**, 373–386 (2017)
19. Tasnimi, A.A.: Strength and deformation of mid-rise shear walls under load reversal. Eng. Struct. **22**(4), 311–322 (2000)
20. Lefas, I.D., Kotsovos, M.D., Ambraseys, N.N.: Behavior of reinforced concrete structural walls. Strength, deformation characteristics, and failure mechanism. ACI Struct. J. **87**(1), 23–31 (1990)
21. Merve, U., Santiago, P., Aishwarya, P., Preethi, S., Cheng, S., Ying, W.: ACI 445B Shear Wall Database. 445B2

Impacts of Web Stiffener Locations on Capacities of Cold-Formed Steel SupaCee Sections

Ngoc Hieu Pham

Abstract This paper investigates the sectional capacities of cold-formed steel SupaCee sections due to the effects of web stiffener locations. SupaCee sections in the form of channel sections are made by creating stiffeners in the sectional webs. This allows such sections to increase their stability and improve their strengths in comparison with the original channel sections. These innovations of the SupaCee sections have been discussed in previous studies on the basis of the commercial sections with the web stiffener locations are regulated in the catalogues from manufacturers. Locations of these stiffeners, therefore, are varied in this paper to find out their influence on the sectional capacities of the investigated SupaCee sections. Full bracings are applied to prevent global buckling failure modes. The sectional capacities are determined using a new design method called the Direct Strength Method as regulated in Australia/New Zealand Standard AS/NZS 4600: 2018. In this method, elastic buckling analysis is compulsory for the design procedure and can be carried out by using commercial software programs. Two cases of applied loads including compression and bending are considered in this investigation. It is found that an increasing trend is obtained for local buckling strengths if the stiffeners reach toward the centroid of the web and there is an opposite trend for distortional buckling strengths.

Keywords Impacts · Web stiffener locations · Capacities · Cold-formed steel · SupaCee sections

1 Introduction

Cold-formed steel channel sections are available worldwide for many decades [1]. These sections in the form of thin-walled sections are sensitive to sectional instability including local and distortional buckling modes. Their stability is subsequently improved by adding stiffeners in the webs to create SupaCee sections. The presence of

N. H. Pham (✉)
Faculty of Civil Engineering, Hanoi Architectural University, Hanoi 100000, Vietnam
e-mail: hieupn@hau.edu.vn

E. Strauss (ed.), *Proceedings of the 7th International Conference on Civil Engineering*, Lecture Notes in Civil Engineering 371, https://doi.org/10.1007/978-981-99-4045-5_23

stiffeners results in significantly improved strengths of SupaCee sections compared to those of channel sections, as discussed in previous papers [2, 3].

For design, the Effective Width Method (EWM) was the conventional method based on the stability of flat plates [4]. This method allowed to account for the local and global interaction buckling modes, but its design procedure was complicated for complex sections or multiple stiffeners. The Direct Strength Method (DSM) has been proposed to solve these limitations and has illustrated its advances in design [5]. The design for complex sections becomes simple, and this method provides an understanding of their buckling behaviors. Elastic buckling analyses are required for the application of this new method with the support of the numerical software program THIN-WALL-2 [6]. The output of this software program is a curve to demonstrate the relationship between the elastic buckling loads and the buckle lengths that can be used for the optimization of sectional capacities. The DSM has been regulated in the Australian/New Zealand Standard AS/NZS 4600: 2018 [7] and will be utilized for the investigation in this paper.

Studies on stiffeners for cold-formed steel sections were carried out by numerous researchers. For compression, an experimental program on cold-formed steel storage racks was performed at the University of Sydney with the variations of edge stiffeners [8–10]. The behaviors of isolated flanges with stiffeners were investigated by Seah and Rhodes [11], and the modifications in the design were proposed on the basis of the Effective Width Method. The behaviors of channel columns with edge stiffeners are studied by Wang and Young, Yan and Young [12–14]. Chen et al. [15] presented an experimental program on cold-formed steel stub channel columns with different stiffeners. For bending, the influence of intermediate and edge stiffeners on the flexural capacities of channel members was studied by Ye et al. [16]. Behaviors of built-up I sections with stiffeners were examined by Manikadan and Arun [17] based on the development of experimental and numerical programs. The impacts of web stiffener locations on the capacities of cold-formed steel SupaCee sections are not reported, and therefore are investigated in this paper.

2 Cross-Sections and Material Properties for Investigation

The SupaCee sections for the investigation are selected from the commercially available sections. The sectional shape and nomenclature of this section are illustrated in Fig. 1. The dimensions of SupaCee sections from the manufacturer's catalogue [18] are given in Table 1. There are two couple of stiffeners in the webs. Based on the catalogue of SupaCee sections provided by the manufacturer [18], it reveals that the distances between single stiffeners in a couple of stiffeners (S) are constant and equal to 42 mm; and the distances from the centroids of each couple of stiffeners to the respective flange sides are also constant for all cross-sections and equal to 44 mm. This may be explained due to the requirements of the manufacturing technology. The distance between the couple stiffeners (GS) is varied to investigate the sectional capacities of SupaCee sections under compression or bending. Sections SC250 are

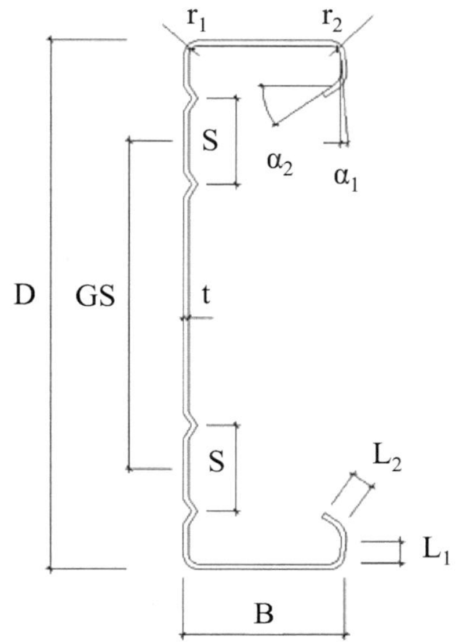

Fig. 1 Nomenclature of SupaCee section

selected for the investigation with different thicknesses, and; the distances (GS) are listed in Table 2.

The material properties of the investigated sections are Grade G450, as regulated in Australian Standard AS 1397 [19]. The stress–strain curve of Grade G450 is demonstrated in Fig. 2, where the yield stress f_y is taken as the 0.2% proof stress. This stress–strain curve is rounded with a low ratio of tensile strength and yield stress due to the cold-forming process. The material properties of Grade G450 include Young's modulus E = 200,000 MPa; the yield strength $f_y = 450$ MPa, and the Poisson ratio $\nu = 0.3$.

3 Determination of Sectional Capacities of Cold-Formed Steel Sections According to AS/NZS 4600

The design of cold-formed steel sections is presented in Clauses 7.2.1 and 7.2.2 of the Australian/New Zealand Standard AS/NZS 4600: 2018 [7] corresponding to compression and bending. Only sectional capacities are considered in this paper; the global buckling strengths (N_{ce}; M_{be}) can be taken as yield strengths (N_y; M_y) as reported in Pham [3].

For compression, the nominal strength of a cold-formed steel section is the lesser of the local buckling (N_{cl}) and distortional buckling (N_{cd}) strengths:

$$N_s = \text{Lesser}\ (N_{cl},\ N_{cd})$$

Table 1 The nominal dimensions of the SupaCee sections [18]

Sections	t	D	B	L_1	L_2	GS	S	α_1	α_2
SC15012	1.2	152	64	7.5	7.5	64	42	5	35
SC15015	1.5	152	64	7.5	7.5	64	42	5	35
SC15019	1.9	152	64	7.5	7.5	64	42	5	35
SC15024	2.4	152	64	7.5	7.5	64	42	5	35
SC20012	1.2	203	76	10	10	115	42	5	35
SC20015	1.5	203	76	10	10	115	42	5	35
SC20019	1.9	203	76	10	10	115	42	5	35
SC20024	2.4	203	76	10	10	115	42	5	35
SC25015	1.5	254	76	11	11	166	42	5	35
SC25019	1.9	254	76	11	11	166	42	5	35
SC25024	2.4	254	76	11	11	166	42	5	35
SC30019	1.9	300	96	14	14	212	42	5	35
SC30024	2.4	300	96	14	14	212	42	5	35
SC30030	3.0	300	96	14	14	212	42	5	35
SC35019	1.9	350	125	15	15	262	42	5	35
SC35024	2.4	350	125	15	15	262	42	5	35
SC35030	3.0	125	125	15	15	262	42	5	35
SC40019	1.9	400	125	15	15	312	42	5	35
SC40024	2.4	400	125	15	15	312	42	5	35
SC40030	3.0	400	125	15	15	312	42	5	35

Note The inner radius $r_1 = r_2 = 5$ mm; t, D, B, L1, L2, GS, S (mm); α_1, α_2 (^{0})

Table 2 The variation of distance *GS* for SupaCee 250 sections

Sections	t	D	B	L_1	L_2	GS	S	α_1	α_2
SC25015	1.5	254	76	11	11	166	42	5	35
SC25019	1.9	254	76	11	11	166	42	5	35
SC25024	2.4	254	76	11	11	166	42	5	35
SC25015	1.5	254	76	11	11	146	42	5	35
SC25019	1.9	254	76	11	11	146	42	5	35
SC25024	2.4	254	76	11	11	146	42	5	35
SC25015	1.5	254	76	11	11	126	42	5	35
SC25019	1.9	254	76	11	11	126	42	5	35
SC25024	2.4	254	76	11	11	126	42	5	35
SC25015	1.5	254	76	11	11	106	42	5	35
SC25019	1.9	254	76	11	11	106	42	5	35
SC25024	2.4	254	76	11	11	106	42	5	35
SC25015	1.5	254	76	11	11	86	42	5	35
SC25019	1.9	254	76	11	11	86	42	5	35
SC25024	2.4	254	76	11	11	86	42	5	35

Note The inner radius $r_1 = r_2 = 5$ mm; t, D, B, L1, L2, GS, S (mm); α_1, α_2 (^{0})

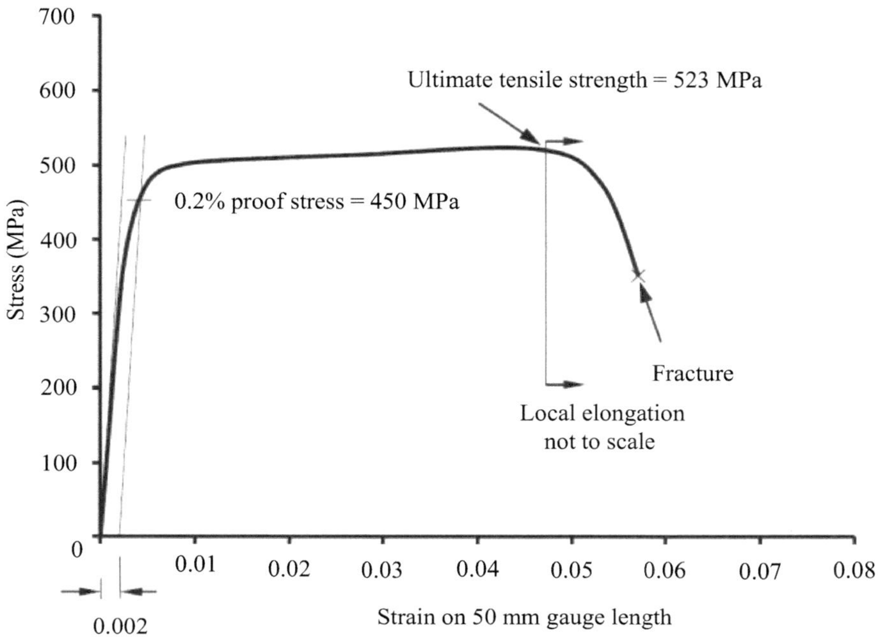

Fig. 2 Stress–strain curve of G450 steel [1]

$$N_{cl} = \begin{cases} N_y & \text{for } \lambda_l \leq 0.776 \\ \left[1 - 0.15\left(\frac{N_{ol}}{N_y}\right)^{0.4}\right]\left(\frac{N_{ol}}{N_y}\right)^{0.4} N_y & \text{for } \lambda_l > 0.776 \end{cases} \tag{1}$$

$$N_{cd} = \begin{cases} N_y & \text{for } \lambda_l \leq 0.561 \\ \left[1 - 0.25\left(\frac{N_{od}}{N_y}\right)^{0.6}\right]\left(\frac{N_{od}}{N_y}\right)^{0.6} N_y & \text{for } \lambda_l > 0.561 \end{cases} \tag{2}$$

where $\lambda_l = \sqrt{N_y/N_{ol}}$; $\lambda_d = \sqrt{N_y/N_{od}}$; N_y is the yield strength in compression; N_{ol} is the elastic local buckling load; N_{od} is the elastic distortional buckling load.

For bending, the nominal moment is taken as the smaller between the local buckling moment (M_{bl}) and distortional buckling moment (M_{bd}).

$$M_s = \text{Lesser } (M_{bl}, M_{bd})$$

$$M_{bl} = \begin{cases} M_y & \text{for } \lambda_l \leq 0.776 \\ \left[1 - 0.15\left(\frac{M_{ol}}{M_y}\right)^{0.4}\right]\left(\frac{M_{ol}}{M_y}\right)^{0.4} M_y & \text{for } \lambda_l > 0.776 \end{cases} \tag{3}$$

$$M_{bd} = \begin{cases} M_y & \text{for } \lambda_l \leq 0.673 \\ \left[1 - 0.22\left(\frac{M_{od}}{M_y}\right)^{0.5}\right]\left(\frac{M_{od}}{M_y}\right)^{0.5} M_y & \text{for } \lambda_l > 0.673 \end{cases} \quad (4)$$

where $\lambda_l = \sqrt{M_y/M_{ol}}$; $\lambda_d = \sqrt{M_y/M_{od}}$; M_y is the yield moment; M_{ol} and M_{od} correspond to the elastic local and distortional buckling moments. Elastic local and distortional buckling loads are determined using the THIN-WALL-2 software program [6].

4 The Effects of Distance Between Couple of Web Stiffeners on the Sectional Capacities of SupaCee Sections

As presented in Sect. 3, only sectional strengths of the investigated sections are considered in this paper; the global buckling modes, therefore, are prevented by using the lateral bracing systems, as illustrated in Fig. 3. In these systems, numerous restraints are installed along the specimen lengths to reduce their effective lengths. The sectional capacities are subsequently determined according to the design equations in Sect. 3.

In terms of buckling analyses, the elastic local and distortional buckling stresses under compression and bending are determined using the commercial software program THIN-WALL-2 [6] as listed in Table 3. Two opposite trends are seen for sectional buckling stresses with the increase in the distance between stiffeners. Local buckling stresses witness an upward trend whereas this is the downward trend for

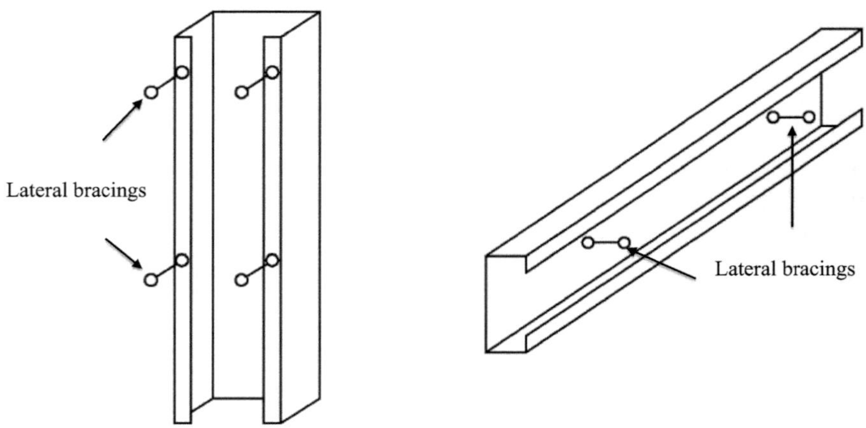

Fig. 3 The lateral bracing systems to prevent global buckling modes

Table 3 Local and distortional buckling stresses of SupaCee 250 sections

Sections	*GS*	Compression				Bending			
		f_{ol}	HWL	f_{od}	HWL	f_{ol}	HWL	f_{od}	HWL
SC25015	166	50.57	230	85.33	700	278.64	200	279.15	700
	146	53.17	230	84.26	700	287.28	200	277.27	700
	126	54.93	270	83.17	700	296.27	200	276.09	700
	106	56.77	270	82.16	700	303.23	200	275.52	700
	86	58.73	270	81.31	700	306.66	200	275.44	700
SC25019	166	74.02	230	116.12	700	411.48	180	371.43	700
	146	76.29	230	114.71	700	421.21	180	369.19	700
	126	78.75	230	113.31	700	431.68	180	367.75	700
	106	80.89	270	112.01	700	439.97	180	367	700
	86	82.52	270	110.9	700	444.29	180	366.81	700
SC25024	166	110.68	200	159.82	700	579.62	160	471.02	600
	146	112.67	230	157.96	700	587.00	180	468.11	600
	126	114.57	230	156.12	700	595.81	180	466.26	600
	106	116.65	230	154.42	700	604.34	180	465.32	600
	86	118.84	230	152.95	700	610.69	180	465.1	600

Note HWL stands for half-wavelength as defined by the investigated length corresponding to the critical buckling stress

distortional buckling stresses in both compression and bending. These sectional buckling stresses are subsequently utilized for the determination of sectional capacities of SupaCee sections as given in Tables 4 and 5 corresponding to compression and bending, where Δ % is the sectional capacity deviation of SupaCee sections in comparison with the original SupaCee sections (GS = 166 mm).

Tables 4 and 5 also show two opposite trends for local and distortional buckling strengths for both compression and bending. The sectional capacities are subsequently determined as the lesser between these two sectional buckling strengths. It is found that sectional capacities are governed by distortional buckling modes for both compression and bending. The sectional capacities are found to undergo decreasing trends with the variations of the distances (GS) between stiffeners although these reductions are less than 3%. The distances between couples of stiffeners, therefore, should not be reached toward the centroid of the webs as the proposed sections in the catalogue from the manufacturer [18].

Table 4 Sectional capacities of SupaCee 250 sections under compression

Sections	GS	Compression					
		N_{cl}	Δ %	N_{cd}	Δ %	N_S	Δ %
SC25015	166	113.11	–	96.83	–	96.83	–
	146	115.24	1.89	96.17	−0.68	96.17	−0.68
	126	116.65	3.13	95.49	−1.38	95.49	−1.38
	106	118.09	4.40	94.87	−2.02	94.87	−2.02
	86	119.59	5.73	94.34	−2.57	94.34	−2.57
SC25019	166	165.57	–	144.99	–	144.99	–
	146	167.42	1.12	144.07	−0.63	144.07	−0.63
	126	169.39	2.31	143.14	−1.28	143.14	−1.28
	106	171.06	3.32	142.27	−1.88	142.27	−1.88
	86	172.32	4.08	141.52	−2.39	141.52	−2.39
SC25024	166	241.18	–	215.01	–	215.01	–
	146	242.74	0.65	213.73	−0.60	213.73	−0.60
	126	244.22	1.26	212.47	−1.18	212.47	−1.18
	106	245.81	1.92	211.29	−1.73	211.29	−1.73
	86	247.47	2.61	210.26	−2.21	210.26	−2.21

Note: Δ % is the sectional capacity deviation of SupaCee sections in comparison with the original SupaCee sections (GS = 166 mm)

5 Conclusions

The paper investigates the effects of web stiffener locations on the sectional capacities of cold-formed steel SupaCee sections under compression or bending. SupaCee sections were selected from the commercial sections, and the distances between stiffeners were varied for the investigation. The sectional capacities were determined according to the Australian/New Zealand Standard AS/NZS 4600: 2018 with the support of the commercial software program THIN-WALL-2. Several remarks are given as follows: (1) As the stiffeners reach toward the centroid of sectional webs, local buckling strengths become higher, but they are lower for distortional buckling strengths. In general, sectional capacities are seen as downward trends for both compression and bending as failures are governed by distortional buckling modes; (2) The web stiffeners should be kept far from the centroids of sectional webs as the original sections in the catalogue.

Table 5 Sectional capacities of SupaCee 250 sections under bending

Sections	GS	Bending					
		M_{bl}	Δ %	M_{bd}		M_S	
SC25015	166	15.33	–	13.81	–	13.81	–
	146	15.45	0.79	13.73	−0.58%	13.73	−0.58%
	126	15.58	1.63	13.67	−1.01%	13.67	−1.01%
	106	15.67	2.24	13.64	−1.23%	13.64	−1.23%
	86	15.71	2.46	13.61	−1.45%	13.61	−1.45%
SC25019	166	22.23	–	19.58	–	19.58	–
	146	22.35	0.52	19.49	−-0.46%	19.49	−0.46%
	126	22.48	1.11	19.42	−0.82%	19.42	−0.82%
	106	22.58	1.55	19.36	−1.12%	19.36	−1.12%
	86	22.61	1.71	19.33	−1.28%	19.33	−1.28%
SC25024	166	31.08	–	26.71	–	26.71	–
	146	31.12	0.14	26.57	−0.52%	26.57	−0.52%
	126	31.20	0.39	26.47	−0.90%	26.47	−0.90%
	106	31.28	0.65	26.41	−1.12%	26.41	−1.12%
	86	31.34	0.82	26.35	−1.35%	26.35	−1.35%

Note Δ % is the sectional capacity deviation of SupaCee sections in comparison with the original SupaCee sections (GS = 166 mm)

References

1. Hancock, G.J., Pham, C.H.: New section shapes using high-strength steels in cold-formed steel structures in Australia. Elsevier Ltd (2016). https://doi.org/10.1016/B978-0-08-100160-8.00011-6
2. Pham, N.H., Vu, Q.A.: Effects of stiffeners on the capacities of cold-formed steel channel members. Steel Constr. **14**(4), 270–278 (2021). https://doi.org/10.1002/stco.202100003
3. Pham, N.H.: Investigation of Sectional Capacities of Cold-Formed Steel SupaCee Sections. In: Proceedings of the 8th International Conference on Civil Engineering. ICCE 2021. Lecture Notes in Civil Engineering, Springer Singapore, vol. 213, pp. 82–94. (2022). https://doi.org/10.1007/978-981-19-1260-3_8
4. Saint-Venant, M.: Discussion in Theorie De L'elasticite Des Corp Solids. (1883)
5. Schafer, B.W., Peköz, T.: Direct strength prediction of cold-formed members using numerical elastic buckling solutions. In: Fourteenth International Specialty Conference on Cold-Formed Steel Structures (1998)
6. Nguyen, V.V., Hancock, G.J., Pham, C.H.: Development of the thin-wall-2 for buckling analysis of thin-walled sections under generalised loading. In: Proceeding of 8th International Conference on Advances in Steel Structures (2015)
7. AS/NZS 4600–2018.: Australian/New Zealand Standard TM Cold-formed steel structures. The Council of Standards Australia (2018)
8. Hancock, G.J.: Distortional buckling of steel storage rack columns. J. Struct. Eng. **111**(12), 2770–2783 (1985). https://doi.org/10.1061/(ASCE)0733-9445(1985)111:12(2770)
9. Hancock, G.J., Kwon, Y.B., Stefan, B.E.: Strength design curves for thin-walled sections undergoing distortional buckling. J. Constr. Steel Res. **31**(2–3), 169–186 (1994)

10. Kwon, Y.B., Hancock, G.J.: Tests of cold-formed channels with local and distortional buckling. J. Struct. Eng. ASCE **118**(7), 1786–1803 (1992)
11. Seah, L.K., Rhodes, J.: Simplified buckling analysis of plate with compound edge stiffeners. J. Eng. Mech. **119**, 19–38 (1993). https://doi.org/10.1061/(ASCE)0733-9399(1993)119:1(19)
12. Wang, L., Young, B.: Cold-formed steel channel sections with web stiffeners subjected to local and distortional buckling - Part I: Tests and finite element analysis. In: The 22nd International Specialty Conference on Recent Research and Developments in Cold-Formed Steel Design and Construction, pp. 229–242. (2014)
13. Yan, J., Young, B.: Column tests of cold-formed steel channels with complex stiffeners. J. Struct. Eng. **128**(6), 737–745 (2002). https://doi.org/10.1061/(ASCE)0733-9445(2002)128:6(737)
14. Wang, L., Young, B.: Design of cold-formed steel channels with stiffened webs subjected to bending. Thin-Walled Struct. **85**, 81–92 (2014). https://doi.org/10.1016/j.tws.2014.08.002
15. Chen, J., Chen, M.T., Young, B.: Compression Tests of Cold-Formed Steel C- and Z-Sections with Different Stiffeners. J. Struct. Eng. **145**(5) (2019). https://doi.org/10.1061/(ASCE)ST.1943-541X.0002305
16. Ye, J., Hajirasouliha, I., Becque, J., Pilakoutas, K. Development of more efficient cold-formed steel channel sections in bending. Thin-Walled Struct. **101** (2016). https://doi.org/10.1016/j.tws.2015.12.021
17. Manikandan, P., Arun, N.: Behaviour of partially closed stiffened cold-formed steel compression member. Arab. J. Sci. Eng. **41**(10), 3865–3875 (2016). https://doi.org/10.1007/s13369-015-2015-0
18. BlueScope Lysaght.: Supapurlins Supazeds and Supacees. Blue Scope Lysaghts (2014)
19. AS1397:2011.: Continuous hot-dip metalic coated steel sheet and strip—coating of zinc and zinc alloyed with aluminium and magnesium. Standards Australia (2011)

Study on the Early Shrinkage Behavior of Coral Aggregate Concrete Reinforced with Ultra-Fine Cement

Guosong Hu, Zhuolin Xie, Jianmin Hua, Lepeng Huang, Songxiao Huang, and Qiming Luo

Abstract The shrinkage behavior of coral aggregate concrete reinforced by ultra-fine cement was studied. In this paper, the ultra-fine cement slurry with different water-slurry ratios was used to prepare the reinforced coral aggregate. Three kinds of concrete water -binder ratio (0.21, 0.27 and 0.33) and five kinds of aggregate treatment methods were considered as variables to study the free shrinkage of concrete under the condition of dry curing. In addition, for the reinforced coral aggregate concrete. Three kinds of reinforcement ratios (1.14, 2.05 and 3.24%) were considered as variables to study the shrinkage performance under the constraint of reinforcement. The experimental results show that at the same age, the shrinkage value of coral aggregate concrete strengthened by ultra-fine cement slurry is less than that of coral aggregate concrete after pickling and less than that of natural coral aggregate concrete. The shrinkage of coral aggregate concrete decreases with the decrease of the water-slurry ratio of ultra-fine cement slurry. The higher the ratio of restraint reinforcement, the larger the pore structure and the smaller the shrinkage of coral aggregate concrete.

Keywords Shrinkage · Coral concrete · Reinforced coral coarse aggregate · Reinforcement · Pore structure

G. Hu · Z. Xie · J. Hua (✉) · L. Huang · S. Huang
School of Civil Engineering, Chongqing University, Chongqing 400045, China
e-mail: cqu2022@cqu.edu.cn

J. Hua · L. Huang
Key Laboratory of New Technology for Construction of Cities in Mountain Area, Ministry of Education, Chongqing University, Chongqing 400045, China

Q. Luo
China Railway 11th Bureau Group Co., Ltd., No. 277 Zhongshan Street, Wuhan 430061, Hubei, China

E. Strauss (ed.), *Proceedings of the 7th International Conference on Civil Engineering*, Lecture Notes in Civil Engineering 371, https://doi.org/10.1007/978-981-99-4045-5_24

1 Introduction

Recently, with the continuous development of marine resources by human beings, the construction of distant islands and reefs has gradually become the norm [1, 2]. In this context, the demand for offshore projects is increasing, and with it, the demand for marine concrete [3], and aggregate is an essential raw material for concrete pouring. When using ordinary concrete, these concrete materials need to be transported by land and sea from the origin of the mainland to the project site, which will incur huge transportation costs and will also take a long time and may delay the construction of the project [4]. Large amounts of coral debris usually accumulate on the beaches of island reefs, and large amounts of coral waste are also dredged up during harbor construction and channel dredging [5]. Coral debris is therefore an ideal material for the production of marine concrete. In fact, coral was used to produce concrete in the 1950s and 1960s [6]. In recent years, many scholars have used coral debris as coarse aggregate to prepare coral aggregate concrete (CAC) and carried out corresponding mechanical properties tests and durability tests and observed its microstructure [7–10], while the shrinkage properties tests for coral concrete have been rarely reported.

When the water in concrete is gradually consumed by hydration and drying, pore stress will occur in the pores in concrete [11]. Capillary stress will lead to the continuous reduction of the distance between the particles inside the concrete, thereby reducing the volume of concrete on a macro level. This phenomenon is called shrinkage [12]. In the shrinkage analysis and prediction of concrete, the capillary tension theory has become an important theory used by scholars [13]. According to previous studies, steel bar has a significant influence on concrete shrinkage [14], and it becomes an important direction to study the influence of steel bar on the shrinkage behavior of concrete. The effects of reinforcement material, diameter, shape and structure type on the shrinkage of ordinary aggregate concrete have been studied and a lot of useful results have been obtained.

Considering that the coral debris has a low cylinder compressive strength, irregular shape and more internal holes, resulting in a correspondingly low strength grade of the prepared coral concrete [15], this study innovatively used ultra-fine cement slurry to modify and enhance the coral coarse aggregate. The test results have shown that the cylinder compressive strength of the reinforced coral aggregate was significantly improved, and the corresponding strength index of the prepared reinforced CAC was also improved. And further studied the CAC free shrinkage and shrinkage behavior under the constraints of reinforcement, and pore structure of coral concrete at different ages under various treatments.

2 Experimental Program

2.1 Coral Aggregates

The study used natural coral aggregate, coral aggregate after pickling and coral aggregate enhanced with ultra-fine cement slurry after pickling. In the pickling process, the coral aggregate is initially cleaned with tap water, then soaked in 3% acetic acid solution for 60 min [16], and then rinsed with tap water for later use.

For the reinforced coral aggregate, the ultra-fine cement with water-slurry ratio of 1.2, 1.4 and 1.6 is used as the reinforcing material in this paper. In the preparation of reinforced coral aggregate, the ultra-fine cement and water were first put into the mixing cylinder according to the requirements of water-slurry ratio, and then coarse aggregate was added. After fully mixing the coarse aggregate, ultra-fine cement and water, the coarse aggregate was moved to the pressurizing device for pressuring. The pressure was set at 0.2 MPa and the pressurizing time was 30 min. After the pressurization was completed, the enhanced coral aggregate was extracted with a fine mesh fishing net and then treated with water for 7 days.

2.2 Mix Proportion and Specimen Production

The binding materials used to assemble concrete specimens contain Portland cement, fly ash and mineral powder. The coarse aggregate particle size range of 5–16 mm, and the fineness modulus of river sand is 2.7. Polycarboxylic acid superplasticizers are used as admixtures. Table 1 shows the mix ratio.

In this study, a total of 5 types of coral aggregate and 3 kinds of reinforcement ratios were designed to study the influence of the reinforced coral aggregate treatment on the shrinkage performance of CAC. See Table 2 for details of the specimens. The reinforcement mode of concrete is divided into centralized reinforcement, and the reinforcement is located in the center of the specimen section to meet the requirements of reinforcement ratio.

Table 1 Coral concrete mix ratio

No	W/B	Cement	Coarse aggregate	Sand	Water	Fly ash	Mineral powder
C1	0.33	270	650	1090	150	90	90
C2	0.27	330	630	1060	150	110	110
C3	0.21	420	615	1030	150	140	140

Table 2 Specimen specifics

No	ρ (%)	D (mm)
C1-N	0	–
C1-A	0	–
C1-R1	0	–
C1-R2	0	–
C1-R3	0	–
C1-R1-D1	1.14	12
C1-R1-D2	2.05	16
C1-R1-D3	3.24	20
C2-N	0	–
C2-A	0	–
C2-R1	0	–
C2-R2	0	–
C2-R3	0	–
C2-R1-D1	1.14	12
C2-R1-D2	2.05	16
C2-R1-D3	3.24	20
C3-N	0	–
C3-A	0	–
C3-R1	0	–
C3-R2	0	–
C3-R3	0	–
C3-R1-D1	1.14	12
C3-R1-D2	2.05	16
C3-R1-D3	3.24	20

Note X-Y-D, X stands for the type of W/B, C1, C2 and C3, respectively, represent the W/B of concrete of 0.21, 0.27 and 0.33. Y is the type of coral aggregate used. N stands for the use of natural coral aggregate, A represents the acid-washed coral aggregate. R stands for aggregate reinforced with ultra-fine cement slurry, R1, R2, R3 represent the ultra-fine cement reinforced slurry ratio of 1.2, 1.4, 1.6. D stands for reinforcement, D1 stands for reinforcement ratio of 1.14%, D2 stands for reinforcement ratio of 2.05% and D3 stands for reinforcement ratio of 3.24%

Table 3 Physical and mechanical properties of coral aggregate

Type	Apparent density (kg/m^3)	Water absorption (%) 1 h	Tube compressive strength (MPa)
CA	1887	12.1	3.0
ACA	1883	13.5	2.7
RCA-1.2	2167	8.2	6.9
RCA-1.4	2111	8.7	6.4
RCA-1.6	2046	10.4	4.6

Note CA stands for natural coral aggregate, ACA stands for acid-washed coral aggregate and R stands for coral aggregate strengthened with ultra-fine cement slurry. 1.2, 1.4 and 1.6, respectively, represent the water-slurry ratio of ultra-fine cement reinforcement

3 Results and Discussion

3.1 Physical and Mechanical Properties of Coral Aggregate

Table 3 shows the natural coral aggregate, the acid-washed coral aggregate and the coral aggregate enhanced by ultra-fine cement slurry. From the table, it can be found that all indexes of coral aggregate after different treatments have changed. Compared with natural coral aggregate, the strength of coral aggregate after acid treatment decreased by about 10%, and the water absorption rate was improved. The apparent density of coral aggregate increased and the water absorption decreased after the treatment of ultra-fine cement slurry with a water-slurry ratio of 1.2. The compressive strength of the cylinder was increased from 3.0 to 6.9 MPa, an increase of 130%.

3.2 Mechanical Properties of Coral Concrete

Tables 4 and 5 show the basic mechanical performance test results at 28 days of age as well as the elastic modulus and Poisson's ratio at different ages. As can be seen from the table, when the W/B is 0.21 (group C3), the compressive strength, axial compressive strength, split tensile strength and elastic modulus of the concrete cube using natural aggregate are 53.4 MPa, 48.2 MPa, 4.31 MPa and 3.04 × 104 MPa, respectively. When the water-slurry ratio of 1.2 is used, the above indexes were increased to 65.3 MPa, 60.5 MPa, 4.98 MPa and 3.34 × 104 MPa, respectively. It increased by 22.3%, 25.5%, 13.5% and 9.9%, respectively.

As shown in the table that the strength of concrete significantly improved with the decrease of the water-slurry ratio of ultra-fine cement slurry. For example, under the W/B of 0.21 (group C3), the cube compressive strength of concrete increased

Table 4 Mechanical test results at 28d age

Concrete type	Cube compressive strength (MPa)	Axial compressive strength (MPa)	Splitting tensile strength (MPa)
C1-N	34.8	30.5	3.33
C1-A	33.7	30.3	3.35
C1-R1	43.2	37.6	3.87
C1-R2	41.5	36.3	3.81
C1-R3	38.1	33.1	3.65
C2-N	41.6	34.8	3.78
C2-A	40.8	33.9	3.71
C2-R1	52.1	43.6	4.35
C2-R2	50.6	41.2	4.26
C2-R3	48.9	38.9	4.18
C3-N	53.4	48.2	4.31
C3-A	52.5	49.1	4.29
C3-R1	65.3	60.5	4.98
C3-R2	63.3	59.1	4.88
C3-R3	57.8	53.1	4.53

Table 5 Elastic modulus and Poisson's ratio at different ages

Concrete type	Elastic modulus ($\times 10^4$ MPa)			Poisson's ratio		
	3d	7d	28d	3d	7d	28d
C1-N	1.78	2.35	2.66	0.252	0.261	0.271
C1-A	1.69	2.33	2.65	0.253	0.263	0.278
C1-R1	1.85	2.46	2.81	0.251	0.257	0.262
C1-R2	1.79	2.36	2.80	0.248	0.258	0.263
C1-R3	1.83	2.38	2.73	0.249	0.255	0.266
C2-N	2.05	2.43	2.89	0.235	0.244	0.251
C2-A	2.03	2.49	2.84	0.234	0.245	0.248
C2-R1	2.19	2.60	3.06	0.219	0.227	0.227
C2-R2	2.15	2.51	3.03	0.221	0.228	0.232
C2-R3	2.16	2.53	2.95	0.238	0.243	0.246
C3-N	2.17	2.61	3.04	0.219	0.234	0.241
C3-A	2.19	2.59	3.02	0.213	0.225	0.232
C3-R1	2.42	2.85	3.34	0.186	0.195	0.202
C3-R2	2.39	2.83	3.27	0.183	0.196	0.201
C3-R3	2.34	2.83	3.16	0.205	0.221	0.232

by 9.5% when the water-slurry ratio of ultra-fine cement slurry was reduced from 1.6 to 1.4, and by 3.2% when the water-slurry ratio of ultra-fine cement slurry was reduced from 1.4 to 1.2. Under the W/B of 0.27 (group C2), the cube compressive strength of concrete increased by 3.5% when the water-slurry ratio of reinforcement was reduced from 1.6 to 1.4 and by 3.0% when the water-slurry ratio of reinforcement was reduced from 1.4 to 1.2. Under the W/B of 0.33 (group C1), the cube compressive strength of concrete increased by 8.9% when the water-slurry ratio of reinforcement was reduced from 1.6 to 1.4, and the strength of concrete increased by 4.1% when the water-slurry ratio of reinforcement was reduced from 1.4 to 1.2.

3.3 Early Humidity Change of CAC

The change curve of concrete internal humidity with age is shown in Fig. 1, and the change of indoor temperature and humidity is shown in Fig. 2. The change of internal humidity of concrete can be divided into two stages: Stage 1, humidity saturation stage (RH = 100%), in this stage, the internal humidity of concrete remains unchanged. Stage 2: Humidity decline stage (RH < 100%). During this phase, the humidity in the concrete continues to decrease. At the beginning of concrete poured, the cement particles are coated with water film, and the holes are filled with liquid water and connected with each other. At this time, the relative humidity measured by the test is 100%. As the hydration reaction progresses, the water is reduced by consumption. On the one hand, the cement continues to hydrate and consume water. While at the same time the water vapor in the macropore continues to diffuse outward through the channels connecting the pore. At this point, the relative humidity inside the concrete is less than 100% and decreases as the process progresses.

3.4 Pore Structure of CAC

Figure 3 shows the main pore structure parameters of coral concrete specimens at different ages, including porosity, average pore diameter, median pore diameter and critical diameter of capillary. From Fig. 3, we can find the following rules:

The pore structure of concrete decreases with the increase of age. For example, under the W/B of 0.21 (group C3), the porosity, average pore diameter, median pore diameter and critical diameter of capillary of natural CAC at the age of 3 days are 18.31%, 38.4 nm, 13.3 nm and 44.4 nm, respectively. At the age of 7 days, the data decreased to 17.77%, 27.6 nm, 11.5 nm and 20.1 nm, respectively. Further decreased to 12.96%, 18.6 nm, 6.0 nm and 11.1 nm at 28 days. This phenomenon can also be found in other groups of specimens.

The pore structure of the concrete using natural coral aggregate is greater than that of the CAC after pickling treatment, and the pore structure of the CAC enhanced by ultra-fine cement slurry is greater than that of the concrete using natural coral

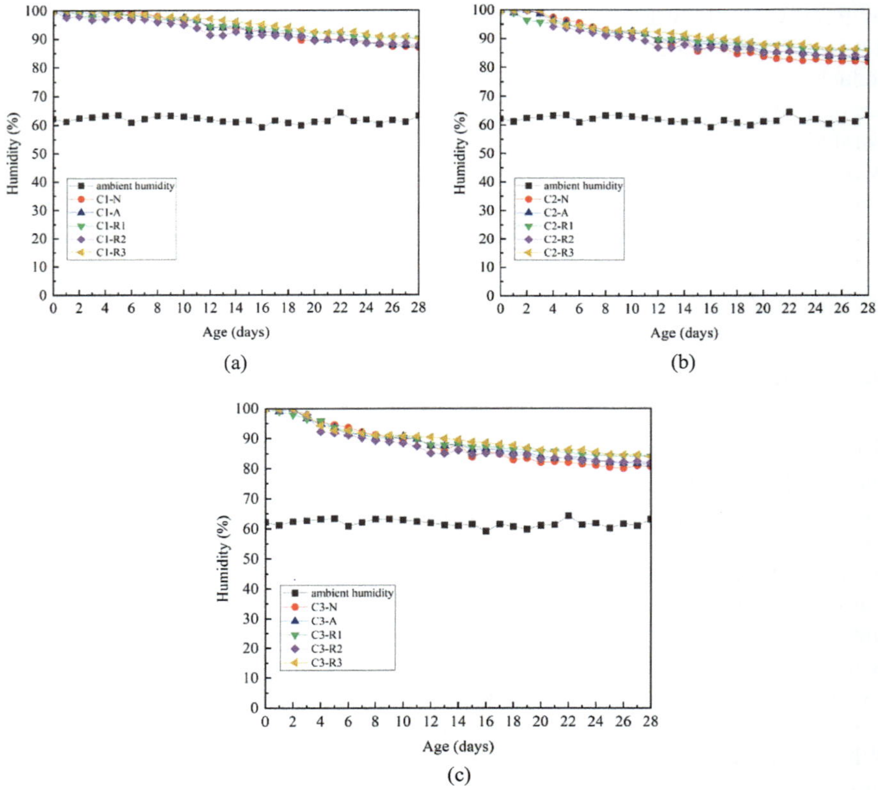

Fig. 1 Humidity variation of different concrete specimens. **a** C1 concrete specimen; **b** C2 concrete specimens; **c** C3 concrete specimen

aggregate. For example, at the W/B of 0.21 (group C3), the porosity, average pore diameter, median pore diameter and critical diameter of capillary of concrete specimens using natural coral aggregate at the age of 28 days are 12.96%, 18.6 nm, 6.0 nm and 11.1 nm, respectively. The porosity, average pore diameter, median pore diameter and critical diameter of capillary of CAC specimens after pickling treatment were 11.91%, 18.2 nm, 5.8 nm and 10.8 nm, respectively. However, the corresponding pore structure parameters of CAC C3-R1 reinforced by ultra-fine cement slurry with a water-slurry ratio of 1.2 were 15.83%, 19.2 nm, 7.4 nm and 17.0 nm. Similarly, the corresponding pore structure parameters of C3-R2 were 15.06%, 17.5 nm, 7.6 nm and 15.4 nm. The corresponding pore structure parameters of C3-R3 are 13.81%, 19.6 nm, 5.9 nm and 13.1 nm.

The resultant force of concrete will decrease with the increase of reinforcement ratio in the specimen constrained by steel bars. In plain concrete specimens, the filling effect of hydration products on pores and the stress acting on the capillary wall lead to the decline of concrete pore structure. However, in the reinforced specimen, the

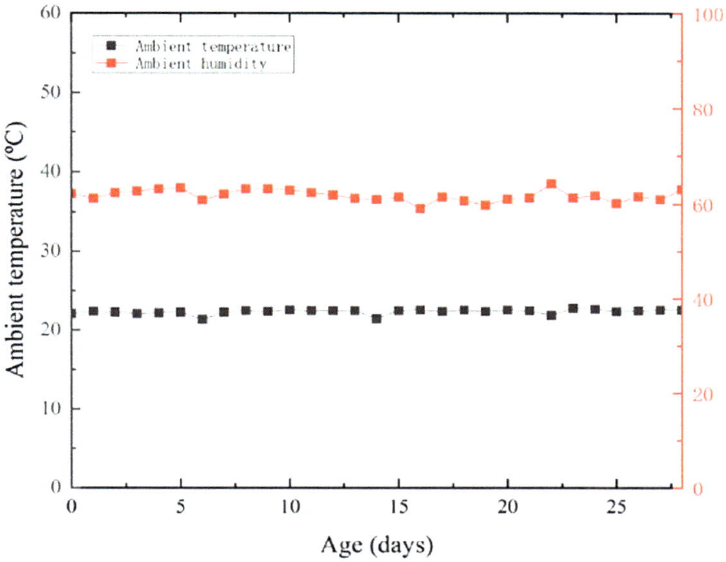

Fig. 2 Variation of indoor temperature and humidity

presence of steel bars leads to the reduction of the force of concrete in the contraction direction, that is, the effect of capillary stress is reduced, and the reduction degree of concrete pore structure caused by capillary stress will be reduced. For example, at 28 days, the porosity, average pore diameter, median pore diameter and critical diameter of capillary of C3-R1-D1 specimens were 16.48%, 16.4 nm, 6.1 nm and 20.8 nm, respectively. The porosity, average pore diameter, median pore diameter and critical diameter of capillary of C3-R1-D2 specimens were 16.73%, 26.5 nm, 8.7 nm and 22.3 nm, respectively. The porosity, average pore diameter, median pore diameter and critical diameter of capillary of C3-R1-D3 specimens were 17.88%, 29.9 nm, 8.8 nm and 25.3 nm, respectively. This is consistent with the test data showing that the pore structure size of the specimen increases with the increase of reinforcement ratio and elastic modulus of steel bars.

3.5 Shrinkage of Concrete

Figure 4 shows the variation of the free shrinkage strain of each coral concrete specimen in the study from 0 to 28 days. It can be found from Fig. 4 that no matter what kind of aggregate, the shrinkage of CAC can be separated into three stages. In the first stage (0–7 days), the shrinkage of concrete in this stage rises rapidly, reaching about 50–80% of the 28 days age. The main reason for the rapid increase of coral concrete shrinkage in this stage is that the moisture in the concrete is consumed

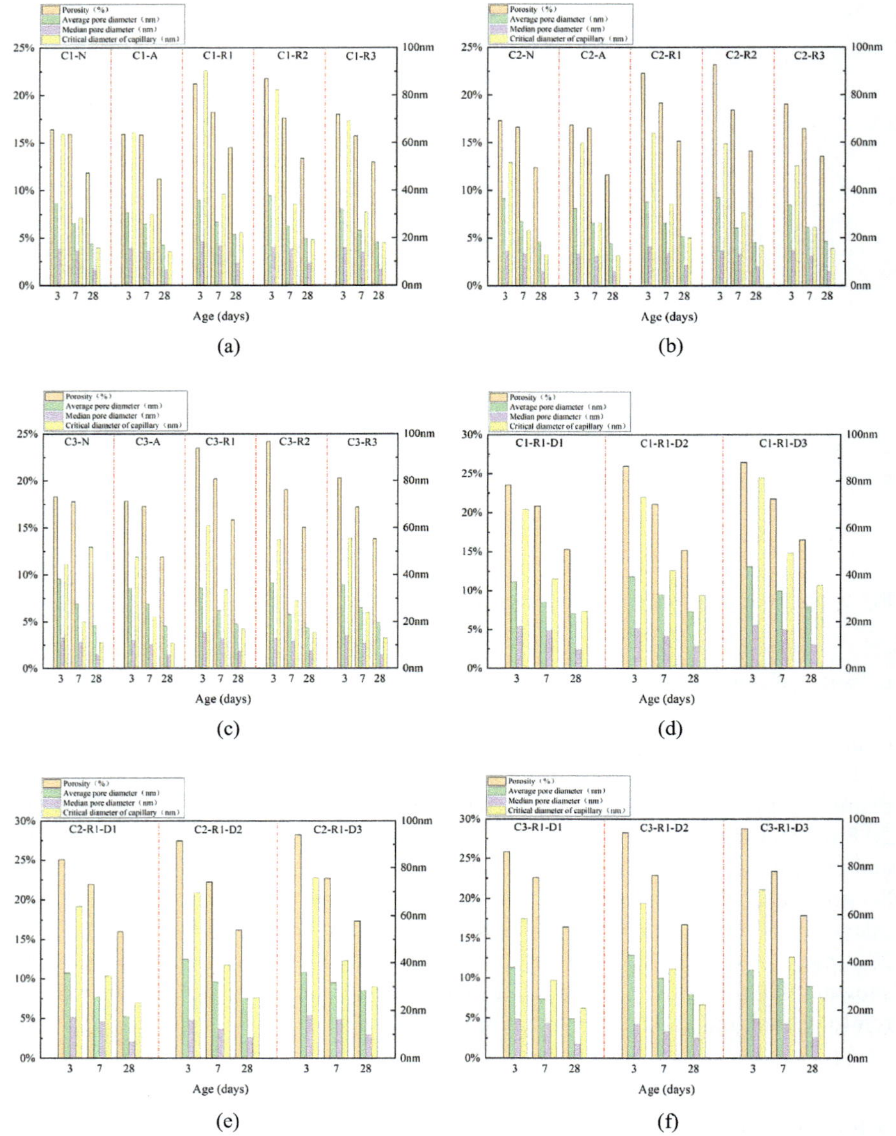

Fig. 3 Pore structure distribution of each specimen at each age. **a** C1 concrete specimen; **b** C2 concrete specimens; **c** C3 concrete specimens; **d** C1-R1 concrete specimens; **e** C2-R1 concrete specimens; **f** C3-R1 concrete specimens

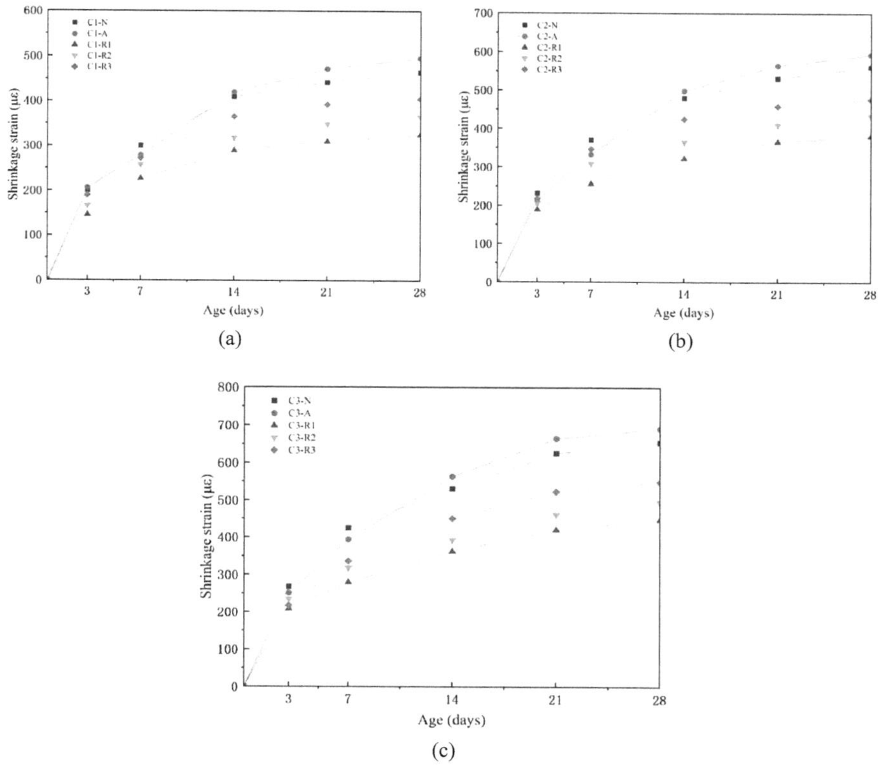

Fig. 4 Shrinkage of concrete specimens without rebar restraint. **a** C1 concrete specimen; **b** C2 concrete specimens; **c** C3 concrete specimen

by the intense hydration of cement in this stage. In the second stage (7–14 days), as the hydration of cement in the concrete slows down, the consumption of water in the concrete decreases, and the shrinkage rate of the concrete also decreases. In the third stage (14–28 days), with the further decrease of cement hydration rate, the shrinkage of concrete gradually maintains a stable state.

The aggregate used in coral concrete has a significant effect on the shrinkage of concrete. The shrinkage of reinforced CAC is less than that of natural CAC at the same proportion and age. For example, at 28 days of age, the shrinkage of C3-R1, C3-R2 and C3-R3 using enhanced CAC was 448 με, 496 με and 530 με, respectively, while the shrinkage of C3-N using natural coral as aggregate was 655 με under the same conditions. The shrinkage of CAC C3-A after pickling with acetic acid was 692 με. That is, at the W/B of C3, the shrinkage strain of CAC reinforced with water-slurry ratio of 1.2, 1.4 and 1.6 decreased by 31.6%, 24.3% and 19.1%, respectively, compared with that of natural CAC at the age of 28 days. The dry shrinkage strain of CAC after acetic acid pickling was 5.6% higher than that of natural CAC. The same phenomenon can be observed under the W/B of C1 and C2. The shrinkage strain of

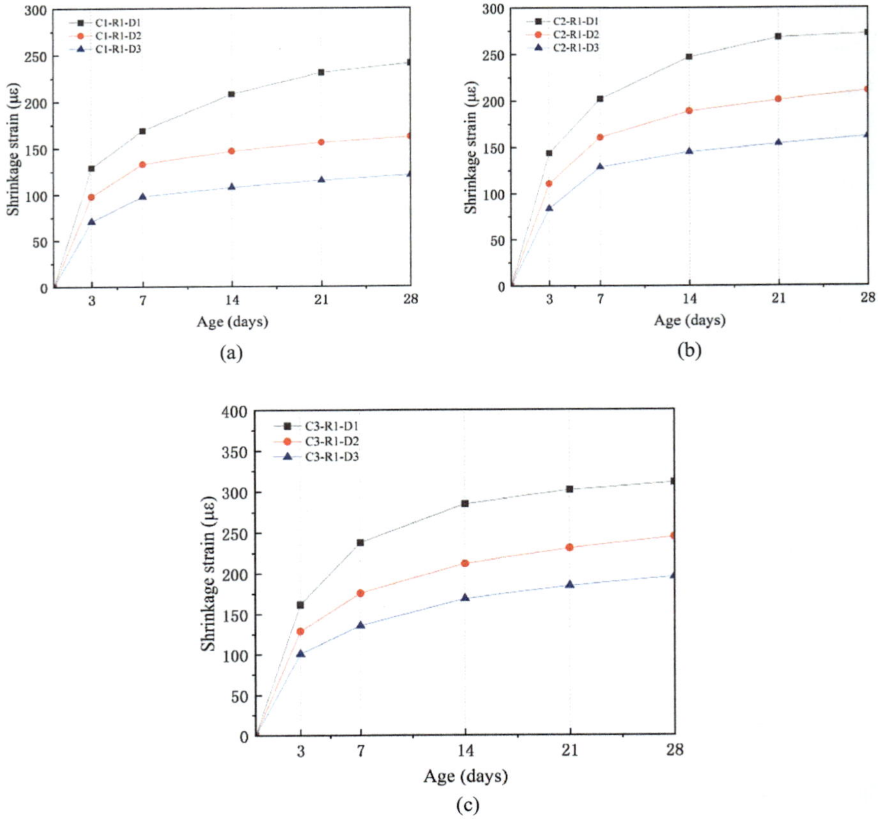

Fig. 5 Shrinkage of concrete specimens constrained by steel bars. **a** C1-R1 concrete specimens; **b** C2-R1 concrete specimens; **c** C3-R1 concrete specimens

C1-R1, C1-R2 and C1-R3 decreased by 30.0, 21.1 and 13.4% compared with C1-N at 28 days of age, while the shrinkage strain of C1-A increased by 9.1% compared with C1-N. Compared with C2-N, the shrinkage strain of C2-R1, C2-R2 and C2-R3 decreased by 35.8%, 29.7% and 20.5%, respectively, at 28 days of age, while the shrinkage strain of C2-A increased by 7.7% compared with C2-N.

When the W/B of concrete and the water-slurry ratio of the ultra-fine cement reinforcement are the same, the reinforcement configuration has a significant impact on the shrinkage of concrete (Fig. 5). Previous studies have shown that the configuration of steel bars in concrete will produce constraint tensile stresses in the concrete in the opposite direction of shrinkage, which will constrain the shrinkage of concrete. This is consistent with what was observed in the experiment. For example, at the age of 28 days, the shrinkage of C3-R1-D1, C3-R1-D2 and C3-R1-D3 specimens with reinforcement ratios of 1.14, 2.05 and 3.24% were 311 με, 245 με and 196 με, respectively, which decreased with the increase of reinforcement ratio. And the shrinkage rate of the unreinforced specimen C3-R1 decreased by 30.6, 45.3 and 56.3%, respectively. This shows that the steel bars have a significant inhibitory effect on the shrinkage of concrete, and this inhibitory effect increases with the increase of the reinforcement ratio.

4 Conclusions

In this study, the shrinkage performance of CAC reinforced with ultra-fine cement slurry was studied. The shrinkage of coral concrete under various aggregate treatments and the corresponding pore structure were tested using linear displacement sensors. The conclusions are as follows:

(1) The pore structure of coral coarse aggregate after pickling treatment is more loose than that of natural coral coarse aggregate, and the coral coarse aggregate enhanced by ultra-fine cement slurry is denser than that of natural coral coarse aggregate. The strength and modulus of elasticity of concrete with reinforced coral aggregate were higher than those with natural coral aggregate.
(2) The shrinkage strain of concrete used ultra-fine cement reinforced coral aggregate is lower than that of concrete used natural coral aggregate is lower than that of coral coarse aggregate after pickling treatment, and the shrinkage strain of coral concrete decreases with the decrease of the water-slurry ratio of ultra-fine cement slurry, and the pore structure parameters of coral concrete also increases with the decrease of the water-slurry ratio of reinforcement material.
(3) Due to the restraint effect of steel bars, the shrinkage of coral concrete is significantly reduced after the reinforcement of ultra-fine cement, and gradually decreases with the increase of reinforcement ratio. The parameters of the pore structure in the steel bar constrained specimen also change with the change of the reinforcement ratio.

References

1. Chen, L., et al.: Green construction for low-carbon cities: a review. Environ. Chem. Lett. (2023)
2. Chen, L., et al.: Strategies to achieve a carbon neutral society: a review. Environ. Chem. Lett. **20**(4), 2277–2310 (2022)
3. Wang, A., et al.: The development of coral concretes and their upgrading technologies: a critical review. Constr. Build. Mater. **187**, 1004–1019 (2018)
4. Liu, J., et al.: Literature review of coral concrete. Arab. J. Sci. Eng. **43**(4), 1529–1541 (2018)
5. Smith, S.V.: Coral-reef area and the contributions of reefs to processes and resources of the world's oceans. Nature **273**(5659), 225–226 (1978)
6. John, G.D.: Coral and salt water as concrete materials. ACI J. Proc. **48**(10)
7. Da, B., et al.: Experimental investigation of whole stress-strain curves of coral concrete. Constr. Build. Mater. **122**, 81–89 (2016)

8. Wu, Z., et al.: Physical and mechanical properties of coral aggregates in the South China Sea. J. Build. Eng. **63**, 105478 (2023)
9. Zhang, J., et al.: Mesoscopic characteristics and macroscopic mechanical properties of coral aggregates. Constr. Build. Mater. **309**, 125125 (2021)
10. Zhou, W., Feng, P., Lin, H.: Constitutive relations of coral aggregate concrete under uniaxial and triaxial compression. Constr. Build. Mater. **251**, 118957 (2020)
11. Kovler, K., Zhutovsky, S.: Overview and future trends of shrinkage research. Mater. Struct. **39**(9), 827–847 (2006)
12. He, Y.X.: Experimental research on pore structure of RCA and its impact on drying shrinkage. In: Advanced Materials Research (2011). Trans Tech Publ
13. Huang, L., et al.: Capillary tension theory for predicting shrinkage of concrete restrained by reinforcement bar in early age. Constr. Build. Mater. **210**, 63–70 (2019)
14. Yoo, D.-Y., et al.: Early age setting, shrinkage and tensile characteristics of ultra high performance fiber reinforced concrete. Constr. Build. Mater. **41**, 427–438 (2013)
15. Cai, Y., et al.: Comparison study on the impact compression mechanical properties of coral aggregate concrete and ordinary Portland concrete. Structures **44**, 1403–1415 (2022)
16. Wang, A., et al.: A gentle acid-wash and pre-coating treatment of coral aggregate to manufacture high-strength geopolymer concrete. Constr. Build. Mater. **274**, 121780 (2021)

Evaluation of Soil-Structure Interaction on RC Framed Irregular Building Under Varying Ground Conditions

Arnab Chatterjee and Heleena Sengupta

Abstract During an earthquake the behaviour of any structure is influenced not only by the response of the superstructure, but also by the response of the soil beneath the structure. Structural failures in past have shown the significance of considering the impact of Soil-Structure Interaction (SSI) during analysis and modelling of structure in high seismic zones. The present study focuses on the influence of SSI in the analysis and design of an irregular 12-storey reinforced concrete frame building with and without shear wall. For integrating the effect of SSI, Models simulating three different soil-foundation conditions based on shear wave velocity with supported underneath raft, pile and raft-pile foundation using spring-dashpot mechanism and a fixed base condition for comparison purposes is carried out. Earthquake motion in response spectrum corresponding to zone V of IS 1893:2016 design spectrum is used to excite the finite element model of soil-foundation structure system. Responses in terms of variation in time period, base shear and storey displacement obtained from the analysis of the SSI models are compared with that obtained from conventional method assuming rigidity at the base of the structure. The results show that SSI effects in the dynamic behaviour of the structure are significant in altering the seismic response with a major increase in the vibration period as well as increase in the system damping and other structural parameters in comparison with the fixed base model, which does not consider the supporting soil-foundation.

Keywords Soil-structure interaction (SSI) · Soil flexibility · Ground conditions · Base shear

---

A. Chatterjee
M.Tech Scholar-Civil Engineering Department, Techno India University, Em Block Sector V, Kolkata, India

H. Sengupta
H.O.D.-Civil Engineering Department, Techno India University, Em Block Sector-V, Kolkata, India
e-mail: arnab6138@gmail.com

E. Strauss (ed.), *Proceedings of the 7th International Conference on Civil Engineering*, Lecture Notes in Civil Engineering 371, https://doi.org/10.1007/978-981-99-4045-5_25

1 Introduction

Failures of structures encountered during 2001 Bhuj earthquake and in Kobe earthquake in 1995 demonstrated the importance of considering SSI, since seismic behaviour of any structure is highly influenced not only by the response of the superstructure, but also by the response of the relative stiffness of the soil-foundation medium underneath. Hence, considering the physical property of the soil-foundation medium during analysis is an important factor in the earthquake response of structures supported over it. Raychowdhury [1] highlighted that considering the foundation nonlinearity, there is a possibility of differential settlement arising out of soil flexibility for low-rise steel moment resisting building frames where SSI needs to be carefully applied for heavily loaded footings owing to high inertial effects. Recent study by Saha et al. [2] primarily attempted to examine the influence of SSI on distribution of seismic design forces on a one storey system using soil-pile raft-structure system clearly indicating that relative acceleration of heavy raft and upper part of the pile with respect to the neighbouring soil attracts extra lateral force and lengthening of fundamental period which may lead to considerable increase in pile head shear. Hora [3] showed that the forces and moments get transferred from the exterior columns towards the interior ones due to elasto-plastic interaction analysis while the soil remains in elastic state, although the soil mass below the outer edges has fully yielded. Dey et al. [4] highlighted that RC frame with shear wall modifies the natural vibrational characteristics resulting in the inelastic global and local response of the frame system to be further modified. The main objective of the current study within this framework is to carry out Response Spectrum Analysis on a G +11 building with irregular plan (L-shaped) satisfying the irregularity provisions listed in IS: 1893–2016 (part 1) resting on three different soil-foundation parameters. The effect of ground motion after taking Soil-Structure Interaction (SSI) into consideration to find out the variation of base shear, storey displacement and time period of the structure against conventional fixed base results.

2 Methodology

2.1 Material Properties of Structure

A three-dimensional model of a twelve-storey residential building with plan irregularity of L-shaped satisfying the plan irregularity provisions as per IS:1893–2016 (part 1) Cl.7.1 3B (Re-Entrant Corners) with $A/L_2 > 0.15$ shown in Fig. 1a and Open Ground Storey (OGS) is considered for this study using SAP2000 finite element software for analysis. The frames are composed of 800 joints, 2700 beam elements (columns and beams) and 732 shell elements(slabs) as shown in Fig. 1a, b.

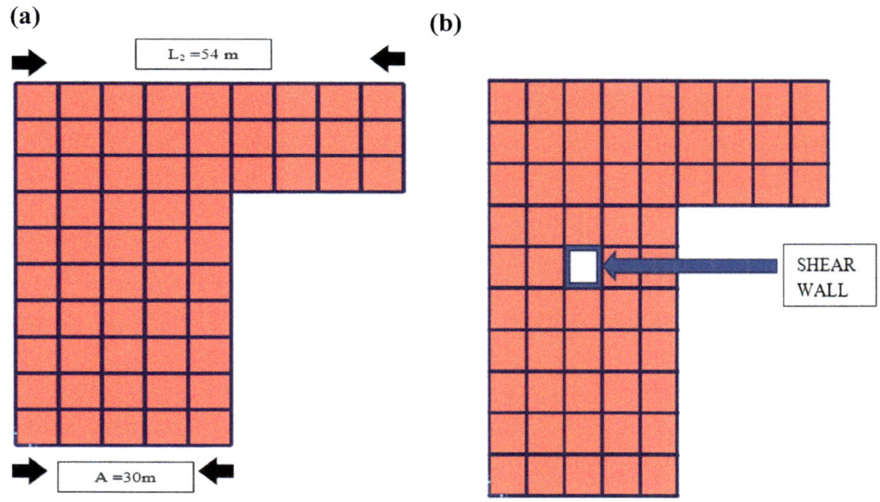

Fig. 1 **a** Plan view of bare frame. **b** Plan view of shear wall frame

2.2 Soil-Foundation Characteristics

Dynamic analysis of the structure and its corresponding interaction with the foundation and soil under the structure affects the response of structure. The interaction between foundation and soil depends on the elastic properties of foundation soil and foundation dimensions. In this study, the soil-foundation flexibility in the analysis is considered by means of replacing the foundation by statically equivalent spring-dashpot system. Three types of soil classified as per IS:1893–2016(part 1) which are Stiff, Medium-stiff and soft soil are given the due consideration for this study. Table 1 discussed about the classification of various soil parameters which will be exercised for further analysis. The soil-foundation model of raft-soil interaction of thickness 650 mm modelled with springs and dampers as shown in Fig. 2b and c replicates the bi-axial translational and rocking stiffness and damping parameters according to the formulation from Richart et al. [5]. While soil-pile interaction depicted in Fig. 2a is modelled using the equations laid down by Nakazawa Method [6]. Table 2 presents the equations of stiffness and damping used to simulate the soil-foundation characteristics incorporating the values derived from calculating the foundation results of raft foundation, raft-pile foundation.

Table 1 Classification of soil parameters

Type of soil	N (standard penetration test value)	Mass density (KN/m^3) (ρ)	Shear wave velocity (m/s) (V_s)	Poisson's ratio υ	SBC KN/m^2	Shear modulus (G)
Stiff (Type-I)	40	21	112	0.25	500	260,000
Medium-Stiff (Type-II)	20	18	84	0.33	230	131,576
Soft (Type-III)	9	17	56	0.48	150	50,674

3 Results and Discussions

In the present work, a comparison is made between three-dimensional building frames resting on fixed base and flexible base considering the interaction of various types of soil-foundation parameters to understand the response of considering SSI compared to fixed base. The buildings with flexible base condition are further studied by incorporating the shear wall to evaluate its effectiveness to control the SSI effect and its variation of results without shear wall. The variation of dynamic parameters such as Time Period, Base Shear and structural behaviour like Storey Displacement using Response Spectrum Analysis is analysed as per IS 1893(Part 1): 2016 [7] code. The results of Response Spectrum Analysis for various models analysed are discussed to highlight the effect of shear wall structure and without shear wall structure and presented thereafter.

3.1 *Time Period*

The time period of a structure is one of the most important factors affecting the seismic response of the building frame. The values of time period obtained for the bare frame and frame-shear wall buildings from the Response Spectrum Analysis of 3D finite element models are as tabulated in Tables 3 and 4.

Inclusion of shear wall results in decline in value of time period in the range of 14%–26% as compared to bare frames with soil flexibility considered with least decrease for raft-pile-soil interaction with stiff and medium-stiff soil parameters whereas it diminishes the most under pile-soil interaction with soft soil parameters. Percentage decrease in time period under fixed base condition by addition of shear wall remains same for the three types of soil. But, values of natural period obtained by considering only soil-foundation interaction is more than considering the combined effect of soil-foundation and shear wall with highest being for building over pile foundation at soft soil condition at 27.8%. It is also noticed that value of time period in case of raft-soil interaction at three types of soil is close to fixed base condition, signifying that raft foundation shows somewhat close to fixed base behaviour.

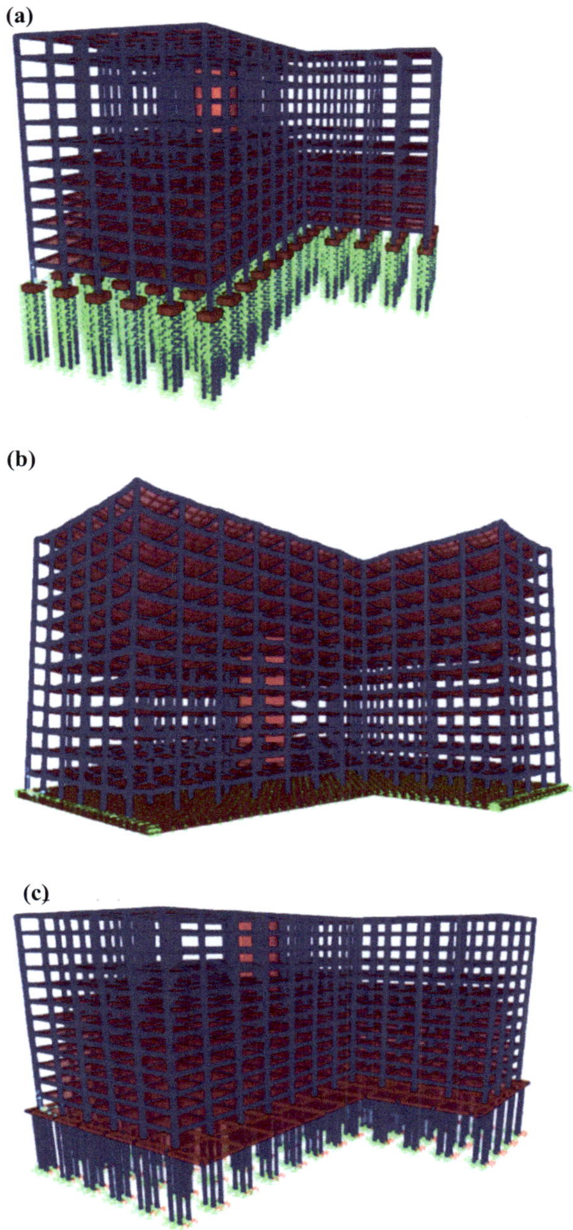

Fig. 2 **a** 3D model of the structure with pile foundation. **b** 3D model of the structure with raft foundation. **c** 3D model of the structure with raft-pile foundation

Table 2 Equations of soil springs and damping as per Richart and Lysmer model

Stiffness equations	Equivalent radius	Remarks
$K_z = 4Gr_z/(1-\mu)$	$r_z = \sqrt{LB \div \pi}$	This is in vertical Z direction
$K_x = (32(1-\mu)\,G\,r_x)/(7-8\mu)$	$r_x = \sqrt{LB \div \pi}$	This induces sliding in horizontal X or Y direction
$K_{\varphi x} = 8Gr^3_{\varphi x}/3(1-\mu)$	$r_{\varphi x} = \sqrt[4]{LB^3 \div 3\pi}$	This induces rocking about X axis
$K_{\varphi y} = 8Gr^3_{\varphi y}/3(1-\mu)$	$r_{\varphi y} = \sqrt[4]{LB^3 \div 3\pi}$	This induces rocking about Y axis
$K_{\varphi z} = 16G\,r^3_{\varphi z}/3$	$r_{\varphi z} = \sqrt[4]{(LB^3 + BL^3) \div 6\pi}$	This induces twisting around vertical Z axis
Damping equations	Equivalent radius	Remarks
$C_z = 0.85K_zR \cdot \sqrt{\rho \div G}$	$r_z = \sqrt{LB \div \pi}$	This is in vertical Z direction
$C_x = 0.576K_xR \cdot \sqrt{\rho \div G}$	$r_x = \sqrt{LB \div \pi}$	This induces sliding in horizontal X or Y direction
$C_{\varphi x} = 0.3K_{\varphi x}R \cdot \sqrt{\rho \div G}/(1 + B_\varphi)$	$B_\varphi = (3(1-\upsilon)\,M_{m0})/(8\rho R^5)$ $r\varphi x = LB^3 \div 3\pi$	This induces rocking about Y axis
$C_{\varphi y} = 0.3K_{\varphi y}R \cdot \sqrt{\rho \div G}/(1 + B_\varphi)$	$r_{\varphi y} = \sqrt[4]{LB^3 \div 3\pi}$	This induces rocking about X axis
$C_{\varphi z} = \sqrt{(K_{\varphi z}/I_z)}/1 + (2\,I_z/\rho R^5)$	$r_{\varphi z} = \sqrt[4]{(LB^3 + BL^3) \div 6\pi}$	This induces twisting around vertical Z axis

3.2 Base Shear

The seismic lateral vulnerability of structures is reflected by the seismic base shear and is one of the main parameters in seismic design of structures obtained from the expressions given in IS1893:2016 code for design spectra of 5% critical damping. The values of Base Shear along X and Y directions for different SSI models analysed are depicted in Fig. 3a and b with the variation in the values of Base Shear of bare frame and with shear wall buildings resting on different categories of soil-foundation medium compared to fixed base are as tabulated in Tables 5 and 6.

Tables 5 and 6 depict the percentage increment in base shear value of shear wall structure compared to bare frame for varying soil-foundation types besides comparing with fixed base condition. As the soil parameter tends to become flexible at pile foundation, so the values of base shears obtained become larger. When the effect of SSI (building-foundation-soil system) is observed it is seen that as flexibility of soil increases the value of base shear increases, since base shear is dependent on the primary factor, natural period (T). With the increase in flexibility of soil, the natural period of the building increases and base shear increases due to the descending curve

Table 3 Time period of various models without shear wall

Considering SSI (sec)			Considering fixed base (sec)		% Increase compared to fixed base
Raft-soil interaction	Stiff soil (Model 1)	1.85	Stiff soil (Model 19)	1.85	No variation
	Medium-stiff soil (Model 2)	1.85			No variation
	Soft soil (Model 3)	1.85			No variation
Pile-soil interaction	Stiff soil (Model 7)	1.855	Medium-stiff soil (Model 20)	1.85	0.05
	Medium-stiff soil (Model 8)	1.86			0.32
	Soft soil (Model 9)	2.6			27.8
Raft-pile-soil interaction	Stiff soil (Model 13)	1.86	Soft soil (Model 21)	1.85	0.32
	Medium-stiff soil (Model 14)	1.86			0.36
	Soft soil (Model 15)	1.931			4

of design response spectrum of design acceleration coefficient corresponding to 5% damping. However, the values of base shear obtained as per conventional method for buildings with fixed base assumed to be constructed over different soil sites shows an increase in the value of base shear with increasing flexibility of soil.

3.3 Storey Displacement

Storey Displacement represents the average value of maximum elastic lateral deflection of each storey with respect to ground level. The deflection is seen to be significantly influenced by the soil flexibility as depicted in Figs. 4 and 5 with the variation in storey displacement values at storey level 11 along X direction is tabulated in Table 7 and along Y direction is tabulated in Table 8. All Storey Displacement values plotted in chart in Figs. 4 and 5 are in 10^{-4} m.

Variation in the storey displacement values at storey level 11 for different building configurations is tabulated in Tables 7 and 8 which highlights that bare frame structure shows more displacement values as compared to shear wall buildings where the deflection values are within the permissible limits in all the soil types. This is due to the additional mass concentration at core due to the inclusion of shear wall thereby showing the merits of including shear wall to reduce the lateral deflection of buildings.

Table 4 Time period of various models with shear wall

Considering SSI with shear wall (sec)			% Decrease compared to Table 3	Considering fixed base with shear wall (sec)		% Decrease compared to Table 3	% Increase of SSI values compared to fixed base
Raft-soil interaction	Stiff soil (Model 4)	1.535	17.2	Stiff soil (Model 22)	1.534	17.25	0.06
Raft-soil interaction	Medium-stiff soil (Model 5)	1.535	17.19	Stiff soil (Model 22)	1.534	17.25	0.072
Raft-soil interaction	Soft soil (Model 6)	1.536	17.16	Stiff soil (Model 22)	1.534	17.25	0.11
Pile-soil interaction	Stiff soil (Model 10)	1.535	17.24	Medium-stiff soil (Model 23)	1.534	17.25	0.072
Pile-soil interaction	Medium-stiff soil (Model 11)	1.535	17.5	Medium-stiff soil (Model 23)	1.534	17.25	0.072
Pile-soil interaction	Soft soil (Model 12)	1.9	25.86	Medium-stiff soil (Model 23)	1.534	17.25	19.43
Raft-pile-soil interaction	Stiff soil (Model 16)	1.5922	14.4	Soft soil (Model 24)	1.534	17.25	3.6
Raft-pile-soil interaction	Medium-stiff soil (Model 17)	1.5925	14.4	Soft soil (Model 24)	1.534	17.25	3.7
Raft-pile-soil interaction	Soft soil (Model 18)	1.593	17.5	Soft soil (Model 24)	1.534	17.25	3.7

Maximum storey displacement is seen for soil-pile interaction while least storey displacement is observed in soil-raft interaction showing that soil-raft foundation interaction displays close to fixed base behaviour. It is also observed from Table 8, that maximum reduction in storey displacement stands at 47.64% for building modelled with shear wall with soil-raft-pile interaction at soft soil condition compared to the same modelling approach with bare frame. With increase in soil flexibility or with decrease in soil shear modulus, structures with shear wall are seen to perform better during seismic activity with lesser storey displacement as compared to bare frame structures. Shear wall insertion during seismic analysis of irregular type of structure helps in reduction of storey displacement in the range of 18–48% depending on soil and foundation parameters thereby making the structural design safe.

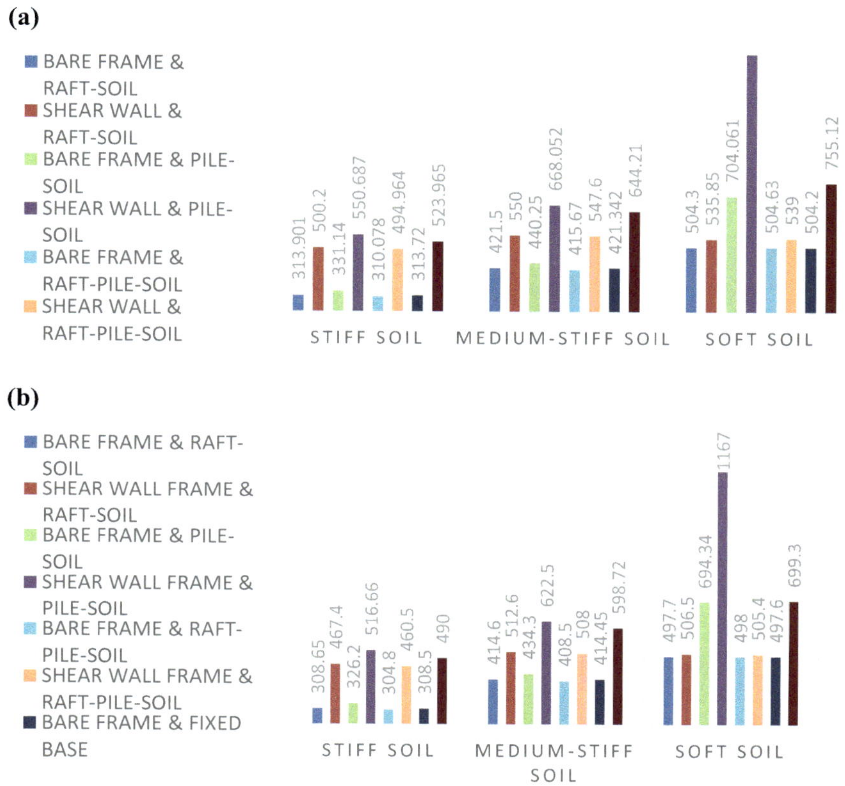

Fig. 3 **a** Base shear along X direction. **b** Base shear along Y direction

Table 5 % Increase in base shear values along X direction

Frame with "shear wall" compared with "bare frame" (Considering SSI)			Frame with "shear wall" compared with "bare frame" (Considering fixed base)	
Raft-soil interaction	Stiff soil	37.24	Stiff soil	40
	Medium-stiff soil	23.4		
	Soft soil	6		
Pile-soil interaction	Stiff soil	39.86	Medium-stiff soil	34.6
	Medium-stiff soil	34		
	Soft soil	44		
Raft-pile-soil interaction	Stiff soil	37.35	Soft soil	33.23
	Medium-stiff soil	24		
	Soft soil	6.4		

Table 6 % Increase in base shear values along Y direction

Frame with "shear wall" compared with "bare frame" (Considering SSI)			Frame with "shear wall" compared with "bare frame" (Considering fixed base)	
Raft-soil interaction	Stiff soil	37.24	Stiff soil	37
	Medium-stiff soil	19		
	Soft soil	1.74		
Pile-soil interaction	Stiff soil	36.87	Medium-stiff soil	31
	Medium-stiff soil	30.23		
	Soft soil	40.5		
Raft-pile-soil interaction	Stiff soil	33.8	Soft soil	28.85
	Medium-stiff soil	19.55		
	Soft soil	1.5		

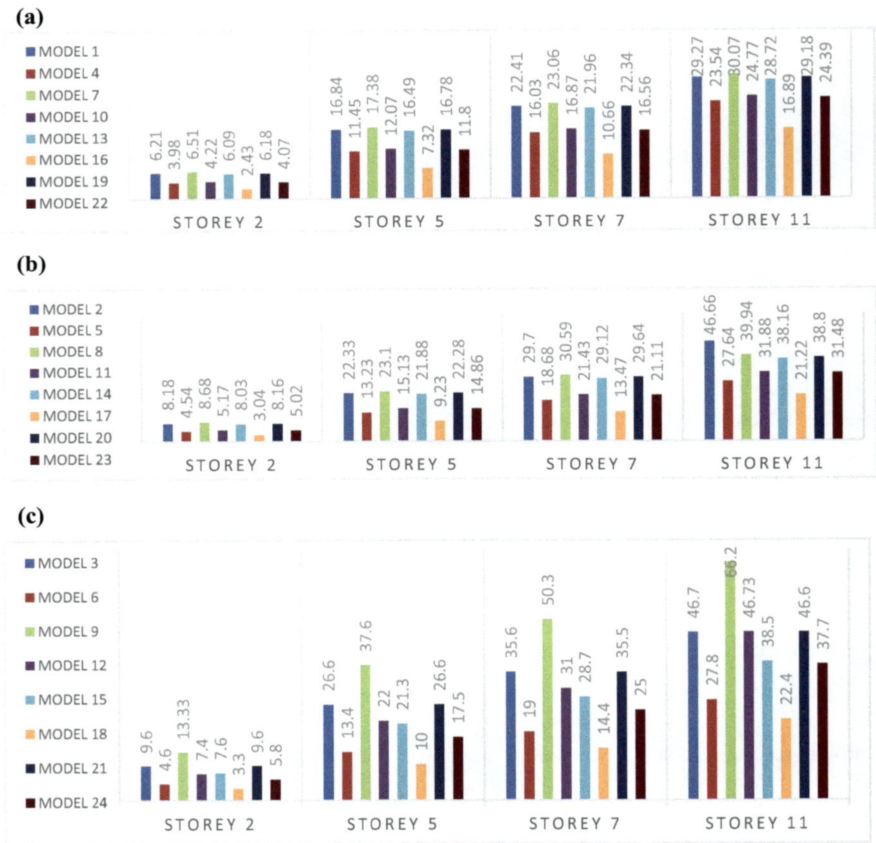

Fig. 4 **a** Storey displacement along X direction on stiff soil. **b** Storey displacement along X direction on medium-stiff soil. **c** Storey displacement along X direction on soft soil

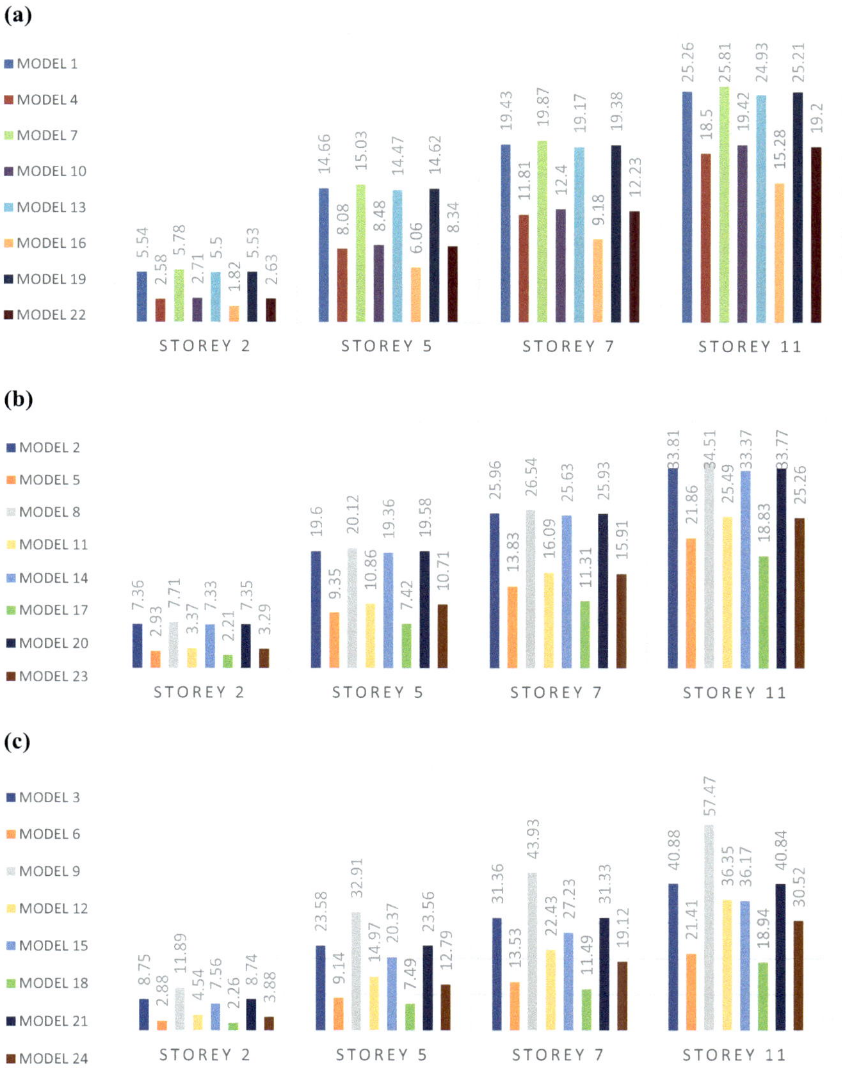

Fig. 5 **a** Storey displacement along Y direction on stiff soil. **b** Storey displacement along Y direction on medium-stiff soil. **c** Storey displacement along Y direction on soft soil

Table 7 % Decrease in storey displacement values at storey level 11 along X direction

Frame with "shear wall" compared with "bare frame" (Considering SSI)			Frame with "shear wall" compared with "bare frame" (Considering fixed base)	
Raft-soil interaction	Stiff soil	19.6	Stiff soil	16.4
	Medium-stiff soil	41		
	Soft soil	40		
Pile-soil interaction	Stiff soil	17.63	Medium-stiff soil	18.8
	Medium-stiff soil	20.2		
	Soft soil	29.4		
Raft-pile-soil interaction	Stiff soil	41.2	Soft soil	19.2
	Medium-stiff soil	44.4		
	Soft soil	42		

Table 8 %Decrease in storey displacement values at storey level 11 along Y direction

Frame with "shear wall" compared with "bare frame" (Considering SSI)			Frame with "shear wall" compared with "bare frame" (Considering fixed base)	
Raft-soil interaction	Stiff soil	26.76	Stiff soil	23.84
	Medium-stiff soil	35.34		
	Soft soil	47.63		
Pile-soil interaction	Stiff soil	24.75	Medium-stiff soil	25.2
	Medium-stiff soil	26		
	Soft soil	36.75		
Raft-pile-soil interaction	Stiff soil	38.7	Soft soil	25.3
	Medium-stiff soil	43.6		
	Soft soil	47.64		

4 Conclusion

Present study attempts to try to analyse and assess the effect of soil-structure interaction on fundamental natural period, base shear and storey displacement on a 12-storey reinforced concrete building frame with shear wall placed at core and bare frame considering the flexibility of supporting soil supported on raft foundation, pile foundation and raft-pile foundation. The results of the study lead to the following conclusions.

(1) Fundamental time periods of the SSI system supported on pile foundation are more than the corresponding values of the same building with raft, raft-pile foundation and fixed base with highest time period in soft soil condition. Hence,

it highlights that with increase in flexibility of soil, time period increases and reduces with the addition of shear wall.

(2) Base shear increases with increase in flexibility of soil with highest value being in pile foundation with the effect of soil-structure interaction whereas compared to conventional fixed base method it decreases but inclusion of shear wall results in an increase of base shear value significantly as compared to bare frame structures with the effect of SSI and fixed base.

(3) The usefulness of shear wall provision is observed to be more significant in case of irregular building as the increase in storey displacement due to SSI is effectively controlled thereby improving the seismic performance of structures.

Finally, it can be concluded that although conventional design procedure omitting SSI is conservative it is required to ensure the structural safety of buildings resting over soft soil due to lateral deflection. However, addition of shear walls counter balances the SSI effect by providing additional stiffness to resist the lateral earthquake forces. Thus, to reduce the SSI effect, structures with shear wall are found to be very useful.

References

1. Ray Chowdhury, P.: Seismic response of low-rise steel moment-resisting frame (SMRF) buildings incorporating nonlinear Soil-Structure Interaction (SSI). Eng. Struct. **33**(3), 958–967 (2011)
2. Rajib, S., Dutta, S.C., Haldar, S.: Seismic response of soil-pile raft structure system. J. Civil Eng. Manage. **21**(2), 144–164 (2015)
3. Hora, M.: Elasto-plastic soil-structure interaction analysis of building frame-soil system. In: 6th International Conference on Case Histories in Geotechnical Engineering, Arlington, VA, Paper No. 1.06, 11–16 Aug 2008
4. Dey, A., Sharma, N., Dasgupta, K.: Influence of shear wall on seismic response of RC frame buildings on pile foundation considering SSI. In 17th World Conference on Earthquake Engineering, Sendai, Japan 13–18 Sept 2020
5. Richart, F.E., Hall, J.R., Jr., Woods, R.D.: Vibrations of soils and foundations. Prentice-Hall Inc., Englewood Cliffs, N.J. (1970)
6. Sosrodarsono, S., Nakazawa, K.: Mekanika Tanah dan Teknik Pondasi. Pt. Pradnya Pramita, Jakarta (2000)
7. IS 1893 (part 1) Indian standard criteria for earthquake resistant design of structures, Bureau of Indian Standards, New Delhi, India (2016)

Electrochemical Technique to Evaluate Carbonation Behavior of Reinforced Concrete

Kulwinder Kaur, Sorabh Saluja, and Shweta Goyal

Abstract The present work postulates the use of electrochemical techniques to evaluate the condition of rebar and concrete subjected to accelerated carbonation front. Potentiodynamic polarization was carried out to find the anodic corrosion current density (i_{corr}), and EIS was used as a non-destructive technique (NDT) to determine the carbonation front progression in concrete. The experimental results reveal that the high-frequency arc of impedance spectra increases with increase in exposure duration. The effect of cement type on the progress of carbonation front is also evaluated. The results show that densification caused due to $CaCO_3$ formation inside the concrete decreased the ingress of carbonation front in OPC concrete. The carbonation depth was further predicted by fitting equivalent circuit on the obtained impedance spectra. Variation in the measured and predicted carbonation depth was found to be below 10% for most of the exposure durations, indicating effectiveness of EIS data to evaluate the carbonation depth of concrete.

Keywords Carbonation · Electrochemical impedance spectroscopy · Equivalent electrical circuits

1 Introduction

The mechanism of rebar corrosion in Reinforced Concrete is a combined function of the properties of concrete and reinforced steel. Concrete provides protection to the rebar firstly, by acting as a physical barrier (presence of cover concrete) against the external aggressive environment and secondly, by developing a passivation layer on the rebar surface because of high pH of concrete [1–3]. However, the ingress of aggressive acidic compounds like, carbon dioxide (CO_2) present in air, sulfuric acid

K. Kaur (✉) · S. Saluja
Department of Civil Engineering, Punjabi University, Patiala 147002, India
e-mail: kulwinder_ucoe@pbi.ac.in

S. Goyal
Department of Civil Engineering, TIET, Patiala 147004, India

E. Strauss (ed.), *Proceedings of the 7th International Conference on Civil Engineering*, Lecture Notes in Civil Engineering 371, https://doi.org/10.1007/978-981-99-4045-5_26

found in industrial environments de-passivates the protective passive layer present around the rebar surface [4].

Carbonation refers to a chemical process that occurs between the atmospheric CO_2 and a hydration product of concrete, i.e. calcium hydroxide, in the presence of moisture [5–7]. It is one of the major causes of deterioration of the reinforced concrete structures situated in places enriched in CO_2 concentration. Increasing generation of CO_2 emission ultimately raised the carbonation rate and carbonation-related cracking in concrete structures [8, 9]. Many concrete structures suffer advanced degree of carbonation because of high level of CO_2 present in capital cities [4].

A variety of destructive methods such as thermo-gravimetric analysis, pH meter test, phenolphthalein indicator, and X-ray diffraction analysis are used as a measure of carbonation front ingress [6, 10–12]. The quantification of the corrosion rate of rebar caused by carbonation can be done by using non-destructive electrochemical techniques also. Electrochemical impedance spectroscopy (EIS) is one of the powerful non-destructive techniques (NDT), which is used to describe the features of various electrochemical mechanisms and to determine the contribution made by the electrode in the system [1, 2]. During EIS technique, an alternating current signal was applied to a sweep of frequencies, generally in the range of mHz to MHz [13].

Various researchers used this technique against chloride-induced corrosion in reinforced concrete to determine the corrosion rate and reaction at steel concrete interface [13–18]. Apart from the corrosion measurement, Dong et al. [6] claimed that carbonation process can be quantitatively accessed in cement mortar by means of EIS. Montemor et al. [19] tried to correlate the concrete resistance with the high-frequency arc of the impedance spectra. Husain et al. [20] used EIS to examine the early age hydration of cement paste made with volcanic ash as additive.

The present study was designed to determine the carbonation depth by using EIS data. The study is performed on reinforced concrete made with two types of cement viz. Ordinary Portland Cement (OPC) and Pozzolanic Portland Cement (PPC). Having the intention to evaluate the corrosion process, potentiodynamic polarization was carried out to determine the anodic corrosion current density (i_{corr}) and EIS to determine the carbonation behavior of concrete. The prediction of carbonation depth was made by obtaining a functional relationship between fitted resistive parameters of high-frequency arc of impedance spectra with exposure duration. The predicted carbonation depth was compared with the depth obtained by using destructive test.

2 Experimental Details

2.1 Materials

Two types of cements were used in the present study viz. Ordinary Portland Cement (OPC) confirming to BIS: 8112–1989 [21], Pozzolana Portland Cement (PPC) confirming to BIS: 1489-2005 [22]. Concrete mixes were prepared by using these

Table 1 Mixture proportions for concrete specimens

Cement[a]	Fine aggregate[a]	Coarse aggregate[a]		Water[a]	Admixture (% weight of cement)
		20 mm	10 mm		
410.00	572.40	836.34	360.40	180	0.30

[a] kg/m^3

cements while keeping the proportion, type and size of coarse and fine aggregates similar in both mixes. The coarse aggregates and river sand confirmed to BIS: 383–2002 [23] were used as fine aggregates. Mix proportions approved for concrete specimen preparation are presented in Table 1. Thermo Mechanically Treated (TMT) steel was used as a rebar confirming to BIS: 1786–1985 [24]. The steel rebar was 360 mm in length and 12 mm in diameter. Rebar was prepared as per ASTM G 109 [25] prior to embedded in concrete prisms. The line diagram and actual photograph of the preconditioned rebar specimen are presented in previous research paper by Kaur et al. [3, 26].

2.2 *Specimen Preparations and Exposure Conditions*

Two types of specimens were cast for current investigation. 100 mm cube specimens were used for carbonation depth measurement, while prism specimens of size 300 × 300 × 150 mm were utilized during electrochemical characterization. All specimens were taken out of moulds after normal curing period of 24 h. Afterward 100 mm cubes were stored in water curing tank for 7 days and prism specimens were kept wet with the help of jute bags till 7 days at temp of 27 ± 3 °C. A total preconditioning period of 15 days was completed by putting all specimens in laboratory environment for further 7 days after completion of 7 days wet curing. In order to achieve unidirectional flow of CO_2, the bottom surface and side surfaces of prism specimens were sealed with sealant.

Thereafter, carbonation depth measurement cubes as well as prism specimens were kept in specially designed carbonation chamber operated at 60–70% relative humidity (RH) and 30 ± 2 °C temperature for 15 days without CO_2 supply. During this period, moisture content inside the specimens is supposed to attain uniform distribution. After the completion of preconditioning period, carbon dioxide is released inside the chamber at a concentration rate of 5% by volume while the temperature and RH range were kept constant as previous.

2.3 Carbonation Depth Measurements by Destructive Method

The carbonation depth was determined at every 15 days of interval to a total exposure period of 90 days. At required test age, 100 mm cubes prepared with both cement types viz. OPC and PPC were split diagonally into two halves. The exposed surface was firstly cleaned with dry cloth and then sprayed with phenolphthalein indicator. The colorless portion reveals carbonated part while purple-red color gives the indication of uncarbonated part [10–12]. The average depth of the carbonated part was found with help of a Vernier caliper by taking measurements from a total of 24 points. The mean of 24 values gives the carbonation depth of the respective specimen.

2.4 Electrochemical Characterization

The electrochemical measurements include half-cell potential (HCP), potentiodynamic polarization, and electrochemical impedance spectroscopy (EIS) performed on reinforced concrete prisms. These measurements were made by using automatic corrosion monitoring (ACM) field machine. ACM machine has provision for the confinement of current with the help of guard ring. The HCP of the top rebar was measured by using Saturated Calomel electrode (SCE) as a reference electrode. A 5 mm thick wet sponge was positioned between prismatic samples and SCE in order to provide ionic conduction (Fig. 1a).

Thereafter, potentiodynamic polarization and EIS were conducted on the top rebar of the prism specimen with the help of guard ring (Fig. 1b). For potentiodynamic polarization measurement, the working electrode was polarized at a sweep rate of 60 mV/min to ± 25 mV from the equilibrium potential. For EIS measurement, alternating current (AC) signal was supplied with a 25 mV RMS for a frequency range of 100 kHz to 10 MHz with five points obtained per decade. Similar frequency range

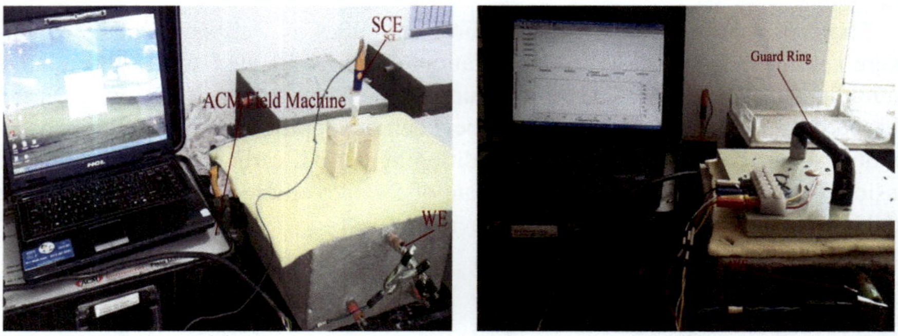

(a) (b)

Fig.1 Experimental setup for **a** half-cell potential **b** electrochemical measurement was used by various practitioners for studying corrosion mechanism in concrete [1, 14, 27, 28]. The experimental setup used for the measurements of half-cell potential and electrochemical measurement is shown in Fig. 1. Further, mathematical analysis of the impedance spectra was carried out by fitting an equivalent circuit using ZMAN software. Since the carbonation process of reinforced concrete system is a complex phenomenon, several models were tried and the best fit equivalent circuit was taken for further analysis.

3 Results and Discussions

3.1 Carbonation Depth Measured by Destructive Method

Figure 2 shows the variation in carbonation depth with exposure duration for both types of concrete mixtures. It was observed from the figure that during the initial exposure duration of 30 days, PPC concrete shows 6% higher carbonation depth as compared to OPC concrete mixture. Carbonation depth behavior goes on increasing with the enhance in exposure duration. At the end of exposure duration of 90 days, PPC concrete shows 43.13% higher carbonation depth on comparison with OPC concrete. Measured higher value of the carbonation depth in PPC concrete can be attributed to the pozzolanic reaction of the fly ash present in PPC concrete. PPC concrete had approximately 25% lower content of CaO and consequently less $Ca(OH)_2$ was formed. Secondly, some part of the $Ca(OH)_2$ formed was consumed by the pozzolanic reaction. These factors lead to lower percentage of $Ca(OH)_2$ in the PPC concrete mix. Hence, during the carbonation reaction, lesser amount of $CaCO_3$ will be formed [19, 29]. $CaCO_3$ gets precipitated as solid matrix in the pores of concrete, owing to its low solubility and leads to densification of the concrete matrix [12, 28].

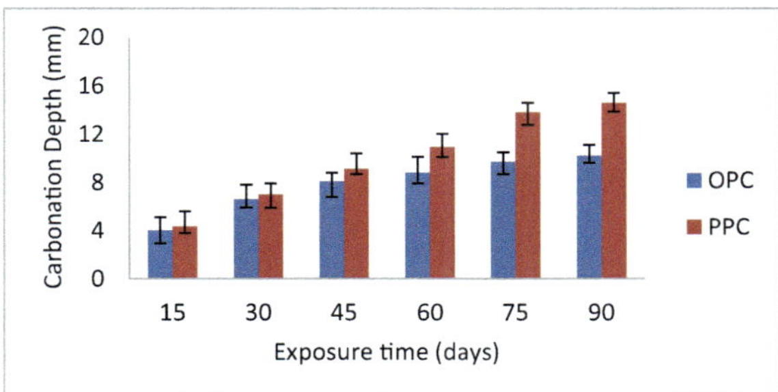

Fig. 2 Carbonation depth variation with exposure duration for both types of concrete mixtures

Table 2 HCP variation in both concrete mixes with exposure duration

Cement type	HCP (mV versus SCE) variation with exposure duration (days)					
	15	30	45	60	75	90
OPC	−149	−158	−167	−180	−206	−223
PPC	−156	−269	−286	−370	−267	−355

3.2 Potential Measurements

Table 2 depicts the evolution of half-cell potential (HCP) with exposure duration. Observed values of HCP for both concrete specimens were found to lie in range specified by RILEM TC 154-EMC [30]. It can be detected that during initial exposure duration both of the concrete specimens show the lower values of HCP, indicating passive behavior for both specimens. The potential readings observed at 30 days of exposure duration for OPC concrete was −158 mV while in PPC concrete was −269 mV indicating intermediate probability of corrosion. Beyond this exposure duration, PPC concrete showed a potential higher than −275 mV, indicating high probability of corrosion. On the other hand, OPC concrete showed the potential value lower than −225 mV at the end of exposure duration.

3.3 Potentiodynamic Polarization Measurements

Polarization curves for both concrete types at variable exposure durations are presented in Fig. 3. The polarization curves clearly indicate that concrete made with different cements performs differently under accelerated carbonation conditions. Observed linear behavior in all polarization curves showed the activation control mechanism in all concrete specimens. During the initial exposure duration, for both OPC and PPC concrete, potential lies in the nobler side, which indicates that rebar embedded in both cement concrete can maintain high corrosion resistance. With the enhance in exposure duration to 90 days, the potential shifts towards active side in case of PPC concrete while, OPC concrete still shows the potential in passive side. Anodic branch of the polarization curve for OPC and PPC concrete starts at −340 mV and −538 mV, respectively, with no explicit passive behavior at 90 days of exposure time.

In order to get anodic corrosion current density (i_{corr}), Tafel extrapolation of these polarization curves was performed [3]. The Tafel slope for the anodic and cathodic reactions was obtained by the extrapolation of linear regions of the polarization curve. The point observed by extrapolating back both of the anodic and cathodic regions corresponds to the corrosion current density (i_{corr}) and the potential at which it lies refer as the corrosion potential (E_{corr}). The corrosion current density observed at 90 days of exposure duration was 0.0694 $\mu A/cm^2$ in OPC concrete and 0.1103 $\mu A/cm^2$ in PPC concrete.

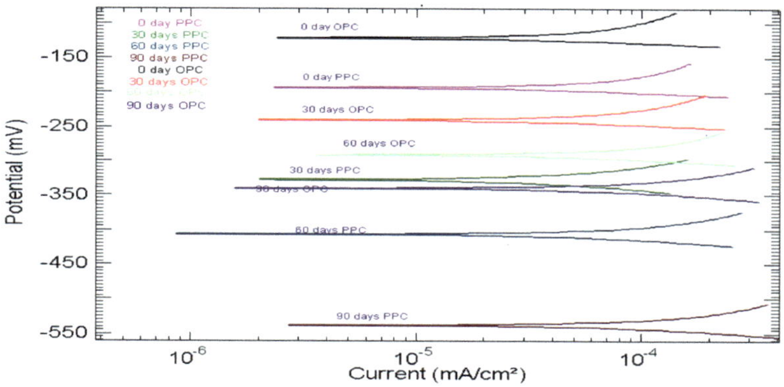

Fig. 3 Polarization curves variation with exposure duration in OPC and PPC concrete

3.4 EIS Measurements

EIS generates plots showing the change in shape and magnitude attributed to the bulk and at steel/concrete interfaces, which can be understood with the help of equivalent electrical circuits [8]. In this present study, ZMAN software was used to interpret the obtained impedance spectra. Basic Randle circuit cannot be utilized to interpret the observed impedance curves because of complex internal structure of reinforced concrete. Previous practitioners [1, 6, 13–16] have suggested lot of equivalent circuits to study the corrosion mechanism occur in concrete. In this current study, several models were tried on the experimentally obtained impedance curves and circuit that fitted best for the experimental data is presented in Fig. 4. In the equivalent circuit, R_s refers for electrolytic resistance of pore solution in concrete, constant phase element (Q_1) refers to the capacitance between the solid/liquid phases of concrete and resistance (R_1) refers to the ion transfer procedure inside the concrete. Similarly, Q_2 refers to the capacitance existing between concrete and rebar; resistance (R_2) refers to charge transfer procedure on the surface of the rebar and Warburg element (W). The high-frequency elements (Q_1, R_1) correspond to the bulk properties of the concrete, while low-frequency elements (Q_2, R_2, W) correspond to the corrosion reaction (Fe^{+2}/Fe^{+3}) that occurs at the rebar surface.

Figure 5 depicts the experimentally obtained (raw) as well as equivalent circuit fitted Nyquist plots for both OPC and PPC concrete. In both mixes, two distinct capacitive arcs were observed at all exposure duration, except the initial day of exposure. First capacitive arc observed at high frequency region corresponds to the bulk properties of the concrete, while electrode-related behavior corresponds to second capacitive arc, which is detected at low frequency region [31]. The absence of high-frequency arc in Fig. 5a, b during initial day of exposure indicates the presence of highly conductive hydroxyl ions and porous structure at the steel/concrete interface. At small frequency, observed large second capacitive arc shows the presence of an extremely resistive passive layer on the rebar surface because of high pH of the pore

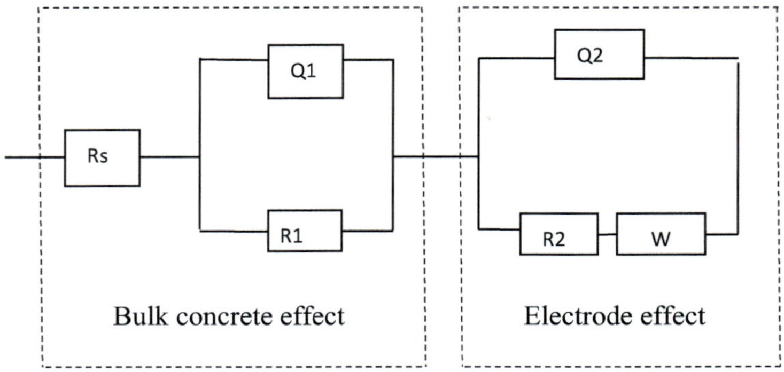

Fig. 4 Equivalent circuit used in the present study to fit the impedance curves

solution [2]. The observed capacitive behavior suggests the presence of homogeneous protective passive layer on the rebar surface [18, 31].

When the CO_2 exposure is augmented from 0 to 60 days, an increase in high-frequency arc with corresponding decay in small frequency arc was obtained for both concrete specimens. Increase in high-frequency arc corresponds to the densification occur inside the bulk concrete, while decrease in small frequency arc indicates the de-passivation of resistive oxide layer present on the rebar surface. Dong et al. [6], Riberio and Abrantes [1] also established the enlargement in radius of high-frequency arc with an increase in carbonation exposure.

Further, curve fitting of the obtained impedance plots was attempted by using ZMAN software with best fit equivalent circuit as shown in Fig. 4. The fitting parameter (R_1), which corresponds to the resistance caused by ions transfer inside the bulk concrete, can be correlated with the properties of bulk concrete and the values of R_1 at different exposure durations are presented in Table 3. Table 3 shows that during initial exposure duration, R_1 value was not observed by equivalent circuit for both types of

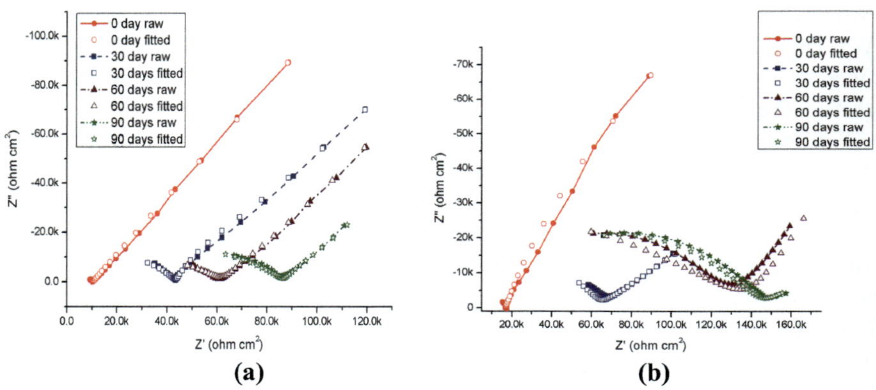

Fig. 5 Nyquist plot obtained at different exposure durations in **a** OPC concrete **b** PPC concrete

Table 3 Fitted resistive parameter (R_1) calculated by using equivalent circuit

Exposure duration (days)		0	15	30	45	60	75	90
R_1 ($k\Omega\ cm^2$)	OPC	–	20.4	38.8	58.7	69.8	84.2	92.3
	PPC	–	19.0	37.9	86.8	106.8	129.4	140.8

concrete because the hydration process of the cement has not been fully developed at this time. At 30 days of exposure time, the R_1 value for both concrete specimens was gradually increased, indicating the progressive hydration process. Thereafter, PPC concrete shows higher value of R_1 on comparison with OPC concrete. PPC concrete shows approximately 47–53% higher value of R_1 as compared to OPC concrete till the end of exposure. Value of R_1 was also supported by the destructive carbonation depth results in which PPC concrete shows approximately 43% higher carbonation depth as compared to OPC concrete.

In order to correlate the results of R_1 obtained by circuit fitting with the carbonation depth obtained by destructive phenolphthalein indicator test, both the values were compared at all exposure durations, and it was observed that the fitted values of resistance R_1 are well supported by the carbonation depth behavior observed for both cement types. Carbonation depth results show that initially there was a smaller variation in carbonation depth but after that variation increases invariably.

3.5 *Prediction of Carbonation Depth from EIS Data*

In Nyquist plot, the high-frequency arc corresponds to the properties of the bulk concrete, which gets altered by the continuous exposure to carbonation. Since carbonation is a bulk concrete phenomenon, so carbonation depth can be predicted with the use of data observed from high-frequency arc of impedance spectra. So, the quantized link between the fitted value of high-frequency arc 'R_1' and carbonation depth was attempted. From carbonation depth measured by phenolphthalein indicator (shown in Fig. 2) and fitted value of R_1 (shown in Table 3), functional relationship between carbonation depth and resistance R_1 was attempted and is presented in Fig. 6. It was established that good correlation can be obtained between carbonation depth and resistance R_1 with correlation coefficient R^2 as 0.957. The carbonation depth was found to be proportional to the square root of R_1. The resultant relationship is presented as:

$$D\,(t) \approx 1.27\{R_1(t)\}^{0.5} - 1.622 \tag{1}$$

where D (t) refers to carbonation depth at time 't', R_1 (t) refers to value of resistance R_1 at time 't' of exposure. Based upon the above relationship, carbonation depth value was predicted as shown in Table 4. The variation in the predicted and obtained carbonation depths is below 10% for most of the exposure durations. As variation

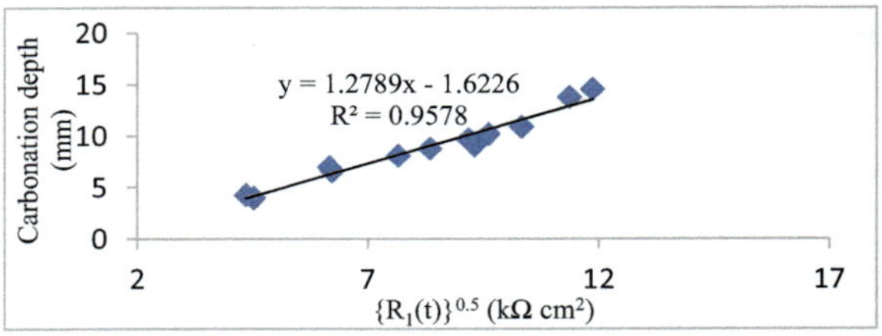

Fig. 6 Established relationship between obtained carbonation depths (destructive) with fitted R_1 values

Table 4 Comparison between obtained and predicted carbonation depth based upon resistance R_1 at different exposure durations

Exposure duration (days)	Specimens	Carbonation depth (mm)		
		Measured by indicator	Predicted by Eq. (1)	Variation (%)
30	OPC	6.6	6.3	4.7
	PPC	7	6.3	11.1
60	OPC	8.8	9.0	2.2
	PPC	10.9	11.6	6.0
90	OPC	10.2	10.6	3.7
	PPC	14.6	13.5	8.1

in carbonation depth value is less so the fitted value of resistance R_1 can be used effectively to assess the carbonation depth of concrete, which could be otherwise measured by performing destructive test on concrete specimen. Hence, EIS can be adopted as an effective NDT technique to measure the carbonation depth of the concrete along with the prediction of corrosion behavior.

4 Conclusions

Based on the experimental test program and modeling analysis of EIS measurements in carbonation rich environment, the following conclusions can be drawn:

1. Carbonation depth results show that type of cement has an important role to resist carbonation-induced corrosion in reinforced concrete.
2. PPC concrete leads to higher carbonation depth as compared to OPC concrete because of lesser formation of $CaCO_3$. The progressive densification of concrete matrix to oppose the ingress of CO_2 into concrete matrix depends upon the amount of $CaCO_3$ precipitation.
3. Polarization curves corroborate the information obtained from the HCP measurement. PPC concrete shows 59% higher corrosion current density as compared to OPC concrete.

4. Equivalent electric circuit used to model the EIS spectra fit the spectra very closely and hence is able to predict the characteristics of concrete and corrosion behavior accurately.
5. A functional relationship correlating fitted high-frequency resistance R_1 obtained from the equivalent circuit and carbonation depth is established that has high correlation coefficient R^2 of 0.957.
6. Variation in the measured and predicted carbonation depth from fitted R_1 value is below 10% for most of the exposure durations, indicating that the fitted value of resistance R_1 can be used effectively to evaluate the carbonation depth of concrete by means of non-destructive technique such as EIS.

References

1. Ribeiro, D.V., Abrantes, J.C.C.: Application of electrochemical impedance spectroscopy (EIS) to monitor the corrosion of reinforced concrete: a new approach. Constr. Build. Mater. **111**, 98–104 (2016)
2. Shen, D.: Electrochemical impedance spectroscopy study on corrosion inhibitor for reinforced concrete. Int. J. Electrochem. Sci. **12**, 4183–4192 (2017)
3. Kaur, K., Goyal, S., Bhattacharjee, B., Maneek, K.: Efficiency of migratory type organic corrosion inhibitor in carbonated environment. J. Adv. Concr. Technol. **14**, 548–558 (2016)
4. Aperador, W., Vera, R., Carvajal, A.M.: Industrial byproduct—based concrete subjected to carbonation, electrochemical behaviour of steel reinforcement. Int. J. Electrochem. Sci. **7**, 12870–12882 (2012)
5. Sisomphon, K., Franke, L.: Carbonation rates of concretes containing high volume of pozzolanic materials. Cem. Concr. Res. **37**, 1647–1653 (2007)
6. Dong, B.Q., Qiu, Q., Xiang, J., Huang, C., Xing, F., Han, N., Lu, Y.: Electrochemical impedance measurement and modeling analysis of the carbonation behaviour for cementitious materials. Constr. Build. Mater. **54**, 558–565 (2014)
7. Aguiar, J.B., Junior, C.: Carbonation of surface protected concrete. Constr. Build. Mater. **49**, 478–483 (2013)
8. Ulloa, E.C., Chab, R.C., Baz, M.S., Borges, P.C., Lopez, T.P.: Corrosion process of reinforced concrete by carbonation in a natural environment and an accelerated test chamber. Int. J. Electrochem. Sci. **8**, 9015–9029 (2013)
9. Bautista, A., Alvarez, S.M., Paredes, E.C., Velasco, F., Guzman, S.: Corrugated stainless steels embedded in carbonated mortars with and without chlorides: 9 year corrosion results. Constr. Build. Mater. **95**, 186–196 (2015)
10. Khunthongkeaw, J., Tangtermsirik, S., Leelawat, A.: A study on carbonation depth prediction for fly ash concrete. Constr. Build. Mater. **20**, 744–753 (2006)
11. Chang, C.F., Chen, J.W.: The experimental investigation of concrete carbonation depth. Cem. Concr. Res. **36**, 1760–1767 (2006)
12. Fukushima, T., Yoshizaki, Y., Tomosawa, F., Takahashi, K.: Relationship between neutralization depth and concentration distribution of $CaCO_3$–$Ca(OH)_2$ in carbonated concrete. Adv. Concrete Technol. ACI SP-179, 347–363 (1998)
13. Braganca, M.O.G.P., Poretella, K.F., Bonato, M.M., Marino, C.E.B.: Electrochemical impedance behavior of mortar subjected to a sulfate environment—a comparison with chloride exposure models. Constr. Build. Mater. **68**, 650–658 (2014)
14. Dhouibi, L., Triki, E., Raharinaivo, A.: The application of electrochemical impedance spectroscopy to determine the long-term effectiveness of corrosion inhibitors for steel in concrete. Cement Concr. Compos. **24**, 35–43 (2002)

15. Ismail, M., Ohtsu, M.: Corrosion rate of ordinary and high performance concrete subjected to chloride attack by AC impedance spectroscopy. Constr. Build. Mater. **20**, 458–469 (2006)
16. Duarte, R.G., Castela, A.S., Neves, R., Freire, L., Montemor, M.F.: Corrosion behavior of stainless steel rebars embedded in concrete: an electrochemical impedance spectroscopy study. Electrochem. Acta **124**, 218–224 (2014)
17. Montemor, M.F., Simoes, A.M.P., Salta, M.M.: Effect of fly ash on concrete reinforcement corrosion studied by EIS. Cement Concr. Compos. **22**, 175–185 (2000)
18. Jamil, H.E., Shriri, A., Boulif, R., Bastos, C., Montemor, M.F., Ferreira, M.G.S.: Electrochemical behavior of amino alcohol based inhibitors used to control corrosion of reinforcing steel. Electrochim. Acta **49**, 2753–2760 (2004)
19. Montemor, M.F., Cunha, M.P., Ferreira, M.G., Simoes, A.M.: Corrosion behaviour of rebars in fly ash mortar exposed to carbon dioxide and chlorides. Cement Concr. Compos. **24**, 45–53 (2002)
20. Hussain, R.R., Ishida, T.: Critical carbonation depth for initiation of steel corrosion in fully carbonated concrete and development of electrochemical carbonation induced corrosion model. Int. J. Electrochem. Sci. **4**, 1178–1195 (2009)
21. BIS 8112: Specification for 43 grade ordinary portland cement. Bureau of Indian Standards (1989)
22. BIS 1489 (part 1)-1991: Specification for Portland Pozzolana Cement. Bureau of Indian Standards 1489-1991 (Reconfirmed 2005)
23. BIS 383.: Specification for coarse and fine aggregates from natural sources for concrete. Bureau of Indian Standards 383 (Reconfirmed 2002)
24. BIS 1786–1985: Specification for high strength deformed steel bars and wires for concrete reinforcement. Bureau of Indian Standards 1786-1985 (Reconfirmed 2004)
25. ASTM G 109-99a: Standard test method for determining the effects of chemical admixtures on the corrosion of embedded steel reinforcement in concrete exposed to chloride environments. American Society of Testing and Materials, G109-1999 (Reapproved 2005)
26. Kaur, K., Saluja, S., Goyal, S., Bhattacharjee, B.: Corrosion inhibitor for preventing carbonation induced rebar corrosion. Indian Concrete J. **94**(11), 29–36 (2020)
27. Liu, M., Cheng, X., Li, X., Zhou, C., Tan, H.: Effect of carbonation on the electrochemical behaviour of corrosion resistance low alloy steel rebars in cement extract solution. Constr. Build. Mater. **130**, 193–201 (2017)
28. Dong, B.Q., Qiu, Q., Gu, Z., Xiang, J., Huang, C., Fang, Y., Xing, F., Liu, W.: Characterization of carbonation behaviour of fly ash blended cement materials by the electrochemical impedance spectroscopy method. Cement Concr. Compos. **65**, 118–127 (2016)
29. Rea, S.P.A., Higuera, R.C., Soberon, J.M.G., Gonzalez, J.H.C., Carmona, V.O., Sanchez, J.L.A.: Carbonation rate and reinforcing steel corrosion of concretes with recycled aggregates and supplementary cementing materials. Int. J. Electrochem. Sci. **7**, 1602–1610 (2012)
30. Elsener, B., Andrade, C., Gulikers, J., Polder, R., Raupach, M.: Electrochemical techniques for measuring metallic corrosion, RILEM TC 154-EMC. Mater. Struct. **36**, 461–471 (2003)
31. Ford, S.J., Shane, J.D., Mason, T.O.: Assignment of features in impedance spectra of the cement-paste/steel system. Cement Concrete Res. 28 (12), 1737–1751 (1998)

Performance Improvement of Reinforced Concrete Beams Strengthened with GFRP Sheet

Achmad Zultan Mansur, Rudy Djamaluddin, Herman Parung, and Rita Irmawaty

Abstract Concrete has a strong compression strength but low tensile strength. As a result, a steel bar is used in concrete construction to increase tensile strength and prevent cracking from occurring due to loading on top of the structure. However, one significant disadvantage of using steel bars in concrete is that it is prone to corrosion. Among the efforts to restore the function of the RC concrete structure when undergoing spalling is by making repairs with grouting and using GFRP sheets. This study aims to increase the bending capacity of reinforcement concrete (RC) beams that have undergone spalling by repairs using sika grout and strengthening the structure with the GFRP Sheet. The results showed that by using grouting improvements in the spalling beam, supported by the GFRP sheet layer, the bending capacity of the reinforced beam increased by 8.9% compared to the control beam (CB). Tensile strain measured in bonded sheets at peak loads for BGRST1, BGRST2, and BGRST3 is 13210.5 $\mu\varepsilon$, 13,797.1 $\mu\varepsilon$, and 15466 $\mu\varepsilon$, respectively. The results of the RC control beam test showed that everything failed to bend while for variation beams, it experienced debonding failure and the concrete decking was peeled off.

Keywords RC · Spalling · Grouting · GFRP sheet

1 Introduction

Chloride attack and carbonation are the two most common causes of rebar corrosion. Typical chloride assault and carbonation do not attack or degrade the concrete cover as chemical deterioration does. Instead, the aggressive chemical penetrates concrete pores and directly corrodes the Electronic Current Steel Reinforcement [1]. Concrete was spotted spalling off the deck, which was thought to be caused by the steel expanding owing to rust produced by the corrosion process. Seawater attacks on concrete include surface erosion brought on by tides, swelling brought on by salt

A. Z. Mansur (✉) · R. Djamaluddin · H. Parung · R. Irmawaty
Department of Civil Engineering, Faculty of Engineering, Hasanuddin University, Makassar, Indonesia
e-mail: achmadzultan@gmail.com

E. Strauss (ed.), *Proceedings of the 7th International Conference on Civil Engineering*, Lecture Notes in Civil Engineering 371, https://doi.org/10.1007/978-981-99-4045-5_27

crystallization, and chemical attack by salt dissolved in the water. Circular drying and capillary suction take place in the concrete just above sea level, and water transports dissolved salts into the concrete [2]. The salts precipitated in the pores due to evaporation, creating tensions that can result in microcracking, which explains why the structure's concrete spalled. Wind depositing salt aerosols on concrete structures near but not exposed to seawater has a similar effect. Corrosion occurs due to minor physical and chemical changes in metals and the environment [3].

If a structure is expected to be exposed to extreme weather conditions or needs to have a long service life, the appropriate choice is to use reinforcing bars that have better corrosion resistance than standard carbon steel rebars as a preventive measure. The removal of concrete to protect the reinforcement following the repair work does not have to be limited to the areas where it is weak, cracked, or damaged. If carbonation or chloride contamination is likely to cause damage to the structure, it is typically necessary to remove structurally sound concrete. Common practice, on the other hand, often overlooks the importance of removing intact concrete [4]. The bonding strength of the repair material against the surface of the removed concrete is a significant issue arising from traditional repair techniques. The substrate's strength and integrity, as well as the surface's cleanliness and roughness, can affect the binding. The surface should ideally be devoid of dust or incoherent residue that could compromise the bonding strength. Bonding chemicals such as cement paste or fine mortar, polymer latex, and epoxy system may be helpful in some circumstances to improve the repair mortar's adhesion [5].

Repair or strengthening the damaged structure is necessary to prevent it from collapsing. Structural restoration seeks to restore and improve the structural elements' strength so that they can resist the load as specified in the design. A patching approach utilizing Sika grout 215 is one option to repair such block fissures. Sika grout 215 is a ready-to-use grouting cement that does not shrink, has a working duration that is appropriate for local temperatures, and may be used to fill cavities, gaps, and temporary stops with high compressive strength [6]. The application of external epoxy-bonded steel plates has been of the most popular ways of strengthening RC beams. Still, it has a major problem with the bond failure at the steel and concrete interphase owing to steel corrosion. Steel plates were replaced with corrosion-resistant and lightweight FRP composite plates to solve these issues. FRPCs aid in increasing strength and elasticity without increasing stiffness excessively. Beams are critical structural components in all constructions exposed to bending, torsion, and shear. Similarly, columns are utilized in a variety of structures as various key elements subjected to axial load combined with/without bending [7]. Externally bonded FRP composites can now be used to strengthen concrete elements quickly and effectively. Due to its high strength, low weight, great fatigue resistance, easy and quick installation, and minimal structural geometry change, retrofitting concrete structures with FRP sheets can often be more economical and technically superior than traditional methods [8]. Traditional approaches would be impracticable in regions with limited access, so Fiber Reinforced Polymer (FRP) systems can be applied. The purpose of this study is to increase the bending strength of the beam in spalling conditions using sika grout 215 as an improvement on concrete and GFRP sheet for strengthening the structure of reinforced concrete (RC) beams.

2 Specimens and Test Setup

2.1 Specimens

Figure 1a is an RC Beam in the condition of the structure is not grouting and strengthening the GFRP so that the tensile reinforcement used is deformed bar D13 and stirrup ϕ8 while Fig. 1b shows the condition of the beam assuming it has undergone spalling in its structure so that there is a reduction in the rebar as a result of which the tensile bone turns into a reinforcement steel bar ϕ 8. Fig. b shows the grouting and strengthening of the GFRP for the lower side and side beams so that the beam structure function can be increased.

There are two variations of specimen made in this study and each variation consists of three specimens, namely, three control beams and three variation beams. Normal beams as control beams are given naming with the symbols CB1, CB2, and CB2. At the same time, the variation beam is a beam that is repaired grouting at the bottom and reinforced by a GFRP sheet on the entire bottom side and GFRP U Wrap on the reinforcement of the reinforced beam fulcrum symbolized by BGRST1, BGRST2, and BGRST3. Figure 2 shows dimension 6 of the RC beam specimen, which is 3.300 mm in length and has dimensions of 150.0 mm × 200.0 mm.

2.2 Materials

This concrete's average cylinder compressive strength is 28 days at 20 MPa. The results of the tensile strength test for plain reinforcement of 8 mm showed that the average melting stress (Fy) was 375.93 MPa. For the presentation of 13 mm threaded reinforcement, the average melting stress (Fy) is 316.02 MPa. For the repair of concrete that undergoes spalling, the Sika Grout 215 product is used @25 kg. In contrast, the GFRP sheet is used to strengthen the beam structure, which is a product of the Fyfe Company, namely, Tyfo® The Fiberwrap Composite System SEH-51A. Epoxy resin is used to adhere the unidirectional GFRP sheets to the concrete surface.

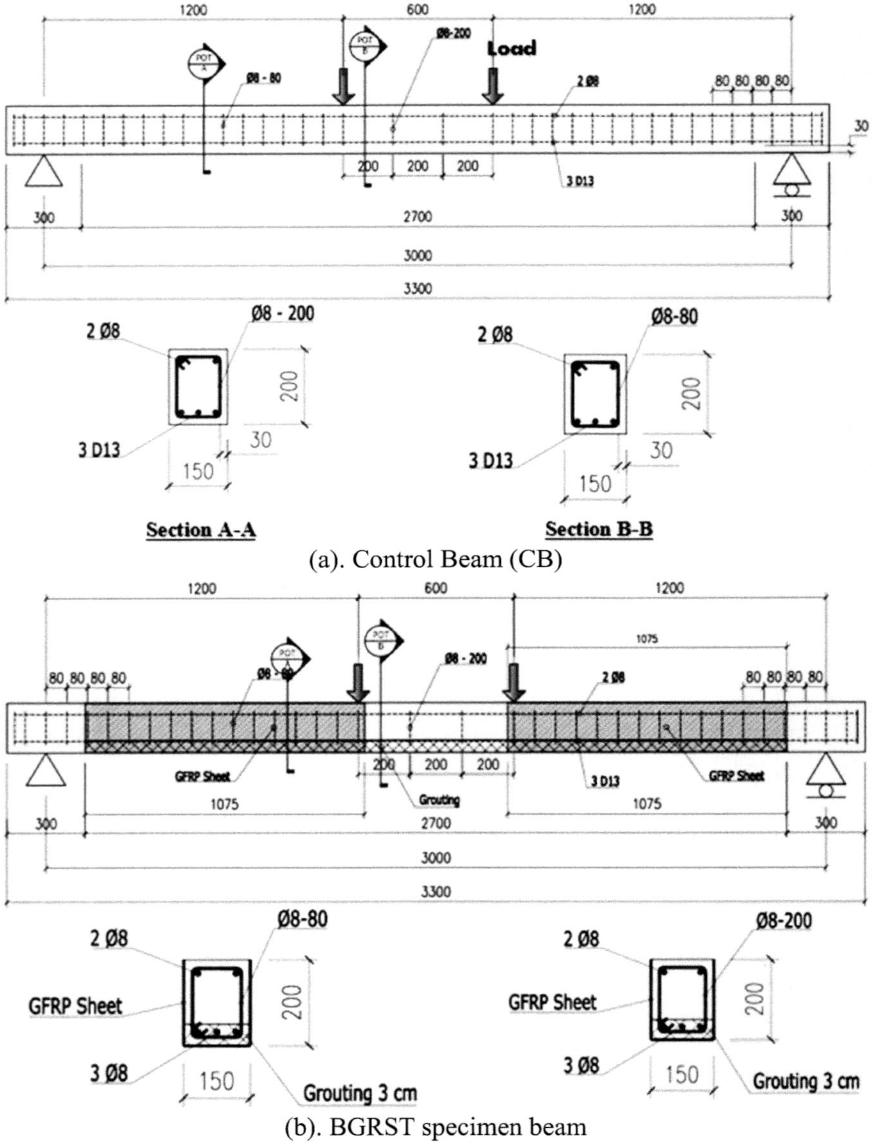

Fig. 1 Detail of specimen

(a) Shows the process of preparing formwork before casting starts. (b) The result of casting by separating 5 cm from the surface so that the reinforcing iron appears visible. (c) Steel reinforcement has undergone corrosion so that repairs are made to the surface of the beams using sika grout 215. (d) Preparation for the installation of GFRP sheets on the beams. (e) Installation of GFRP sheet along the bottom of the beams. (f) The result of installing a GFRP sheet for the fulcrum area of the beams.

Fig. 2 Fabrication of test material

2.3 Fabrication and Set Up Specimen

This research conducted experimental testing and was carried out at the Materials and Structures Laboratory of the Department of Civil Engineering, Hasanuddin University (Fig. 3). The main object of the study was the structure of standard reinforced concrete beams, which were assumed to be subject to corrosion (spalling) in the reinforcement, so repairs were carried out by grouting and retrofitting using GFRP sheet.

Static loading is used with a constant actuator ramp speed of 0.05 mm/s until the beam fails. Data readings in the data logger are taken at each 1 kN load increase under normal conditions, while for certain conditions, such as in first cracking, yield, and ultimate load data are taken more tightly. Meanwhile, observations of the test beam continue to be monitored visually, especially for the development of cracks that occur due to increased load and the behavior of failures. Loading is carried out until it reaches the ultimate load for recording vertical deflection, installed LVDT 100 mm on some parts of the test object. For strains on longitudinal reinforcement, concrete, and GFRP, several strain gauges are installed at certain positions, as in Fig. 3.

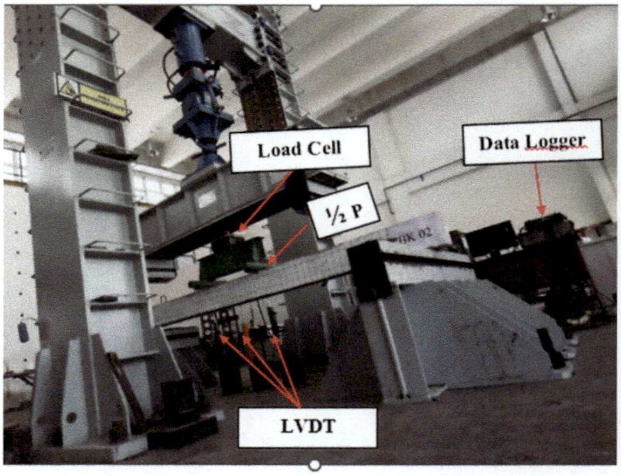

Fig. 3 Setup specimen

3 Results and Discussions

3.1 Maximum Flexural Capacity of RC Beams

The results of the flexural test (Table 1), it was seen that the ultimate loads of the control beams CB1, CB2, and CB3 were 29,74 kN, 30.57 kN, and 30.65 kN, respectively, with the average of 29,74 kN. Meanwhile, the ultimate load of strengthened beams BGA1, BGA2, and BGA3 was 32.85 kN, 30.25 kN, and 34.12 kN with the average of ultimate load of 32.41 kN. During the testing, it was observed that all the six beams tested were almost similar behavior in the initial loading stage. It is because all the beams' failure occurred in flexural failure.

In the case of the control beam (CB), load of first crack appeared in the middle span of the beam at the largest load among the three existing specimens of 2.13 kN. With increasing load, some cracks were observed. Eventually, the beam failed due to wide crack propagation from the bottom fibers and the destruction of

Table 1 Maximum load of RC beams

Beams		Ultimate load (kN)	Average maximum load (kN)	Percentage increase (%)
Control beams	CB 1	28,12	29,74	–
	CB 2	30,45		
	CB 3	30,65		
Variation beams	BGRST 1	32,85	32,41	8,98%
	BGRST 2	30,25		
	BGRST 3	34,12		

concrete in the upper fibers of the beam at the largest load of the existing three specimens of 30.38 kN. Please note that the first paragraph of a section or subsection is not indented. The first paragraph that follows a table, figure, equation etc. does not have an indent, either.

The first loading cracks appeared in the BGRST beam of the three variation beams, i.e., at a load of 9.1 kN. More flexural cracks were seen as the load gradually increased, and the earlier flexural cracks expanded and spread vertically upward. At the ultimate load of 32.41 kN, the beam's flexural failure with concrete crushing in the bottom fiber is occurred. The results showed that by using grouting improvements in the spalling beam, which was then strengthened by the GFRP sheet layer, the bending capacity of the reinforced beam increased by 8.9% compared to the normal beam (CB). The function of the grouting is to cover the spalling on the beam so that the dimensions of the beam return to perfection then the influence of the GFRP sheet on the bottom of the beam keeps the connection of the grouting mortar with the old concrete from easily collapsing due to increased load. Adding GFRP U-Wrap reinforcement to the fulcrum area can increase the beam's shear strength but cannot increase the bending strength. Therefore, most beams fail due to bending failure.

3.2 Maximum Load Versus Deflection Behavior

Table 2 shows the relationship of load and maximum displacement for all beams. The displacement referred to here is the displacement in the midspan of the beam. The results showed that the repair of RC concrete using sika grout 215 made the composition of the concrete no longer homogeneous, thus affecting the displacement of the concrete. The effect of the reinforcement treatment with the GFRP sheet affects the bending strength but not for the deflection behavior of the mid-span beam.

Table 2 Summary of experimental result

Beams		Ultimate load (kN)	Mid-span displacement (mm)	Failure mode
Control beams	CB 1	28,12	51,86	Flexural failure
	CB 2	30,45	54,38	
	CB 3	30,65	49,74	
Average		**29,74**	**51,99**	
Variation beams	BGRST 1	32,85	64,86	Debonding failure
	BGRST 2	30,25	49,99	Debonding failure and the concrete decking was peeled off
	BGRST 3	34,12	59,08	Debonding failure and the concrete decking was peeled off
Average		**32,41**	**57,98**	

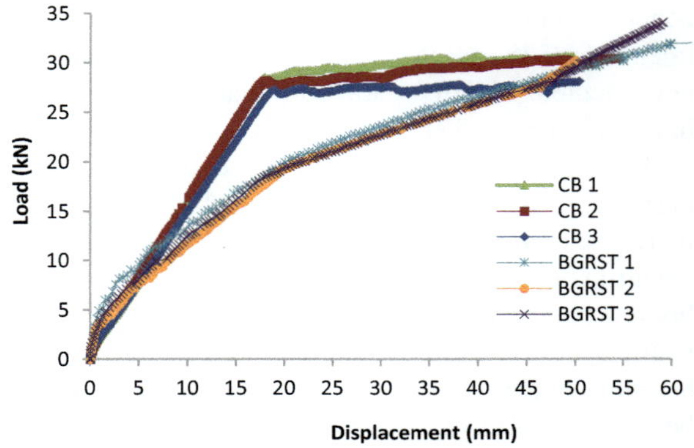

Fig. 4 Load–displacement curves

The load–displacement curve is shown in Fig. 4. It can be seen that control beams (CB1, CB2, and CB3) exhibit the same behavior as strengthened beams (BGRST1, BGRST2, and BGRST1) to the resulting load. Deformation of CB is quite large than BGRST in Fig. 4, CB has not been strengthened so that when receiving the ultimate load, the rebar will slowly melt and crack the RC beam. Beyond the resulting load, the rigidity of the reinforced beam is stiffer than the control one. Thus, using the GFRP sheet at the bottom of the beam, the post-yield bending rigidity of the reinforced beam is significantly increased from the normal beam. The effect of repairs with sika grout makes the strength of concrete not optimal so that it has an impact on a large displacement value but with the use of GFRP it can increase the maximum load up to 8.98%. The average control beam load value is 29.74 kN with a deflection of 64.86 mm; as the load increased, the reinforcing steel bar began to yield, which is indicated by a significant increase in deflection without a substantial load increase; the nonlinear relationship curve is now much flatter than it was previously. This occurs until the variation beam (BGRST) reaches a maximum load of average 32.41 kN and a deflection of 57.98 mm.

3.3 Maximum Load Versus Strain Behavior

It is observed in Fig. 5 that the main compressive strains of concrete at CB1, CB2, and CB3 are 2063.81 $\mu\varepsilon$, 3282.86 $\mu\varepsilon$, and 3039.05 $\mu\varepsilon$, respectively. Meanwhile, the ultimate compressive concrete strains for variation beams in BGRST1, BGRST2, and BGRST3 were 2220 $\mu\varepsilon$, 2066.67 $\mu\varepsilon$, and 3120.95 $\mu\varepsilon$, respectively. The result implies that a normal beam fails when the concrete reaches the final strain the concrete (assumed to be equal to 3000 $\mu\varepsilon$). The nature of concrete that has become composite and has been strengthened greatly affects the strain of concrete; as can be seen from the graph, the average value of concrete strain is reduced by 11.66% of the control beam. The relationship between load and tensile strain in a GFRP sheet is depicted in Fig. 6. It can be noted that for BGRST1, BGRST2, and BGRST3, the tensile strain

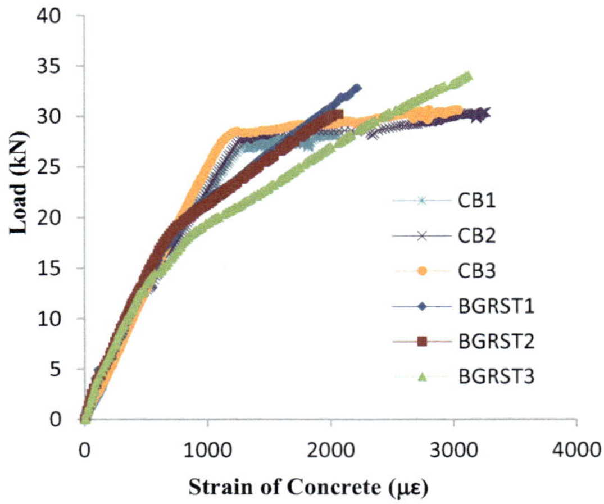

Fig. 5 Load-strain of concrete

measured in bonded sheets at peak loads is 13210.5 με, 13,797.1 με, and 15,466 με, respectively. The ultimate strain (fu) of the GFRP sheet (equivalent to 20,000 με) is less than these values. This shows that GFRP sheet debonding caused crack failure in reinforced beams. As the load increases, the graph tends to reach a linear shape. At a certain load level, the strain distribution curve becomes linear, which means the connection begins to fail. This corresponds to the bonding capacity along the part of the FRP surface that receives the loading. With increasing loading, this means that the load transfer on the FRP is shifted along the FRP surface with concrete until the debonding process occurs.

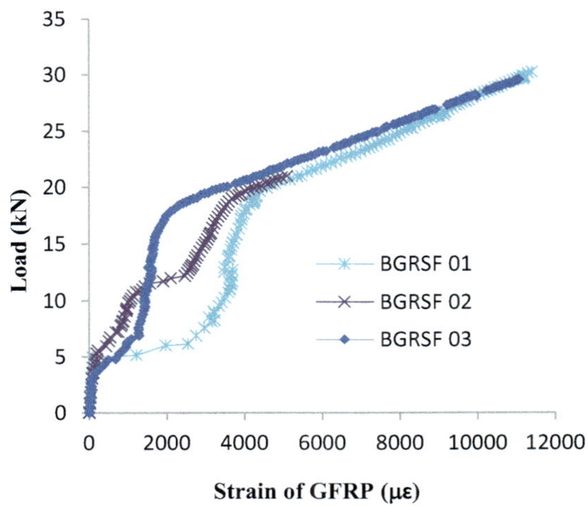

Fig. 6. Load-strain of GFRP

3.4 Crack Pattern

The test results in Fig. 7 show that all beams have a bending failure; this failure begins with the inability of the beam to accept loads that exceed the strength. The initial crack occurs in 1/3 of the central span of the beam, which continues to undergo a propagation of cracks that move intensively from the pull side to the compressive side of the beam and the type of crack that occurs is a bending crack. The tensile stress at the bottom of the concrete part is caused by a bending moment in part under consideration. This process continues until the peak load is reached, where the load no longer increases. Still, the deflection continues to grow, especially in cracks that are wide enough and then suddenly decrease dramatically.

From Fig. 7, the failure of the specimen ranges from cracking fractures to variation objects undergoing debonding. Debonding failure occurs when there are visible loose bonds between the FRP-S and the beam. All types of beams experience a bending failure that begins more previously with an initial cracking signal and gradually breaks a loud sound. FRP-S debonding in Fig. 7a–c is characterized by the sound of cracking in the specimen until it appears FRP-S has detached from the concrete; the pattern of occurrence of FRP-S discharge starts from the position of the tip to the center of the specimen. GFRP wrapping positively affects the highest strength, limiting the spread of splicing so that cracks increase. In this study, the failure model was a debonding failure in FRP-S. FRP-S has a higher quality than concrete and

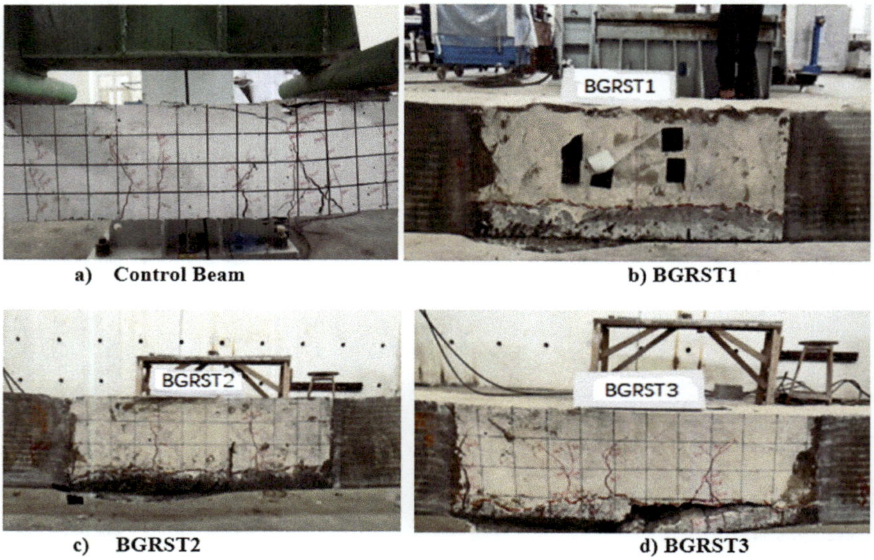

Fig. 7 Failure mode of specimen

reinforced steel; therefore, when the load exceeds the material capacity, reinforced steel and concrete will be destroyed earlier. Because the concrete and reinforced steel have been eliminated first, the mixed reaction between the concrete surface and FRP-S is reduced, so the bond between FRP-S and concrete is no longer fused.

4 Conclusions

Based on the findings and discussions, the following conclusions can be drawn:

(a) Using GFRP sheets as a reinforcement for corrosion concrete could increase the bending capacity. The strengthened GFRP can increase the load maximum of 8.98% compared to the control beams.
(b) For all beams show bending failure, where cracks propagating in the vertical direction from the bottom side to the compressive side.
(c) The average ultimate load magnitude from the results of repair and retrofitting methods for RC beams at the time of debonding is 32.41 kN.

Acknowledgements Thank you to Indonesia Endowment Funds for Education (LPDP) and The Center for Education Financial Service (PUSLABDIK) for providing research funding assistance for the Doctoral Program (S3), so that it can be used to finance this research. The authors wish to acknowledge PT. Fyfe Fiberwrap Indonesia for supplying GFRP sheets and epoxy materials.

References

1. Hamedi, G.H.: The laboratory study of the effect of using liquid anti-stripping materials on reducing moisture damage of HMA. AUT J. Civ. Eng. **1**(1), 23–30 (2017). https://doi.org/10.22060/CEEJ.2017.12240.5156
2. Dermawan, A.S., Dewi, S.M., Wibowo, A.: Performance evaluation and crack repair in building infrastructure. In: IOP Conference Series: Earth and Environmental Science Papers, vol. 328 (2019). https://doi.org/10.1088/1755-1315/328/1/012007
3. Djamaluddin, R., Hijriah, R., Irmawati, R., Faharuddin, R., Wahyuningsih, R.T.: Delamination mechanism of GFRP sheet bonded on the reinforced concrete beams. In MATEC Web Conference, vol. 258, p. 03009 (2019). https://doi.org/10.1051/matecconf/201925803009
4. Djamaluddin, R., Irmawaty, R.: Relationship model of the moment capacity of GFRP sheet strengthened RC beams to the duration of sea water exposure. Proc. Eng. **180**, 1195–1202 (2017). https://doi.org/10.1016/j.proeng.2017.04.280%0A%0A
5. Hijriah, H., Parung, H., Djamaluddin, R., Irmawaty, R.: Debonding failure of externally bonded GFRP sheet on flexural strengthening of reinforced concrete beams. Int. J. Eng. Sci. Innov. Technol. **8**(2), 16–22 (2019). https://www.ijesit.com/Volume8/Issue 2/IJESIT201902_03.pdf

6. Thanoon, W.A., Jaafar, M.S., Kadir, M.R.A., Noorzaei, J.: Repair and structural performance of initially cracked reinforced concrete slabs. Constr. Build. Mater. **19**, 595–603 (2005). https://doi.org/10.1016/j.conbuildmat.2005.01.011
7. Machmud, H., Tjaronge, M.W., Djamaluddin, R., Irmawaty, D.R.: The capacity of reinforced concrete beams post rebars yielded with Frp sheet strengthening. Int. J. Civ. Eng. Technol. **10**(09), 232–241 (2019)
8. Barton, B., et al.: Characterization of reinforced concrete beams strengthened by steel reinforced polymer and grout (SRP and SRG) composites. Mater. Sci. Eng. **412**, 129–136 (2005). https://doi.org/10.1016/j.msea.2005.08.151

Assessment of the Pressure-Impulse Curves of Reinforced Concrete Panels Considering Full Blast Loading History

Nasser A. Alarfaj and Omar M. Alawad

Abstract The pressure-impulse curve is used as a design technique that allows the assessment of a component's dynamic behavior when subjected to blast loading. The pressure-impulse curve has two asymptotes, which draw the lower limits of a structural component's level of protection. Pressure-impulse curves are typically constructed considering a simplified version of the blast loading history where the positive phase is only considered. This simplification often causes tilting (toward the right) to the pressure-impulse curve in the impulsive region, thus altering the impulse asymptote representation. Therefore, the study aims to assess the angle of tilting for the pressure-impulse curves that are constructed with full blast loading history for a considered range of reinforced concrete panels. The results indicate that all examined cases exhibit angle of tilting, i.e. ranges between 3.5 and 17.1°, which results in an increase of the margin of safety for the impulsive region compared with the pressure-impulse curve of positive phase blast load only.

Keywords Blast · The P-I curve · Negative phase · Impulse asymptote

1 Introduction

When an explosion incident occurs, a rise in pressure above the atmospheric pressure occurs at the onset of an explosion and for a period of time, i.e. denoted as the positive phase, which is then followed by a period under atmospheric pressure, where this period is known as the negative phase. The performance of a structural component subjected to blast loading can be obtained using a simplified single-degree-of-freedom system. During early design stages, the structural system can be quickly assessed for any given blast information using the

N. A. Alarfaj (✉) · O. M. Alawad
Department of Civil Engineering, College of Engineering, Qassim University, Buraydah, Saudi Arabia
e-mail: n.aarfaj@gmail.com

O. M. Alawad
e-mail: omar.awwad@qu.edu.sa

E. Strauss (ed.), *Proceedings of the 7th International Conference on Civil Engineering*, Lecture Notes in Civil Engineering 371, https://doi.org/10.1007/978-981-99-4045-5_28

pressure-impulse curve. The P-I curve records the positive pressure and positive impulse when the maximum displacement for the system achieves a given level of protection. The P-I curve is typically generated only considering the positive phase of the blast history. However, omitting the negative phase from the blast loading history for some structural component cases may result in an inaccurate prediction of the response compared to the full blast load history. This can occur, for instance, to a component that reaches its maximum displacement after the negative period of the blast load is completed, i.e. short period of blast scenarios and a component with a long natural period. Hence, this study aims to assess the P-I curve of full blast loading history for a given range of reinforced concrete panels.

2 Background

2.1 Blast Loading History

An explosion is a release of chemical energy that produces a blast wave radiating from the center of the blast, usually with the production of high temperatures and the release of high-pressure gases. A rise in pressure above the atmospheric pressure occurs at the onset of an explosion and for a period of time, i.e. denoted as the positive phase. This phase is then followed by a period with the pressure drops below atmospheric pressure, which is known as the negative phase Fig. 1. Blast intensity can be measured by scaled distance, Z, using Eq. (1) where W is the charge weight of the explosive material, and R is the distance from the center of the blast to the structural component (standoff distance).

$$Z = \frac{R}{W^{\frac{1}{3}}} \tag{1}$$

Blast loading history can be predicted using empirical method techniques such as Kingery and Bulmash (1984) [1] approach, which is widely used for determining positive blast loads; while the negative phase information can be determined by Granström (1956) approach [2]. This is done by considering the charge weight and the standoff distance parameters. The idealized curve of the blast loading history can then be approximated to a piecewise linear loading history as it is recommended in the UFC-3-340-02 [3] and shown in Fig. 1. The actual decay in the idealized curve is approximated to an equivalent triangular pressure pulse, thus, fictitious positive duration, t_{of}, and fictitious negative duration, $t^{-}{}_{of}$, are used and can be calculated using Eqs. (2) and (3); where, i is the total impulse of positive phase, i^{-} is the total impulse of negative phase, P is peak pressure of positive phase, and P^{-} is peak pressure of negative phase.

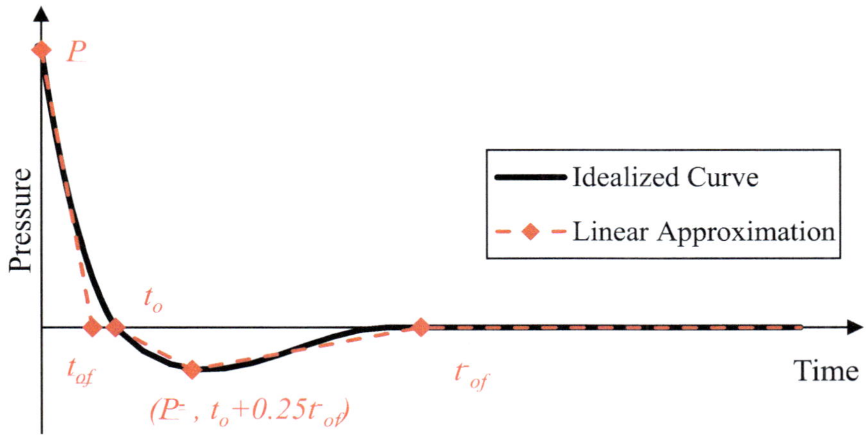

Fig. 1 Blast loading history

$$t_{of} = 2 \times \frac{i}{P} \tag{2}$$

$$t^-_{of} = 2 \times \frac{i^-}{p^-} \tag{3}$$

2.2 *Single Degree of Freedom System (SDOF)*

A real structural system that is subjected to blast loading can be simplified into an equivalent SDOF system by using the transformation factors as shown in Fig. 2 The displacement history, *y(t)*, of the equivalent SDOF can be obtained using Eq. (4); where, K_{LM} is the load-mass transformation factor, *R(y(t))* is the SDOF system resistance function, *M* is the mass of the system and *F(t)* denotes as the history of the blast load. Time-stepping methods, then, can be used to solve Eq. (4) and to find the displacement history *y(t)* as in Fig. 2. The maximum displacement can be evaluated with the required level of protection which can range from superficial damage (i.e. the component has no visible damage) to blow-out (i.e. the component is completely overwhelmed). More details on level of protection can be found in PDC-TR-06-08 [4].

$$K_{LM} M * y''(t) + R(y(t)) = F(t) \tag{4}$$

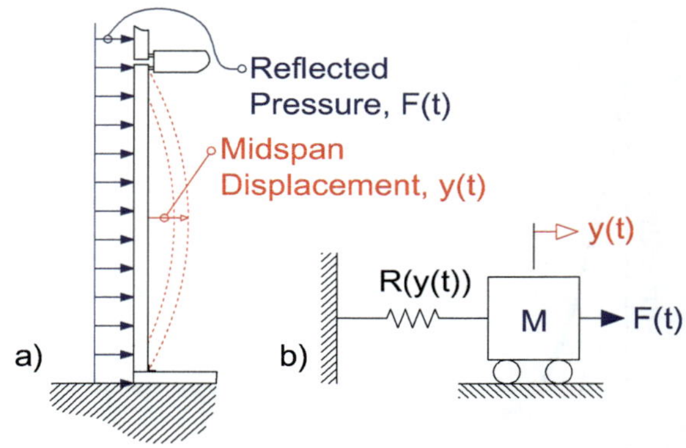

Fig. 2 **a** Real system and **b** Equivalent SDOF system

2.3 *Pressure-Impulse Curve*

The P-I curve is a logarithmic scale diagram that milestones the positive pressure and impulse where the response limit for a structural component is met. The P-I curve is used for preliminary design stages to assess a component's dynamic behavior under various blast loading scenarios. A typical P-I diagram provides information concerning the level of damage of a particular component which ranges from superficial damage (i.e. the component has no visible damage) to blow-out (i.e. the component is completely overwhelmed). The P-I curve can be generated using an SDOF tool, upon which the maximum displacement (or allowable damage level) is obtained.

The P-I curve is typically generated considering the positive phase of the blast history only and has a unique shape with defined asymptotes. The curve vertical and horizontal asymptotes represent the lower limit for impulse and pressure, respectively, for a level of damage. The impulse asymptote can be defined using Eq. (5), where E is strain energy, K_{LM} is the load mass factor, and M is mass of the component. The asymptotes can be used as a normalizer to generate rapidly the P-I curve for wide a range of structural components parameters, rather than iterating the SDOF solver numerous times which can be tedious, especially for preliminary assessment of a structural component at the early design stages [5]. However, omitting the negative phase from blast loading history for some structural components may result in underestimating or overestimating the response compared to the full blast loading history [6]. As a result, the positive phase P-I curve asymptotes may not truly reflect the full loading blast history P-I curve.

$$I_{min} = \sqrt{2EK_{LM}M} \tag{5}$$

3 Literature Review

Several studies investigated the influence of omitting the negative phase [6–11]. For instance, Rigby et al. [6] indicated that there are no approaches that can definitively permit neglecting the negative phase during blast analysis. Krauthammer and Altenberg [7] have observed glazing failure against the loading direction. This suggests that the negative pressure phase increases the rebound, resulting in maximum deflection against the loading direction as also observed in Teich et al. [8].

Syed et al. [9] showed that positive (inbound) displacement did not alter much when the study took into account the negative phase of the blast loading, while negative (rebound) displacement increased. Similar observations were noticed by Gantes et al. [10] where maximum displacements were observed to be in the period of the rebound of the system as a result of its occurrence within the period of the negative phase of the blast load.

Teich and Gebbeken [8] examined the influence of the negative pressure phase. The authors presented a method for modeling the blast wave, taking into account the negative pressure phase. The study presented a new reflection coefficient for the negative pressure phase that was derived from experimental data. The study showed that the negative pressure dominates the structural response significantly for long natural period systems when subjected to weak blast loads; hence, the standard triangular approach with positive only may lead to underestimated results.

Yang and Ahmari [11] developed the P-I charts using the model of a fully restrained ductile beam system for the three different degrees of deformation and negative phase loading. The study showed tilting in the P-I curves in the impulsive region toward the right, i.e. increasing the margin of safety. The severity of tilting was affected by the amount of negative impulse and aspect ratio of length to height of the beam. The study also stated that the impulse asymptote of non-negative simplified blast history would no longer be true for the tilted P-I curves.

Previous studies showed that neglecting the negative phase can lead to inaccurate performance for blast-resistant systems. Since P-I curves are typically constructed considering the positive phase only of a blast load, then it may also fail to provide the required safety measurements or result in an uneconomical choice. Hence, this study aims to assess P-I curves considering the effect of including the negative phase and the resulted tilting in the impulsive region. The study examines a wide range of reinforced concrete panels and evaluates the slope of tilting for future work that interests in defining a new impulsive asymptote equation for the tilted P-I curves.

4 Parametric Study

A SDOF analysis is used to evaluate the amount of tilting of the P-I curves of reinforced concrete panels when considering full blast loading and varying designed parameters. The analysis is carried out for a variety of span lengths, L, (i.e. 8, 12, 15,

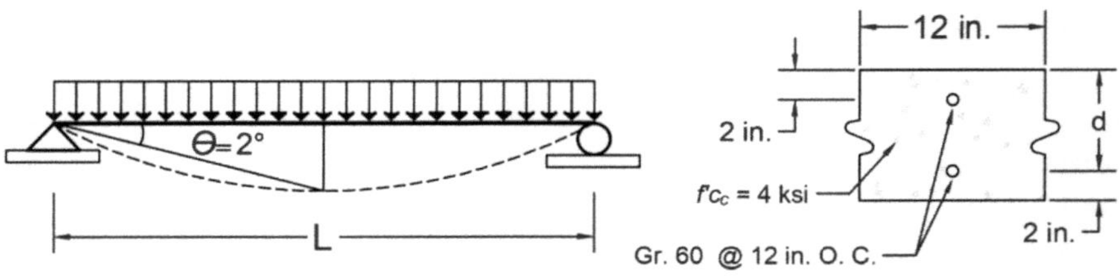

Fig. 3 Panel support and cross-section detailing

20, and 30 feet), as well as a range of areas of grade 60 steel reinforcement (i.e. 0.11, 0.2, 0.31, 0.44, 0.6, and 0.79 square inches), and various range of steel reinforcement depths, d, (i.e. 4, 5, 6, 7, 8, 9, and 10 inches). The concrete cover retained constant for all examined cases, which is equal to 2 inches on both sides of the panels. Concrete compressive strength, *f'c*, is chosen to be equal to 4 ksi. Medium level of protection degrees (i.e. angle of support rotation, θ, is 2°) according to PDC TR 06 08 [4] and a simply supported boundary condition are considered. Figure 3 shows the details of the panel's supporting system and the cross-section.

5 Results

The P-I curves of full blast load are obtained using the SDOF analysis for the set of parameters that were selected in the study. Figure 4 shows a sample of the P-I curve when considering full blast loading history where tilting is observed, compared with the P-I curve of positive phase only. It also can be noticed that tilting occurs in the impulsive region, while no major variation is observed in the quasi-static region.

In order to measure the amount of tilting in the P-I curve of full blast loading, the slope of the curve's tilting is calculated. This is carried out by selecting the first five values of the impulse and pressure vectors. These values are then converted to the logarithmic values where then the slope of its regression line is calculated. This procedure is carried out for the whole P-I curves (i.e. total of 210 curves) in the parametric study. The slope of tilting is considered only for the impulsive region as it possesses the most dramatic change as explained earlier. To better represent the results, the angle of tilting for the regression line, as shown in Fig. 4, is obtained from the slope results.

The variation in the angle of tilting is plotted against the natural period, *Tn*, which is calculated using Eq. (6); where k is the elastic stiffness of the component, K_{LM} is the load-mass factor, and M is mass of the component Fig. 2. The Tn for the examined cases ranges from 20.2 to 567.5 ms. The angles of tilting range between 3.5 and 17.1 deg. Figure 5 shows that significant angle of tilting occurs for all examined cases as a result of including the negative phase. Figure 5 also shows that angle of tilting increases with the increase of Tn, which indicates an increase in the margin of safety of the impulsive region compared with the P-I curve of positive phase blast load only

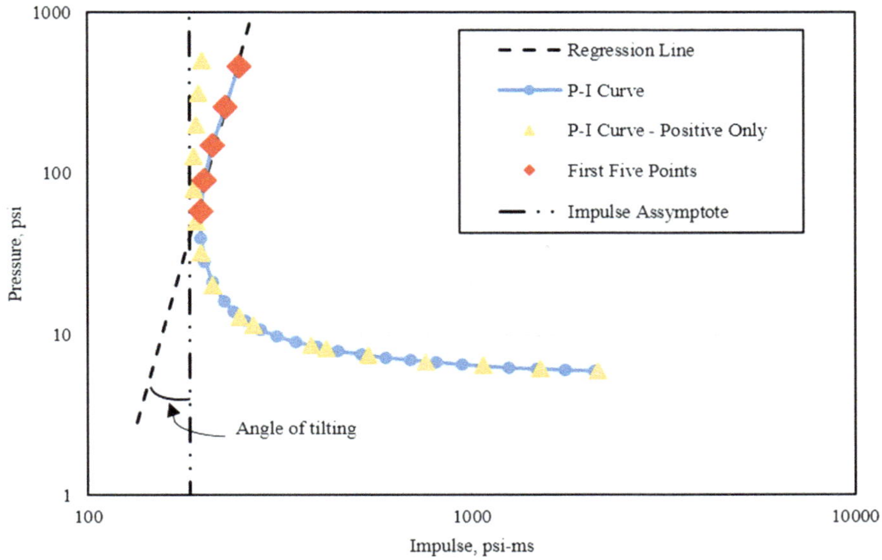

Fig. 4 The P-I curves of full and positive phase blast loadings

(i.e. uneconomical design outputs). The significant angle of tilting, observed in the examined cases, reveals the need of establishing new impulse asymptote formulations. Hence, deriving the equation of the new impulse asymptote for the tilted P-I curve will be the goal of a future study.

$$Tn = \frac{2\pi}{\sqrt{\frac{k}{K_{LM} * M}}} \tag{6}$$

6 Conclusions

The P-I curves with negative phase are assessed for a given range of reinforced concrete panels, simply supported, and medium level of protection (i.e. resulting in 210 P-I curves). The assessment is performed considering the angle of tilting for the regression line of the first five points of the P-I curve in the impulsive region and plotted against the natural period of the examined cases. The results showed the angles of tilting between 3.5 and 17.1°, which indicated that all examined cases exhibited tilting in the P-I curve. The angle of tilting increased with larger Tn which indicates an increase in the margin of safety of the impulsive region compared with the P-I curve of positive phase blast load only (i.e. uneconomical design outputs). The results provided in this research may indicate the necessity for establishing new formulations of the tilted impulse asymptote.

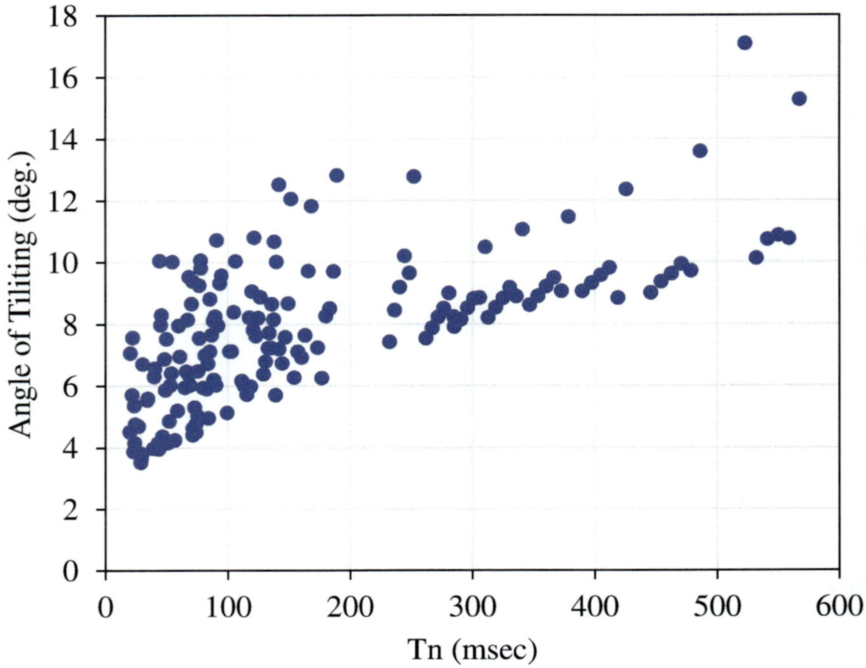

Fig. 5 Angle of tilting of the regression line versus Tn

References

1. Kingery, C.N., Bulmash, G.: Air blast parameters from TNT spherical air burst and hemispherical surface burst. USA, Technical Report (1984)
2. Granström, S.A.: Loading characteristics of air blasts from detonating charges. Stockholm, Technical Report (1956)
3. UFC 3-340-02: Structures to resist the effects of accidental explosions (2005)
4. PDC-TR 06-08: Single degree of freedom structural response limits for anti-terrorism design. U.S. Army Corps of Engineers (2008)
5. Alawad, O.M., Gombeda, M.J., Naito, C.J., Quiel, S.E.: Simplified methodologies for preliminary blast-resistant design of precast concrete wall panels. PCIJ **64**(4) (2019). https://doi.org/10.15554/pcij64.4-03
6. Rigby, S.E., Tyas, A., Bennett, T., Clarke, S.D., Fay, S.D.: The negative phase of the blast load. Int. J. Prot. Struct. **5**(1), 1–19 (2014). https://doi.org/10.1260/2041-4196.5.1.1
7. Krauthammer, T., Altenberg, A.: Negative phase blast effects on glass panels. Int. J. Impact Eng. 17 (2000)
8. Teich, M., Warnstedt, P., Gebbeken, N.: Influence of negative phase loading on cable net facade response. J. Archit. Eng. **18**(4), 276–284 (2012). https://doi.org/10.1061/(ASCE)AE.1943-5568.0000083

9. Syed, Z.I., Mendis, P., Rahman, S.A.: Effect of large negative phase of blast loading on structural response of RC elements. In: MATEC Web of Conferences, vol. 47, p. 02015 (2016). https://doi.org/10.1051/matecconf/20164702015
10. Gantes, C.J., Pnevmatikos, N.G.: Elastic–plastic response spectra for exponential blast loading. Int. J. Impact Eng. **30**(3), 323–343 (2004). https://doi.org/10.1016/S0734-743X(03)00077-0
11. Yang, M., Ahmari, S.: Investigation of effect of negative phase of blast loading on cable net curtain walls through the linearized stiffness matrix method. Int. J. Impact Eng. **61**, 36–47 (2013). https://doi.org/10.1016/j.ijimpeng.2013.06.004

Effect of Steel Fiber Volume Ratio on Bending Moment Transfer Coefficient of SFRC Shield Tunnel Under Staggered Assembling

Shuo Yu, Huajun Sun, Miaofeng Cao, and Changbao Liu

Abstract The transfer coefficient of bending moment is one of the key parameters of shield tunnel design. It is an important index to measure the force transfer property of shield tunnel under staggered assembling. In this paper, the mechanical parameters of segments with different steel fiber volume ratios are determined by a series of standard tests. The stiffness of joint of the SFRC shield tunnel is analyzed by numerical calculation. Based on the results, the model of the three-ring SFRC shield tunnel is established. Finally, the effect of steel fiber volume ratio on bending moment transfer coefficient of SFRC shield tunnel under staggered assembling is investigated. The result shows that the steel fiber volume ratio of segment is inversely proportional to the bending moment transfer coefficient. When the steel fiber ratios increase, the bending moment transfer coefficient at the position of 101.25° and 168.75° reduces greatly. The location of the maximum moment coefficient is not affected by steel fiber ratio increase, while the location of the minimum bending moment coefficient will shift from the roof to bottom of tunnel.

Keywords Steel fiber volume ratio · SFRC shield tunnel · Staggered assembling segment · Bending moment transfer coefficient

1 Introduction

The shield tunnel method has been widely used in urban rail transit construction, at present, most tunnels are assembled by reinforced concrete (RC) segments [1, 2]. As the tensile and crack resistance of concrete is not well, the segment is easy to be damaged and cracked. Once the segment appeared deep cracking, the

S. Yu (✉) · M. Cao · C. Liu
Powerchina Huadong Engineering Corporation Limited, Hangzhou 310000, China
e-mail: yu_s@hdec.com

H. Sun
Zhejiang Provincial Development and Reform Commission, Hangzhou 310000, China

E. Strauss (ed.), *Proceedings of the 7th International Conference on Civil Engineering*, Lecture Notes in Civil Engineering 371, https://doi.org/10.1007/978-981-99-4045-5_29

external groundwater will corrode the internal reinforcement [3, 4], which will lead to structural deterioration and reduce the service life of the shield tunnel.

In order to remedy the above shortcomings of traditional reinforced concrete, the steel fiber reinforced concrete (SFRC) has been proposed for the production of shield tunnel segment. The SFRC is a composite material that mixing short steel fiber in common concrete, these randomly distributed steel fiber can effectively hinder the expansion of micro-cracks and the formation of macro-cracks. It can significantly improve the tensile, flexural, impact, and fatigue resistance. Some researchers have studied the mechanical properties of steel fiber tunnel segments. Meng [5] found that steel fiber reinforcement delayed the initial impact of segment cracking and increased the limit of proportionality. Abbas [6] carried out the full-scale test, experiment shows that the SFRC segments exhibited more stable post-peak cracking behavior compared to that of RC segments. Buratti [7] studied the crack tip opening displacement of SFRC by bending tests, the result shows that the crack-bridging effect of the fiber significantly reduces the crack width opening in the tunnel lining. These papers mainly focused on mechanical properties of the single SFRC segment. There is no research about the mechanical properties of the SFRC shield tunnel.

For the assembled shield tunnel, the joint has a great influence on the mechanical properties of the shield tunnel. Gong [8] used rectangular segment instead of curved segment, compared the mechanical properties of RC and SFRC joints by joint test, the results show that the peak load-bearing capacity of the SFRC joints are higher than that of the RC joints. Avanaki [9] compared the longitudinal seam joint stiffness of SFRC and RC and pointed out that the joint behavior of RC shield tunnel and the joint behavior of SFRC shield tunnel are totally different. Therefore, it is not reasonable to apply the design model of RC shield tunnel to design SFRC shield tunnel.

There are many design models of RC shield tunnel considering joint effect, mainly include uniform rigidity ring model, average uniform rigidity ring model, multi-hinge ring model, beam-spring model. The average uniform rigidity ring model is one of the most popular methods to be used in the shield tunnel design. The model contains one key parameter, which is the transfer coefficient of bending moment ζ, the ζ is an important index to measure the force transfer property of shield tunnel under staggered assembling, and the factor is normally set to be 0.3 in Japanese standards. Koyama [10] provided a formula for calculating the moment transfer coefficient and pointed out that the moment transfer coefficient is closely related to the effective ratio of the bending rigidity. Liu [11] analyzed the inter-ring transfer mechanism by systematic experiments and provided the reference values of bending moment transfer coefficient under different loads. Zhu [12] studied the bending moment transfer capability of special shaped shield tunnels through experiments, the results show that the moment transfer capacity is greatly affected by burial depth. But most studies of force transfer property focus on the RC shield tunnel. Little attention is paid to the SFRC shield tunnel.

In our research, the mechanical parameters of segments with different steel fiber volume ratios are determined by the compressive test, the indirect tensile test, and the tensile flexural test. The stiffness of the longitudinal seam and the circumferential seam joint of the SFRC shield tunnel are analyzed by numerical calculation. Based on the results, the model of the three-ring SFRC shield tunnel is established. Finally, the effect of steel fiber volume ratio on bending moment transfer coefficient of SFRC shield tunnel under staggered assembling is investigated. The reference values of bending moment transfer coefficient with different steel fiber volume ratios are provided.

2 Details of SFRC Shield Tunnel Under Staggered Assembling

Each ring of SFRC shield tunnel consists of six pieces. It includes one key block (K), two adjacent blocks (B1, B2), and three standard blocks (A1, A2, A3). The key block has a center angle of 21.5°, the adjacent blocks have center angles of 67.5°, and the standard blocks have center angles of 68°. The segment rings are assembled according to an "A-B-A" pattern. The two adjacent blocks are connected by bolts, the form of bolt is M30 bending bolt. There are 12 longitudinal seam bolts set for each ring, the circumferential seam bolt is set in the adjacent ring's edge, the total number of circumferential seam bolt is 16. All the circumferential seam joints of the segments contain tenons, but the longitudinal seam joints do not contain tenons. The detail is shown in Figs. 1 and 2.

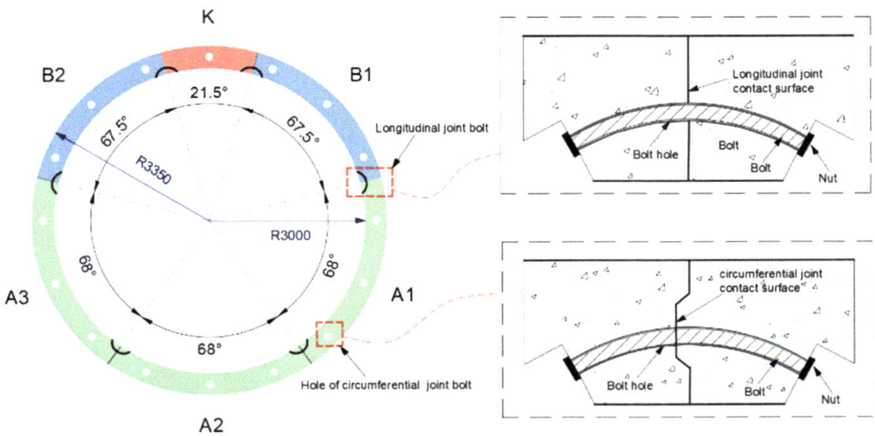

Fig. 1 Joint structure of shield tunnel

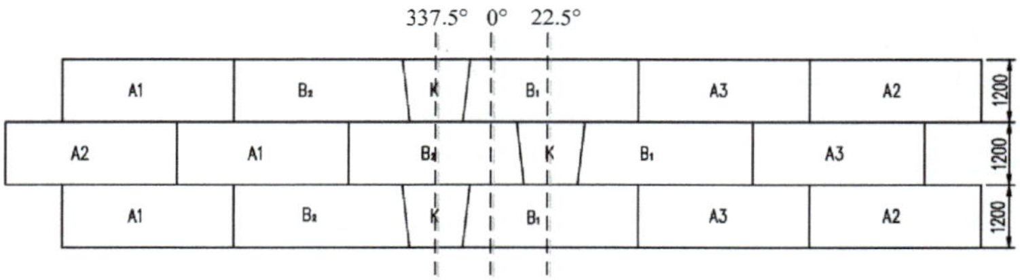

Fig. 2 Staggered assembling form of shield tunnel

3 SFRC Shield Tunnel Calculation Model

3.1 Concrete Parameters of SFRC Shield Tunnel

In order to determine the mechanical parameters of steel fiber-reinforced concrete segments, it is necessary to analyze the constitutive relationship of steel fiber reinforced concrete by a series of standardized experimental tests (3/dosage for each test), all tests according to CECS 13:1989 [13], the dimension of tensile-flexural strength specimens is 150 mm × 150 mm × 550 mm, the dimension of indirect tensile strength specimens is 150 mm × 150 mm × 150 mm. The dimension of compressive strength specimens is 150 mm × 150 mm × 150 mm, as shown in Fig. 3.

The type of steel fiber is hooked shape, the length is 50 mm, the diameter is 0.8 mm, aspect ratio is 62.5, the tensile strength is 1169 MPa, and the elastic modulus is 210 MPa, as shown in Fig. 4. The specimens contain three different steel fiber contents. The relevant parameters of steel fiber reinforced concrete are shown in Table 1. Figure 5 is the stress–strain curve of compressive test. The compressive strength and indirect tensile strength of concrete with different steel fiber ratios are shown in Table 2.

Figure 6 shows the load–deflection curves of tensile flexural test, according to the RILEM TC 162-TDF [14], the residual tensile flexural strengths (f_{RX}) can be

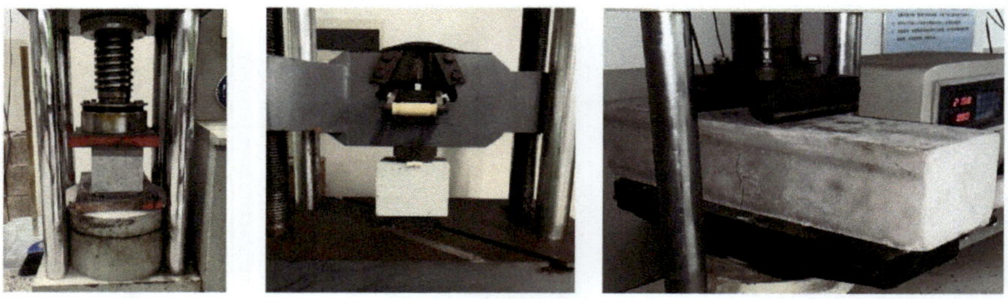

(a) Compressive test (b) Indirect tensile test (c) Tensile flexural test

Fig. 3 Standard test of steel fiber reinforced concrete

Fig. 4 Type and size of steel fiber

Table 1 Composition of the concrete

Steel fiber ratio (%)	Cement (kg/m^3)	Fly ash (kg/m^3)	Water (kg/m^3)	Sand (kg/m^3)	Gravel (kg/m^3)	Water reducing agent (kg/m^3)
0.4	313	78	180	790	1040	9.78
0.6	313	78	180	790	1040	9.78
0.8	313	78	180	790	1040	9.78

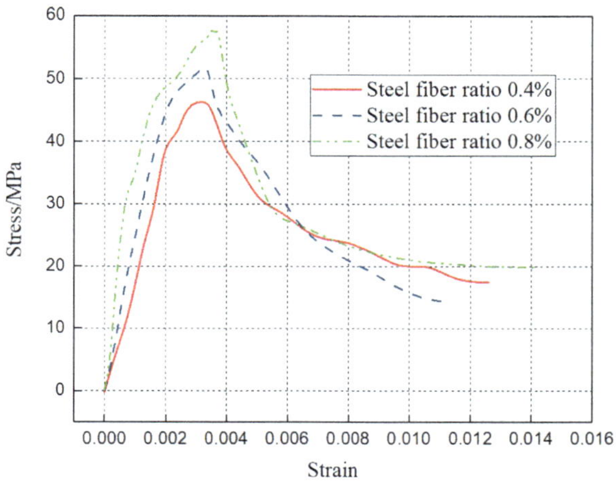

Fig. 5 Stress–strain curve of compressive test

Table 2 Compression and indirect tensile strength of SFRC

Steel fiber ratio (%)	Compression strength f_h (N/mm^2)	Indirect tensile strength f_{ct} (N/mm^2)
0.4	46.2	4.2
0.6	51.3	4.8
0.8	57.6	5.6

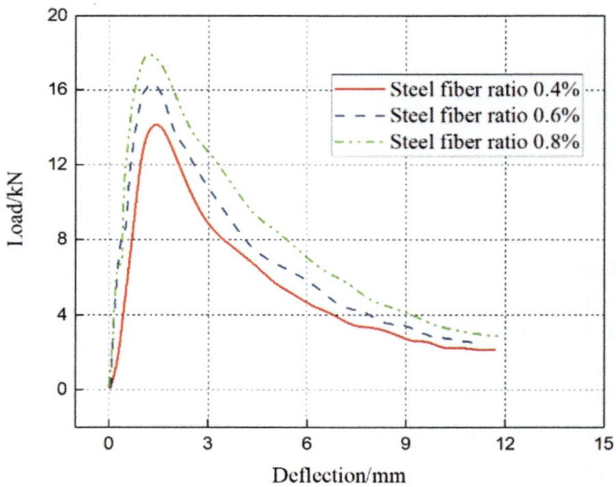

Fig. 6 Load deflection curve of tensile flexural test

calculated, as shown in Table 3, the corresponding crack opening values of f_{R1}, f_{R2}, f_{R3}, f_{R4} are, respectively, 0.5 mm, 1.5 mm, 2.5 mm, 3.5 mm.

In order to simulate the mechanical properties of steel fiber reinforced concrete segments, the damaged plasticity model was used to simulate the segment damage degree, the constitutive relationships of SFRC can be simulated based on the σ-ε diagram according to RILEM TC 162-TDF [14], as shown in Fig. 7 and Table 4.

Table 3 Parameter value of f_{R1}, f_{R2}, f_{R3}, f_{R4}

Steel fiber ratio (%)	f_{R1}(N/mm^2)	f_{R2}(N/mm^2)	f_{R3}(N/mm^2)	f_{R4}(N/mm^2)
0.4	2.8	2.5	2.3	2.1
0.6	3.2	2.9	2.6	2.4
0.8	3.6	3.3	2.9	2.7

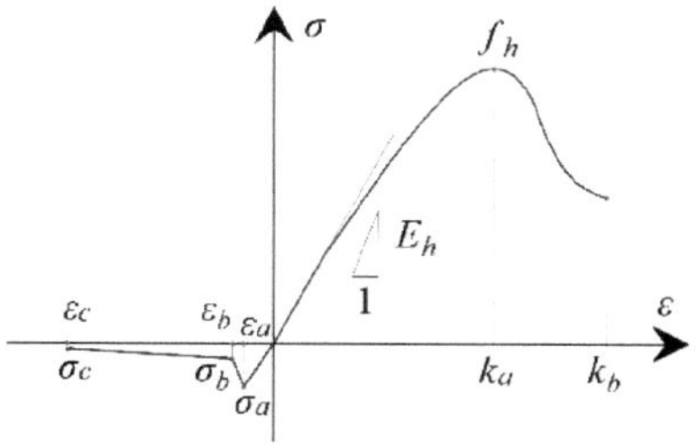

$$\begin{cases}\sigma_a = f_{ct} \\ \sigma_b = 0.45 f_{R1} k_h \\ \sigma_c = 0.37 f_{R4} k_h\end{cases} \qquad \begin{cases}\varepsilon_a = \sigma_a / E_h \\ \varepsilon_b = \varepsilon_a + 0.01\% \\ \varepsilon_c = 2.5\%\end{cases}$$

$$\begin{cases}k_a = -2.0‰ \\ k_b = -3.5‰\end{cases}$$

$$k_h = 1.0 - 0.6\frac{h[\mathrm{cm}] - 12.5}{47.5}\ (12.5 \le h < 60)$$

$$E_h = 9500\sqrt[3]{f_h}$$

Fig. 7 Stress–strain relationship of SFRC

Table 4 Mechanical parameters of SFRC

Steel fiber ratio (%)	σ_a/MPa	σ_b/MPa	σ_c/MPa	ε_a/‰	ε_b/‰	ε_c/‰	E_h/MPa
0.4	4.2	1.22	0.75	0.12	0.22	25	34.04
0.6	4.8	1.40	0.86	0.14	0.24	25	35.23
0.8	5.6	1.57	0.97	0.15	0.25	25	36.69

3.2 *Stiffness of SFRC Tunnel Joints*

Joints are the weak areas of shield tunnel, the bending stiffness and shear stiffness of joints are usually used to express the force transfer and non-linear change characteristics of joint structures after loading. Therefore, the reasonable values of joint stiffness are very important for evaluation performance of multi-ring shield tunnel. In order to analyze the stiffness of joints with different steel fiber ratios, refer to joint test method of Liu [11], different joint numerical models are established by ABAQUS. In the model, Table 4 is used to determine the parameters of the segment concrete, the elastic modulus of bolt is 200 GPa, the yield strength of bolt is 400 MPa, and the tensile strength is 500 MPa. The stress model of joint test is shown in Figs. 8 and 9, the numerical analysis of the joint stress is shown in Figs. 10 and 11.

The two types of joint stiffness curve with different steel fiber ratios are shown in Figs. 12 and 13, it can be seen that the stiffness of joints presents the characteristics of non-linear change. In order to calculate the internal forces of multi-ring tunnels more efficiently and accuracy, the stiffness curve of joints is simplified to three-fold curve, as shown in Figs. 14 and 15. In the model of SFRC shield tunnel, the joints are simulated by rotating spring and shearing spring, the stiffness values at different stages are shown in Table 5.

Fig. 8 Stress model of longitudinal seam joint test

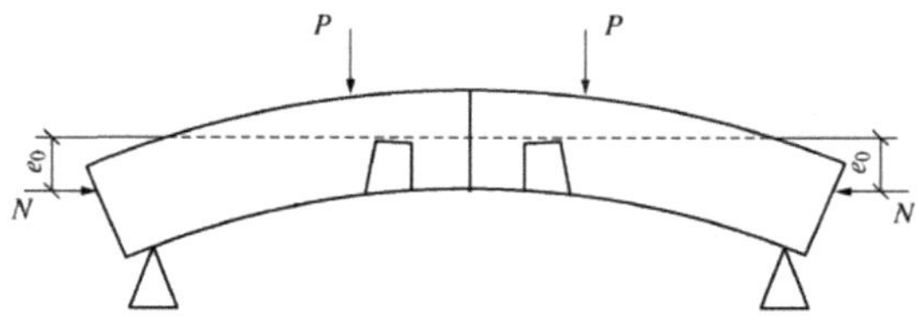

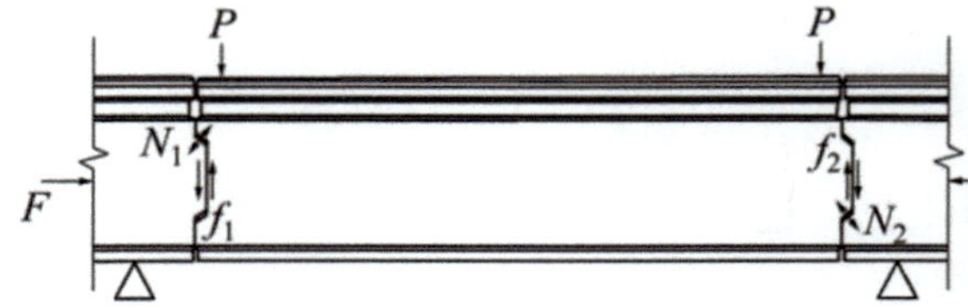

Fig. 9 Stress model of circumferential seam joint test

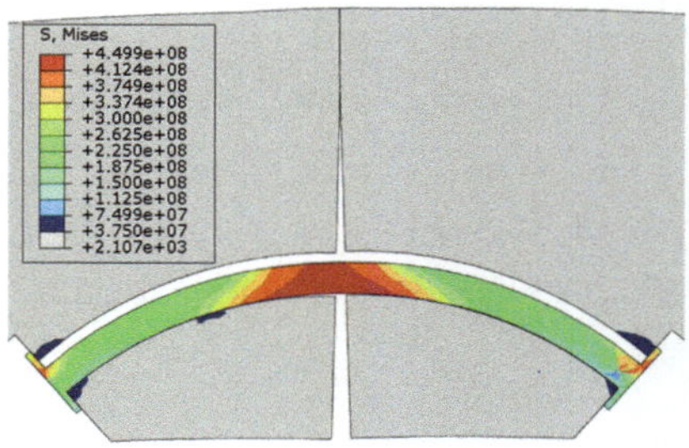

Fig. 10 Analysis of longitudinal seam joint stress

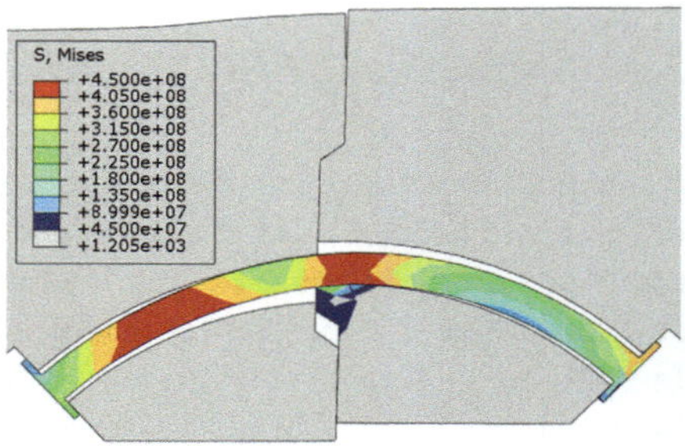

Fig. 11 Analysis of circumferential seam joint stress

$$K_L = \frac{M_L}{\theta_L} \tag{1}$$

$$K_C = \frac{F_C}{S_C} \tag{2}$$

where K_L is the rotational stiffness of the longitudinal seam joint, the M_L is the bending moment, the θ_L is the joint angle. The K_L is the shear stiffness of circumferential seam joint, the F_C is the shear force, S_C is the dislocation.

3.3 Boundary Condition of Calculation Model

Three types of soil layer are selected in the model, as shown in Fig. 16. The thicknesses of three types of soil layers from top to bottom are 15, 6.2, and 15 m. The

Fig. 12 Rotational stiffness of longitudinal seam joint

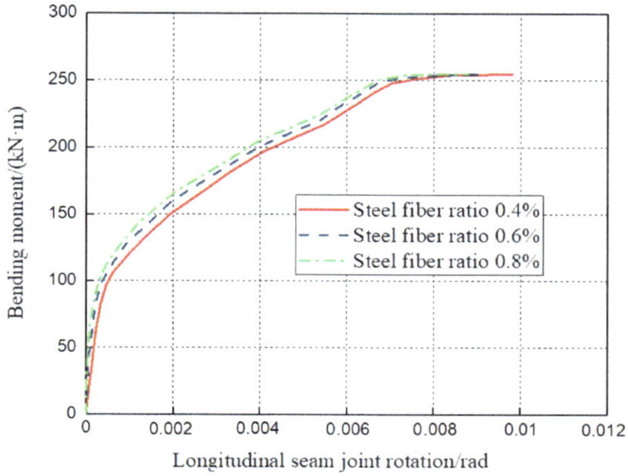

Fig. 13 Shear stiffness of circumferential seam joint

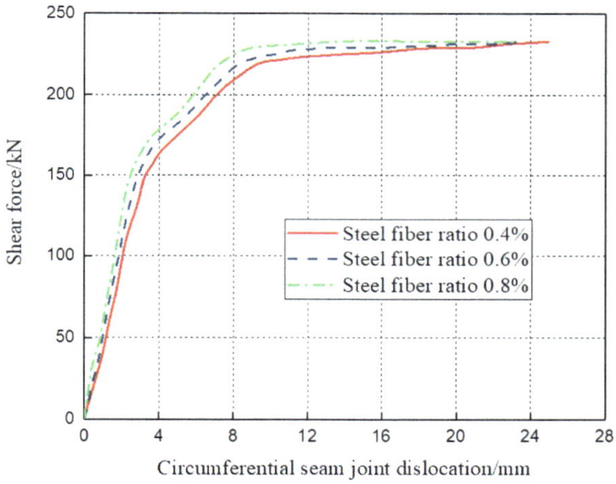

SFRC shield tunnel is located in the second layer of soil, three rings SFRC shield tunnels under staggered assembling are selected for analysis, as shown in Fig. 17. Ignoring ground surface load, three types of soil parameter are shown in Table 6.

4 Result Analysis

In order to study the bending moment transfer coefficient between adjacent rings under staggered assembling, Koyama [10] proposed a formula for calculating moment transfer coefficient, as shown in Fig. 18.

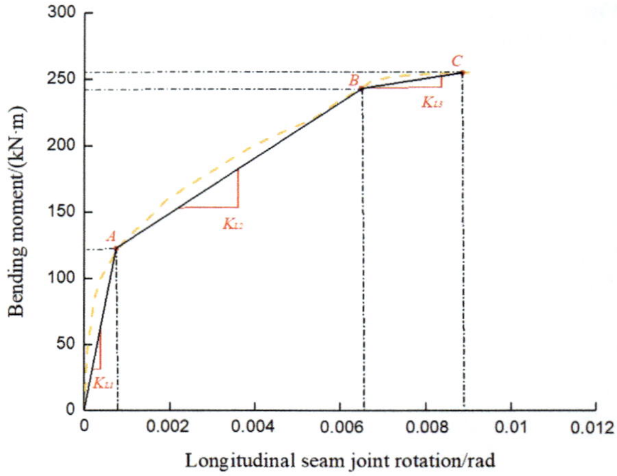

Fig. 14 Triple line rotational stiffness model of joint

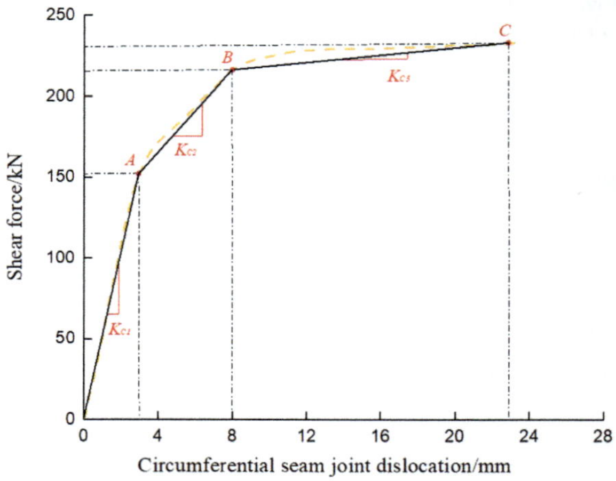

Fig. 15 Triple line shear stiffness model of joint

Table 5 Rotational and shear stiffness value under different steel fiber ratios

Segment steel fiber ratio	K_L (kN·m/rad)			K_C (kN/m)		
	K_{L1}	K_{L2}	K_{L3}	K_{C1}	K_{C2}	K_{C3}
0.4	15,625	20,500	9090	50,620	13,212	1067
0.6	16,142	20,623	8653	51,032	13,434	958
0.8	16,543	20,867	8210	51,531	13,656	921

$$M_O = M_S + M_J \tag{3}$$

$$M_J = (1 - \zeta)\frac{M_O}{2} \tag{4}$$

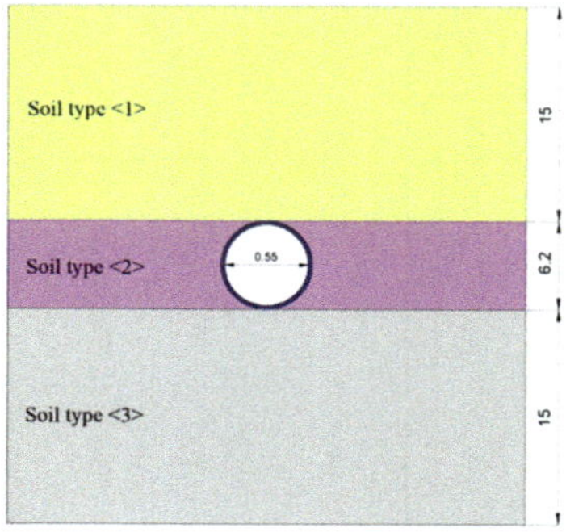

Fig. 16 External condition of shield tunnel model

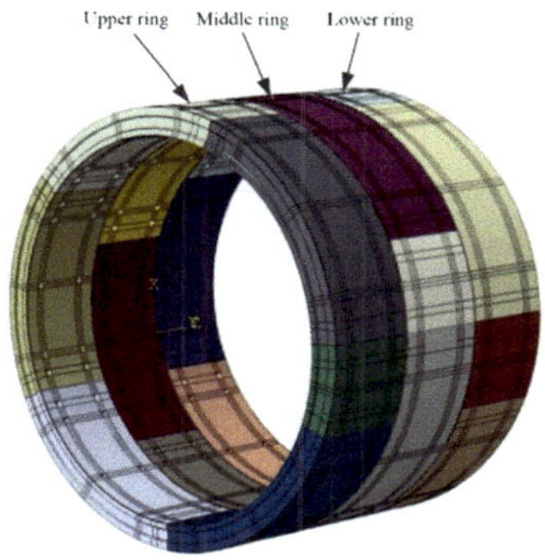

Fig. 17 Three rings' model of SFRC shield tunnels

Table 6 Geotechnical properties of the soil

Soil type	Geotechnical name	Natural density (g/cm^3)	Cohesion (kPa)	Internal friction angle (°)	Compression modulus (MPa)
<1>	Artificial fill layer	1.8~1.9	14~20	20~24	2.6~4.3
<2>	Silt layer	1.6~1.7	15~19	9~13	1.6~3.4
<3>	Rock layer	1.7~2.2	45~60	25~40	5.0~15.0

$$M_S = (1 + \zeta)\frac{M_O}{2} \tag{5}$$

where the M_S is the bending moment in segment section, the M_J is the bending moment at joint, the M_O is the sum of bending moment, the ζ is the moment transfer coefficient.

The moment transfer coefficients at six longitudinal seams' joint section of the middle ring are selected for analysis, as shown in Fig. 19. Figure 20 is the bending moment transfer coefficients with different steel fiber ratios, the sections of #3, #5,

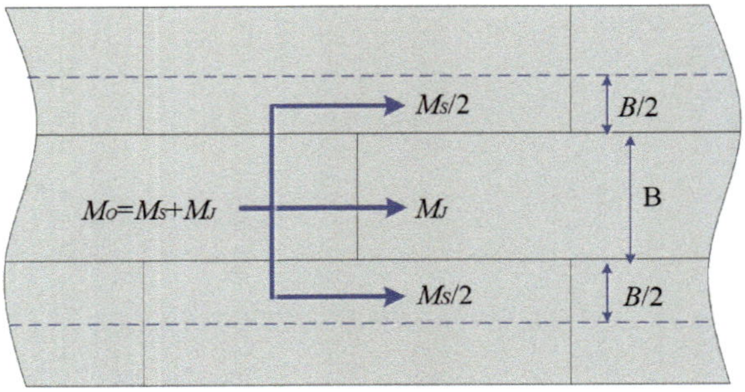

Fig. 18 Bending moment transfer coefficient model of shield tunnel

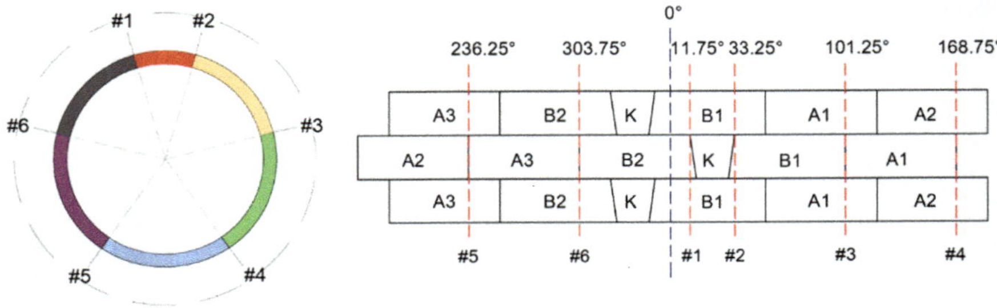

Fig. 19 Schematic map of shield tunnel joint angle

#6 are in the positive moment area, the sections of #1, #2, #4 are in the negative moment area. It can be seen that the steel fiber volume ratio is inversely proportional to the bending moment transfer coefficient. When the ratios of steel fiber increase, the bending moment transfer coefficient at the position of 101.25° and 168.75° reduce greatly.

When the steel fiber ratio is 0.4%, the minimum bending moment coefficient is located at 11.75°, the maximum bending moment coefficient is located at 303.75°. With the increase of steel fiber ratio, the angle of maximum bending moment coefficient remains changed, while the angle of minimum bending moment coefficient shift from to 11.75° to 168.75°.This shows that the location of the maximum moment coefficient is not affected by steel fiber ratio, while the location of the minimum bending moment coefficient will shift from the roof to bottom of tunnel.

Figure 21 is the bending moment transfer coefficients in different bending moment areas, it shows that the bending moment transfer coefficient in the positive bending moment area is larger than that in the negative bending moment area. When the steel fiber ratio is 0.4~0.8%, in the positive bending moment area, the bending moment transfer coefficient is 0.11~0.19, in the negative bending moment area, the bending moment transfer coefficient is 0.05~0.12.

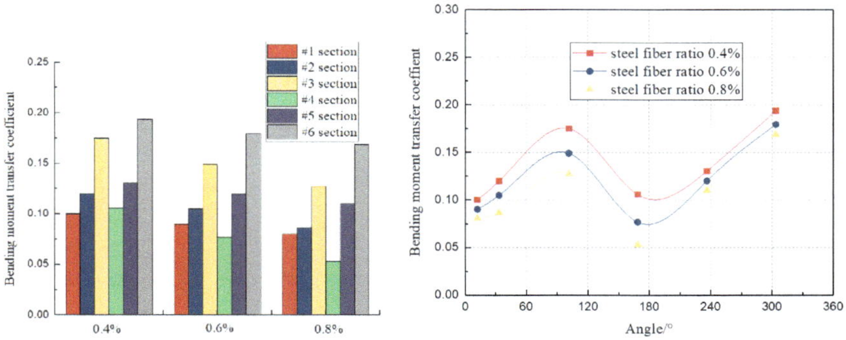

(a) In different sections (b) At different angles

Fig. 20 Bending moment transfer coefficient with different steel fiber ratios

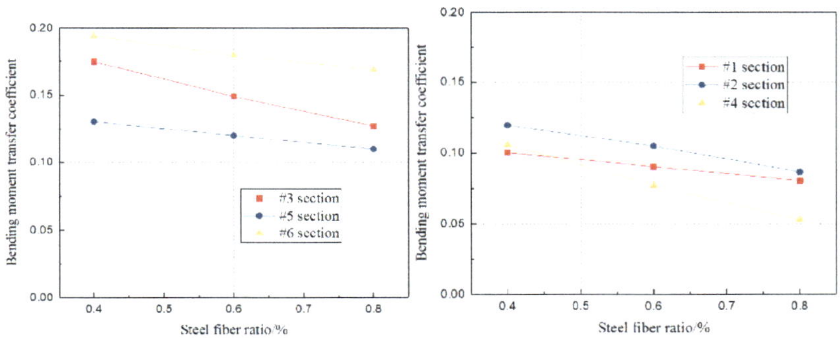

(a) In positive bending moment area (b) In negative bending moment area

Fig. 21 Bending moment transfer coefficient in different bending moment areas

5 Conclusion

In this paper, the effect of steel fiber volume ratio on bending moment transfer coefficient of SFRC shield tunnel under staggered assembling is investigated, and the reference values of bending moment transfer coefficient with different steel fiber volume ratios are provided. The result mainly includes:

(1) The steel fiber volume ratio of segment is inversely proportional to the bending moment transfer coefficient. When the ratios of steel fiber increase, the bending moment transfer coefficient at the position of 101.25° and 168.75° reduces greatly.
(2) With the increase of steel fiber ratio, the location of the maximum moment coefficient is not affected by steel fiber ratio, while the location of the minimum bending moment coefficient will shift from the roof to bottom of tunnel.

(3) When the steel fiber volume ratio is 0.4~0.8%, in the positive bending moment area, the bending moment transfer coefficient is 0.11~0.19, in the negative bending moment area, the bending moment transfer coefficient is 0.05~0.12.

References

1. Jin, H., Tian, Q.R., Li, Z., et al.: Ability of vibration control using rubberized concrete for tunnel invert-filling. Constr. Build. Mater. **317**, 125932 (2022)
2. Jin, H., Yu, S., Zhou, S.H., Xiao, J.H.: Research on mechanics of longitudinal joint in shield tunnel by the nonlinear spring equivalent method. KSCE J. Civ. Eng. **23**(2), 902–913 (2019)
3. Jin, H., Yu, S.: Study on corrosion-induced cracks for the concrete with transverse cracks using an improved CDM-XFEM. Constr. Build. Mater. **318**, 126173 (2022)
4. Jin, H., Tian, Q.R., Li, Z.: Crack development of rebar rust in rubberized concrete using mesoscale model. Constr. Build. Mater. **321**, 126409 (2022)
5. Meng, G., Gao, B., Zhou, J., et al.: Experimental investigation of the mechanical behavior of the steel fiber reinforced concrete tunnel segment. Constr. Build. Mater. **126**, 98–107 (2016)
6. Abbas, S., Soliman, A.M., Nehdi, M.L.: Mechanical performance of RC and SFRC precast tunnel lining segments: a case study. ACI Mater. J. **111**(5), 501–505 (2014)
7. Buratti, N., Ferracuti, B., Savoia, M.: Concrete crack reduction in tunnel linings by steel fibre-reinforced concretes. Constr. Build. Mater. **44**, 249–259 (2013)
8. Gong, C., Ding, W., Mosalam, K.M., et al.: Comparison of the structural behavior of reinforced concrete and steel fiber reinforced concrete tunnel segmental joints. Tunn. Undergr. Space Technol. **68**, 38–57 (2017)
9. Avanaki, M., Hoseini, A., Vahdani, S., et al.: Numerical-aided design of fiber reinforced concrete tunnel segment joints subjected to seismic loads. Constr. Build. Mater. **170**, 40–54 (2018)
10. Koyama, Y.: Present status and technology of shield tunneling method in Japan. Tunn. Undergr. Space Technol. **18**(2), 145–159 (2003)
11. Liu, X., Dong, Z., Song, W., et al.: Investigation of the structural effect induced by stagger joints in segmental tunnel linings: Direct insight from mechanical behaviors of longitudinal and circumferential joints. Tunn. Undergr. Space Technol. **71**, 271–291 (2018)
12. Zhu, Y.: Prototype loading test on bending moment transferring coefficients of special shield tunnels with large cross-section. Tunnel Constr. **37**(10), 1269–1275 (2017). (in Chinese)
13. CECS 13:89: Test methods used for steel fiber reinforced concrete (1989)
14. Colombo, M.D.P., Rilem, T.C.: 162–TDF σ-ε design method-final recommendation. Mater. Struct. **36**(8), 560–570 (2003)

Building Information Technology, Road Condition Monitoring and Construction Management

An Assessment of Road Condition Monitoring Practice and Technologies in the Philippines

John Paul T. Dacanay, Lestelle V. Torio-Kaimo, and Lea B. Bronuela-Ambrocio

Abstract Road construction has been one of the main focuses of the Philippine government in the last 6 years under the Build, Build, Build Program. In addition to the existing road network, the program constructed more than 29,000 km of roads. This continued construction of new roads warrants a road monitoring system capable of keeping up with the growing inventory of road networks. Recently, advancements in the road monitoring system have been prevalent in developing countries, and many developments are geared toward automating road condition monitoring. The tools used in road condition monitoring include sensors, vehicles' onboard devices, and audio and video streams. This paper presents the existing practices and technologies used in road surveying and monitoring in the Philippines. A comprehensive review of existing road monitoring systems used in other countries was consolidated and compared to assess their applicability and capacity to monitor Philippine Road conditions. The current local road condition monitoring system is also discussed, which is necessary for the future development and enhancements of maintaining a sound system that regularly monitors these pavements.

Keywords Pavement management system · Road defects · Philippine roads · Road survey · Road monitoring

J. P. T. Dacanay (✉)
Science Research Specialist II, University of the Philippines National Center for Transportation Studies, U.P. Campus, 1101 Diliman, Quezon City, Philippines
e-mail: jptdacanay@gmail.com

L. V. Torio-Kaimo · L. B. Bronuela-Ambrocio
Assistant Professor, Institute of Civil Engineering, University of the Philippines-Diliman, U.P. Campus, 1101 Diliman, Quezon City, Philippines

E. Strauss (ed.), *Proceedings of the 7th International Conference on Civil Engineering*, Lecture Notes in Civil Engineering 371, https://doi.org/10.1007/978-981-99-4045-5_30

1 Introduction

The Philippine National Road Network has a total road length of 33,119 km. The network is composed of 10,875 km (32.84%) of asphalt pavement, 21,651 km (65.37%) of concrete pavement, and 592 km (1.79%) of the unpaved road, which connect the cities and municipalities of the Philippine archipelago.

The network is managed by the country's Department of Public Works and Highways (DPWH). The department is responsible for programming the budget for the maintenance of existing national roads and the construction of new national roads while maintaining and repairing existing national road networks. The rest of the local roads are maintained by either local government units or private groups.

The Philippine National Road Network is classified into three classes: National Primary Roads, National Secondary Roads, and National Tertiary Roads. The National Primary Roads are roads with "a contiguous length of significant road sections extending linearly without any breaks or forks that connect major cities (at least around 100,000 population) comprising the main trunk line or the backbone of the national road system" [1]. On the other hand, a National Secondary Road directly connects cities, major ports and ferry terminals, major airports, tourist service centers, and major National Government Infrastructure to National Primary Roads. It connects provincial capitals within the same region. Other roads under DPWH that perform a local function are considered under National Tertiary Roads.

In recent years, the previous administration focused on its Build, Build, Build Program as the administration's infrastructure legacy resulting in the construction of several infrastructure projects nationwide. Based on July 2021 data, it was reported that 29,264 km of roads had been constructed, widened, and rehabilitated. This program includes bridges, farm-to-market roads, farm-to-mill roads, missing links, bypasses or diversion roads, airports, seaports, roads to economic zones, and declared tourism destinations [2]. These completed projects were promised to boost the economic growth of the country. However, the costs for these new roads continue after the completion of the project. In order to keep the road network in good service and performance, it will require continuous maintenance and monitoring.

The continued construction of new roads warrants a road monitoring system capable of keeping up with the growing inventory of road networks. This paper focuses on the current Philippine Road condition monitoring system and the enhancements it needs to optimize and maintain regular monitoring on these pavements.

This research aims to compare and assess the current practices and systems used in the Philippines for road monitoring by benchmarking them with the methods and equipment used in other countries. This is achieved by identifying road survey solutions used in developed countries, comparing them to the local system that is currently being implemented, and evaluating and recommending potential technologies that could enhance the local road monitoring system of the Philippines.

This study aims to achieve three objectives:

1. Identify automated road survey solutions in foreign settings and cite their advantages and disadvantages.
2. Characterize the local road survey system currently being implemented.
3. Compare the identified local and foreign systems and find potential technologies to enhance the local road monitoring system.

Consequently, the discussion of this paper is only limited to tools and equipment currently used in road condition monitoring. The programs and software used in the monitoring will not be covered.

Road networks transport millions of passengers daily, and their performance and safety are the primary concern of the local government. With this, constant monitoring and maintenance must be provided to keep the road networks at their optimum performance. Traditionally, visual survey examination of roads is performed by experienced operators by walking down the line and adopting necessary contact surveying tools to tally and measure specific characteristics such as depth, width, and length. This method takes time and resources and considers the operators' safety. Currently, the Bureau of Maintenance, a division under DPWH, monitors and maintains the Philippine national road. This division is a team of 16 engineers and personnel that surveys, monitors, and maintains national roads in the Philippines. Such a limited workforce will significantly benefit the enhancement of their road monitoring system.

2 Pavement Monitoring

Generally, pavement condition monitoring is part of a pavement management system. It is carried out to completely ascertain the road condition by taking its structural and functional evaluation to provide helpful information and data for analysis to arrive at consistent, cost-effective, and justifiable decisions for preserving a pavement network. The functional condition of the pavement mainly considers the characteristics of a highway section's riding quality or safety. In contrast, the structural condition is determined by deflection, layer thickness, and material characteristics. Several methods used to determine the functional and structural condition of a pavement include visual survey, IRI, non-destructive test, and destructive test. The paper focuses on visual condition surveys because they can consider both the functional and structural conditions of a pavement. It is also the most common method used for routine evaluations because it is fast, economical, and conventional.

Visual condition surveys are a qualitative measure of the overall condition of the road. Traditionally, an inspector walks along the road to note the defects on the pavement. Ideally, two or more evaluators would arrive at the same assessment of the section's current condition for any given section of the highway. Still, many aspects of pavement evaluation are highly subjective. Nowadays, many advanced tools and equipment have been developed to address the challenges of visual condition surveying, which helps decrease the subjectivity of the method and standardize the pavement condition assessment.

Current technologies employed in most of the world for visual condition assessment are capture systems such as cameras and lasers that collect the surface features of the pavement. This automation of the survey process itself reduces the time spent by expert personnel on the field. In effect, the sensory systems serve as a visual aid to the expert personnel as the data are inspected in an office setting.

Developments to produce a more efficient survey and assessment process continue. Some developments in assessing the pavement condition incorporate the use of accelerometers. Staniek [3], Souza [4], and Lima [5] utilize the fine variations in acceleration detected as an estimate of the overall roughness of the pavement section. The degree and frequency by which the acceleration varies shall determine the pavement condition index of the pavement section. The locations of points along the pavement are also tracked through GPS and plotted on a map along with the estimated condition index. Singh [6] uses accelerometers to detect pavement distresses such as cracks and potholes and plots them across an online mapping service such as Google Maps. The accuracy of detection of distresses is reported to be 88%.

Another method uses a combination of lasers and cameras [7]. The cameras capture the horizontal texture of the road, while the lasers capture the elevation data of the road. These spatial data are then combined through software into a three-dimensional map of the road section. This map is still subject to the expert's assessment of the road.

A large volume of research found focuses on the use of computer vision. This approach enables the computer to simulate vision by interpreting photographic data and attempts to contextualize the data by identifying pavement distresses such as potholes, cracks, and scaling. This approach improves many traditional processes in interpreting the captured field data by effectively eliminating the time consumed by experts examining and assessing the data and the subjectivity associated with human perception. This approach can make the assessment process quicker and more standardized, as computers are independent of human biases.

2.1 Automated Visual Survey Equipment in Other Countries

Australia. The Australian Road Research Board (ARRB) currently has 10 survey vehicles for pavement surface inspection. These Intelligent Safe Surface Assessment Vehicles (ISSAVes) are deployed on roads to collect pavement data. It has installed cameras for imaging the pavement, sensors for measuring skid resistance, and sensors for detecting pavement deflections. The vehicle collects data at an operating speed of 50 kph [8]. The processed data will then report any detected cracks and map them along the road section. The crack detection process is automated. An overall condition assessment can also be produced.

United Kingdom. The Transport Research Laboratory conducts visual condition surveys for the country's roads. The survey may either be conducted via a walk-through by expert personnel (which is generally avoided) or via a survey vehicle equipped with a forward-facing camera capable of capturing images every 5 m at normal traffic speeds; a downward-facing camera that captures a single line of pixels with a length of 4096; and, a global positioning system accurate to ± 50 cm. The collected data are assessed by expert personnel for any presence of defects and an estimate of its overall condition [9].

South Korea. South Korea's Ministry of Land, Infrastructure, and Transportation (MOLIT) employs an ensemble of vehicles to monitor their roads. They have four vehicle types that collect data on the pavement structure (ground penetrating radar), deflection (falling weight deflectometer), skid resistance (pavement friction tester), and surface features. For surface features, the ministry uses the Automatic Road Analyzer (ARAn) vehicle equipped with six video cameras and sensors for profiling the road for rutting [10]. The collected data are still interpreted by experts for the presence of distress and for assessing the overall condition of the road section. The cameras installed in the vehicles can resolve up to 1/32 of an inch.

United States of America. In the United States, an ensemble of vehicles captures visual and spatial surface data for pavements. Like South Korea, their Department of Transportation uses ARAn vehicles to collect pavement surface data.

Minnesota's Department of Transportation (MnDOT) uses a Digital Inspection Vehicle (DIV) for its pavement surveys. This vehicle is equipped with front and rear lasers for profile and rut measurements, respectively, and five video cameras (2 at the front and rear each, one at the top, to capture video footage of the pavement. The DIV can travel up to 55 mph during surveys [8].

The data collected by both survey vehicles are consolidated and interpreted by field experts to detect any distresses and assess the overall condition of the pavement.

India. India's Central Road Research Institute implemented an automated road survey system of survey vehicles to collect pavement surface data for assessment. The cars are equipped with rear-mounted video cameras for capturing the texture and distress on the pavement, laser sensors for road profiling and measurement of rutting, and GPS for spatial referencing of points of interest. The camera system can capture a pavement area of 3.5 m wide by 1.0 m long. The data collected by the survey vehicles are still consolidated and interpreted by field experts to detect any distresses and assess the overall condition of the pavement.

New Zealand. The New Zealand Transport Agency surveys the condition of its highways through its specialist vehicle, Sideway-force Coefficient Routine Investigation Machine or SCRIM+. The SCRIM+ survey vehicle collects information regarding the skid resistance, texture, roughness, rutting, geometry, and centerline GPS coordinates of the roads. This information is used to monitor highway performance, plan future work programs, analyze trends, and predict changes in road conditions. This survey vehicle is equipped with numerous lasers, accelerometers, inclinometers, gyroscopes, GPS, and tilt sensors to help gather data in evaluating the condition of the road [11].

2.2 Local Pavement Monitoring

DPWH has the Road and Bridge Information Application (RBIA) that has monthly data on the Philippine National Road Network and the Bangsamoro Autonomous Region in Muslim Mindanao (BARMM) road network. The RBIA consolidates road quality, road sections, and corresponding maintenance offices. DPWH has established District Engineering Offices (DEOs) to monitor the National Road Network. DPWH has two groups that are most involved in the road monitoring process.

The DPWH Planning Service (DPWH-PS) is the division responsible for programming road maintenance budgets for each District Engineering Office. It monitors the condition of roads by outsourcing them through a third-party company. This outsourced road survey company is equipped with LIDAR, a laser profiler, and cameras that automatically detect pavement defects that DPWH-PS monitors for their pavement condition metric: the Visual Condition Index (VCI). The VCI is a metric ranging from 0 to 100 that is calculated from the various weights multiplied by a specific unit of measurement designated to pavement defect. A road section free of defects and fully serviceable has a VCI of 100. That value decreases over time as the road is being used by ongoing traffic.

3 Current System in Inspecting Philippine National Road

Philippine roads can be classified either under national roads or local roads. Local roads are subject to the jurisdiction of the local government units, while national roads are under the jurisdiction of the Department of Public Works and Highways.

The national roads are maintained by District Engineering Offices (DEOs) under DPWH. Each DEO manages a certain number of road sections as mandated under their district. Each province may have up to three DEOs, and some cities in metropolitan areas (such as Quezon City in the National Capital Region) can have two DEOs. The total road length under their jurisdiction can range from 25 to 450 km, with an average of 175 km. When a DEO's total road length covered reaches high enough, a department order can be released to split off the DEO into two offices to make the maintenance duties more manageable.

District engineering offices have field inspectors deployed to conduct road condition assessment surveys. Field inspectors can either be from the maintenance or planning divisions. Maintenance division personnel are responsible for identifying defects on the road for rectification by the DEO. Planning division personnel (regarding road inspections) are responsible for validating inspection reports from the maintenance division to determine its qualification for rehabilitation.

Maintenance field inspector teams can range from 10 to 30 personnel depending on the size of the DEO's district. Each field inspector can have a varying number of road sections under their watch. These field inspectors travel to their assigned road sections via a survey vehicle. Once at the road section, field inspectors may traverse the road section on foot or via a vehicle. With a vehicle, they can reach speeds of up to 60 kph to spot any defects. These defects may be on the pavement or the roadside, as maintenance groups are responsible for maintaining the whole road right of way. The defects to be identified are listed in DPWH Department Order (DO) 189 Series of 2022, which updates the former Department Order 41 Series of 2016. The latter department order updates the original DO with one more defect for DEO Maintenance Teams to spot. The defects are (1) potholes, (2) alligator cracks, (3) major scaling, (4) shoving and corrugation, (5) pumping and depression, (6) no/faded road markings, (7) low/inverted shoulders, (8) lush vegetation, (9) clogged drains, (10) open manholes, (11) no/inadequate sealant, (12) cracks, (13) raveling, (14) unmaintained road signages, (15) unmaintained bridges, and (16) unmaintained guardrails.

The maintenance field inspectors encode these 16 defects onto a report that is sent to the DPWH Bureau of Maintenance for review and perusal. These defects are also estimated according to the specified unit of measure by the DO. Measurements can be in number, linear meters, or square meters. These measurements are estimated based on the expert's judgment. It takes 2 to 3 days for each DEO to produce one report of the whole district to relay its current condition to DPWH-BOM. The maintenance divisions conduct at least one to two surveys each week.

Similarly, planning field inspector teams also conduct their own surveys. Their surveys are more precise as they are the group that endorses road sections that require rehabilitation. Roads for rehabilitation require time and approvals for budget allocations for the districts that need it. Per Department Order 120 Series of 2019, the planning field inspectors only evaluate the defects on the carriageway. Table 1 enumerates the defects that this group spots depending on the nature and section of the pavement. Another defect that was not included in the list is vegetation control.

Like maintenance field inspectors, planning field inspectors are deployed to assess the road conditions assigned to them. The difference they have with the maintenance field inspectors, apart from the defects they identify and their limitation to the carriageway only, is that their inspection process is done exclusively on foot. Planning field inspectors precisely measure defects every 100 m and according to the units specified in the department order. Their inspection process of the whole district lasts from 2 to 3 months. Their average pace per day provided an entire team of field inspectors is deployed, reaches 8 to 10 km. Planning division inspection reports are submitted to the DPWH Planning Service group.

In addition, planning division personnel also assess reports of road sections endorsed by the maintenance personnel for rehabilitation. The planning personnel checks if the criteria are met for rehabilitation. They may also conduct an onsite survey to validate those findings.

Surveys are also conducted at the management level to audit the assessments performed by the DEOs. It must be noted that the maintenance assessment system also corresponds to a performance rating system of the DEO, whose rating corresponds to a specific incentive [12]. DPWH-BOM also conducts their own survey to audit the reports of the DEO Maintenance personnel, which is conducted once every two months. The DEO maintenance personnel usually assists surveys with

Table 1 List of defects surveyed by the planning field inspectors

Flexible pavement	Rigid pavement	Unsealed roads	Drainage	Shoulders
Patches	Joint sealant distress	Gravel thickness	Side drain	Unsealed shoulders
Potholes	Faulting at transverse joints	Material quality		Sealed, asphalt surfaced, and concrete shoulders
Surface failures	Spalling at joints	Crown shape		
Road slip/cut	Pavement cracking	Roadside drainage		
Pavement cracking	Shattered slabs	Road slip/cut		
Pavement rutting	Road slip/cut			
Wearing surface (raveling/flushing/polishing)	Wearing surface			
Edge break (horizontal)				

BOM as they traverse the whole district. These surveys usually last for a week. DPWH Regional Offices conduct audit field surveys in the districts they manage. Their surveys typically last a week and are conducted once every quarter of the year. Maintenance field inspectors also assist them as they traverse the districts.

4 Challenges of Road Condition Monitoring in the Philippines

Estimating a pavement's overall condition in the Philippines has primarily depended on the assessment of experts and seasoned personnel. In manual surveying, personnel is deployed on a specified road section to look for various distresses on the road and measure dimensions and apparent severity based on their expert assessment. This process continues until the road section has been completely inspected. Alternatively, DPWH employs cameras to capture the image and video footage of the road sections. The footage is then interpreted and checked by experts for any presence of pavement distress. The amount and severity of these detected pavement distresses will contribute to the final assessment of the condition of the road section. This semi-automated process takes days, given the volume of the inputs taken, and is prone to human error caused by subjective assessments.

Developing a system that can detect road defects similar to those used in developed countries would significantly improve the quality and operation of road monitoring and maintenance in the Philippines. Given the scarcity of an experienced workforce in the planning and maintenance division and the length of the road network to be maintained, the information collected during surveying would significantly impact the entire road management system in the country. However, the cost would play a substantial factor in procuring such equipment since the budget is limited, and several units may need to be acquired, given the geographic constraints. The main goal is to optimize the automation of road condition monitoring by building a system of low-cost sensing solutions that will be sufficient to acquire the necessary information in road surveying.

5 Conclusion and Recommendations

The increasing road infrastructure in the Philippines will need an efficient and reliable pavement management system to protect them upon completion. The data gathered in the field are used to assist decision-makers in developing the most appropriate intervention for existing roads. Thus, it is crucial to create a pavement condition monitoring system that is fast, objective, and consistent.

This research paper investigated the current practices and tools used in road condition monitoring in the Philippines. Recently, many studies have established the use of new monitoring tools and techniques in evaluating road pavement conditions. Latest innovations have shown that combining different sensing technologies can improve the accuracy and quality of collected pavement condition data. Upon reviewing the systems used by transport agencies in other countries, it is evident that the existing Philippine road condition monitoring system needs enhancements to maintain regular and more standard surveys on these pavements. It is also vital that the current system be reviewed and updated to keep pace with the current trends and competence in pavement management systems.

Acknowledgements This work is under the Project PAVE—Prototype Automated Visual Survey Equipment, supported by the Department of Science and Technology, and monitored by the Philippine Council for Industry Energy and Emerging Technology Research and Development. The authors would like to acknowledge the collaborating agency, the Department of Public Works and Highways, and their field experts for making this research possible.

References

1. Department of Public Works and Highways: Department Order No. 41 Series of 2016: amended policy guidelines on the maintenance of national roads and bridges (2016)
2. Patinio, F.: Build, build, build: paving the way to progress. https://www.pna.gov.ph/articles/1147990. Accessed 16 Sept 2021
3. Ragnoli, A., De Blasiis, M.R., Di Benedetto, A.: Pavement distress detection methods: a review. Infrastructures. **3**(4), 58 (2018). https://doi.org/10.3390/infrastructures3040058

4. Souza, V.M., Giusti, R., Batista, A.J.: Asfault: A low-cost system to evaluate pavement conditions in real-time using smartphones and machine learning. Pervasive Mob. Comput. **51**, 121–137 (2018). https://doi.org/10.1016/j.pmcj.2018.10.008
5. Lima, L.C., Amorim, V.J.P., Pereira, I.M., Ribeiro, F.N., Oliveira, R.A.R.: Using crowdsourcing techniques and mobile devices for asphaltic pavement quality recognition. In: 2016 VI Brazilian Symposium on Computing Systems Engineering (SBESC) (2016). https://doi.org/10.1109/sbesc.2016.029
6. Singh, G., Bansal, D., Sofat, S., Aggarwal, N.: Smart patrolling: an efficient road surface monitoring using smartphone sensors and crowdsourcing. Pervasive Mob. Comput. **40**, 71–88 (2017). https://doi.org/10.1016/j.pmcj.2017.06.002
7. Wang, K.: Automated survey of pavement distress based on 2D and 3D laser images MBTC DOT 3023 principal investigator (2011)
8. Australian Road Research Board: Data Collection|Australian Road Research Board. https://www.arrb.com.au/data-collection-services (2021)
9. Gavilán, M., Balcones, D., Marcos, O., Llorca, D.F., Sotelo, M.A., Parra, I., Ocaña, M., Aliseda, P., Yarza, P., Amírola, A.: Adaptive road crack detection system by pavement classification. Sensors **11**(10), 9628–9657 (2011). https://doi.org/10.3390/s111009628
10. Varistha Ltd.: TRL|Visual Condition Surveys (VCS). TRL|The Future of Transport. https://trl.co.uk/solutions/infrastructure-asset-management/surveys---investigations/visual-condition-surveys–vcs (2020)
11. High speed data collection programmes. https://www.nzta.govt.nz/roads-and-rail/road-composition/pavement-condition-surveys/high-speed-data-collection-programmes/. Accessed 11 Feb 2022
12. Comprehensive Policy Guidelines on the Maintenance of National Roads and Bridges. https://www.dpwh.gov.ph/dpwh/sites/default/files/filefield_paths/do_189_s2022.pdf. Accessed 11 Feb 2022

BIM Cost Calculator: Contract Costing of Building Information Modeling Services Using Parametric Estimates for BIM-Based Projects in the Philippines

Nel Ann Beloso and Dante Silva

Abstract This study aims to develop a web-based tool for cost estimation that will simplify the process of estimating contract price for design projects utilizing BIM technologies. It becomes a highly complex matter when financial issues influencing contracts are at stake. This research will concentrate on the BIM aspects of the services, scope, and procurement that are involved in a project and how they affect contract price. This initiative also aims to provide a framework that can be used as a basis in the Philippines for contract pricing of BIM-related services and will focus on BIM design and pre-construction or tender stage costs. A quantitative survey was used to gather data from BIM professionals of local and international companies based in the Philippines which involved in projects that utilized BIM into construction. The findings of this study have concluded the key parameters that are significant contributors to the overall cost estimation and determined the average project duration and manpower for each project type for different Level of Development of BIM services. Considering high level of accuracy and appropriateness, the researchers discovered that parametric cost estimating is commonly used. These parameters were then utilized in the development of web-based costing tool. Similarly, the BIM Cost Calculator produces consistent findings, and the underlying indicators reveal that it is effective. It was suggested that the construction industry in the Philippines may consider the use of the developed framework and the BIM Cost Calculator tool to support for their costing need.

Keywords BIM contract costing · BIM cost calculator tool · BIM services cost

N. A. Beloso (✉) · D. Silva
School of Civil, Environmental, and Geological Engineering Mapúa University, Manila, Philippines
e-mail: belosonelann@gmail.com

D. Silva
e-mail: DLSilva@mapua.edu.ph

E. Strauss (ed.), *Proceedings of the 7th International Conference on Civil Engineering*, Lecture Notes in Civil Engineering 371, https://doi.org/10.1007/978-981-99-4045-5_31

1 Introduction

In the field of architecture, engineering, and construction (AEC), the BIM or also known as "Building Information Modeling" plays a significant role as a digital version of construction methodologies that often used or applied in this particular industry. BIM also be pertained to—"Building Information Modeling" as well as "Building Information Model" and also can refer to "Building Information Management" and how this particular process was related and often associated with an intention to produce sophisticated and well-executed output for intended stakeholders and projects [1]. The use of BIM or "Building Information Modeling" is often portrayed to be used with several modeling software applied with technical engineering concepts and principles and with the needed architectural knowledge and skills for its effective application. These may range from 2 to 3D to clash analysis to the deployment of multiple software plug-ins, etc. The "Building Information Model", or BIM, is the product of modeling efforts; it is the computerized design of a building project that is shown to clients and other important parties. "Building Information Management", on the other hand, refers to the large-scale patterns of data processing and communication that involve working together with a specific team and with different clients and other stakeholders.

In relation to the conduct or launching of a particular project, contracts play a significant role in making this possible. Specifically, contracts are considered as the basis of procurement and scope of works. Through the application or execution of contracts, a team of BIM stakeholders closely work with its intended clients in order to effectively carry out the various requirements of different project properties. An effective BIM contracting is considered to be helpful in making contractors make better pricing costs and in terms of effective streamlining of the construction work and also helps in minimizing the tendency of the emergence of human errors that can surface during design or pre-construction phase. As a result, pricing is primarily determined by the procurement and scope of work, which define the bounds of various project deliverables.

According to a study, technical and financial offers represent only a partial scope of what is actually performed due to differences in engineering received during the bidding process. When budgeting with the traditional method, mistakes can occur, reducing the company's margin. The author also mentioned that preparing a good budget requires three factors: reliable engineering information, an executed plan with a deadline for the construction procedure, and sufficient resources to execute it [2].

In order to better understand the stages of a building project with BIM overlay, a study proposed a framework be built with an outline of the cost components of BIM procurement. To help stakeholders understand more about the BIM services at each step of a project, the costs of procurement are broken down into pre-contract and post-contract phases. Each of these phases is then broken down into more transactional phases [3].

Incorporating BIM and Lean construction principles into a structural model increased project scope, timeliness, and decremental effect on the overall total cost while laying the groundwork for rethinking the risks and challenges inherent in the construction industry [4]. This implied that the model proposed by the author showed a reduction in the project cost which made it become related to the present study. Building Information Modeling (BIM) technology adoption has been shown to increase the accuracy of cost predictions, and a structural equation modeling (SEM) approach has been used to analyze these effects. Analysis confirmed that most respondents anticipate more accurate cost estimates to be developed with BIM visualization, its dependable database, and coordinated data. By using BIM, they believe they can cut down on the time spent on cost estimates [5].

A challenge, however, is that BIM in the Philippines has not yet been fully utilized in the design and construction sector. Following research in 2019 by a Filipino AEC professional organization on the topic of BIM adoption, it was concluded that the lack of widespread acceptance of BIM principles in the Philippines' AEC sector is mostly due to the high cost of BIM software and the scarcity of locally available competent labor. It is estimated that just a third of Philippine construction sector players use BIM, back in 2013. Veterans are more likely to stick with the status quo rather than try new things since they're more comfortable with the status quo. There were a number of issues that made BIM adoption challenging in the Philippines. To assess whether BIM had a good impact on the work of construction businesses in Manila, surveys were sent out to those in the industry. Furthermore, the authors explained that the results revealed that 64.8% of the 110 participants agreed with the findings. 33.6% of those surveyed had no idea what BIM was, according to the results. One of the main reasons why BIM isn't widely used is the high cost of BIM software and the scarcity of BIM experts in the Philippines [6].

Because of the expensive cost of BIM-related software, the scarcity of BIM experts, general satisfaction with existing software (CAD), and lack of knowledge, the Philippines' construction sector and government has been reluctant to adopt the full benefits of BIM. Since then, researchers have been analyzing the effects of costing in the AEC industry, notably on the flow of capital. Given its technological foundation, BIM differs in approach from conventional building methods. There is a lack of knowledge of where to start by potential micro and macro businesses in the country, hence, the research believes to start in cost pricing, as it covers the main components of a project contract: procurement, manpower, services, and project duration. A major objective of this study is to develop a framework for BIM services costing and a web-based cost estimate tool that will simplify the process of estimating contract prices, services, and procurements of design projects using BIM technology.

2 Methods

2.1 Methodology

Quantitative and qualitative data were collected and analyzed using descriptive and case study research methods, respectively, for this study. The research is divided into four parts: Part 1, the evaluation and assessment of costing methodologies and general BIM information gathering; Part 2, the collection, analysis, and interpretation of data; Part 3, development of BIM services costing framework and the web-based cost estimate tool: BIM Cost Calculator; and Part 4, the validation of the BIM Cost Calculator's efficacy, accuracy, and effectivity. To back up the theoretical and technical aspects of the study, the researcher has utilized resources in journals, publications, and other similar studies. The data collected will be analyzed statistically through descriptive analysis using answers from BIM professionals from local and international companies based in the Philippines involved in projects using BIM.

2.2 Data Analysis

32 participants have been surveyed, and data from linked studies and archives have been subjected to content analysis and a focus group. The collected information was stratified for better classification. Since the organizations being assessed are expected to have diverse incomes and project costs, this stratification method will be most appropriate. To accomplish this, we will use a fixed response rate as an anchor, and then calculate the fraction of the whole population that falls within our sample. The required stratified random sample size is calculated by taking this sample size and multiplying it by the size of the strata. For the quantitative part of the study and the survey questionnaire, the researcher used descriptive statistics, such as percentages, frequency counts, weighted means, and Pearson correlation.

3 Results

3.1 Performing Comparative Analysis of Traditional Versus BIM-Based Costing Strategy

Table 1 indicates the comparison between traditional and BIM-based cost strategies to determine the most appropriate method to use for costing and pricing estimation in construction projects. As shown in the table, most of the traditional-based cost

Table 1 Comparative Analysis on Traditional vs BIM-based costing strategy

Traditional-based cost strategy		BIM-based cost strategy	
Activity-based costing (ABC)	Undistorted cost estimates. Each activity's resources and materials are listed. Each resource's cost is determined by its receipt. Calculate the material unit cost and list of job-related activities	Parametric estimating method	This is possible by employing regression analysis on a database of two or more systems that share similar characteristics to establish price estimation correlations (CERs), which figure out prices based on the performance of the system or architecture (such as speed, range, weight, and thrust)
General bidding model	Suited to the markets when costs and bids can be predicted. This is the ratio between a contractor's costs and the bids of their competitors	Analogous estimating method	The estimated project costs are based on past expenses from a similar project. This method combines historical data with the project manager's expertise
de Neufville et al. bidding model	Bidders' risk aversion, which incorporates economic conditions, project size, and contractor risk attitudes	Bottom-up estimating method	A large project is often divided into smaller subtasks. The project manager then prepares an expenditure estimate for each manageable work bundle
Fuerst bidding model	Models that account for aspects including worker productivity, weather, external provider efficiency, and unforeseen site conditions	Three-point Estimating	It gives three project-cost projections. In the first case, work and resources are effectively allocated, in the second case, both time and resources are wasted. The third case depicts usually in the midst of two extremes
LOMARK bidding strategy method	Used by micro-to-medium-sized regional businesses. This method figures out the best markup by looking at the local market model as a whole	Data analysis method	It analyzes the recognized possibilities to determine the options or techniques for executing the project's job
Risk-pricing method	This model evaluates and values project risk. This value can be added to the final bid price to get rid of the risk of loss for the contractor	Project management information system method	A method that uses resource management software to plan, organize, and estimate resources

strategies were designed for the bidding process and the costs associated with it. Furthermore, some of the methods are also utilized to estimate the risk associated with performing and developing the construction.

3.2 Determining the Suitable Cost Estimation Method to Use for BIM-Based Projects

Based on the comparative analysis in the previous table, the present study determines the suitable cost estimation to use for BIM-based projects. In Table 2, it was then discovered that among those strategies that have been mentioned, parametric cost estimating method was utilized for the projects which involved BIM technology as compared to activity-based costing under traditional cost strategy.

Table 2 Comparative analysis on activity-based versus parametric cost estimating method

Traditional-based cost strategy		BIM-based cost strategy	
Activity-based costing (ABC)		Parametric estimating method	
Advantages	Identifying costly and non-value-adding operations It helps you question operating costs It enhances product and customer profitability analysis and supports performance management strategies	Advantages	It's more accurate and reliable than comparable estimates and easy to implement Enable to get useful and accurate data Senior management supports it due to the model's quality; and Estimates can be used for one project and repeated for subsequent projects
Dis-advantages	Data collection and compilation take time and money Accounting reports don't always provide source data ABC's reports aren't acceptable for financial reporting purposes ABC's data may be incompatible with managerial performances Not as relevant for businesses with low operating-to-overhead ratios	Disadvantages	Standardizing and generating estimates take a lot of time, effort, and money There must be similarities between projects and tasks to standardize and generate estimates It's difficult to account for socioeconomic, political, and cultural differences; and It's difficult to produce accurate estimates of every pricing or project event driven by parametric benchmarks

3.3 *Identifying the Current Status and Uses of BIM in the Philippines*

Figure and tables below present the current status, uses, and application of BIM in the Philippines. Figures and tables below represent the data gathered from 32 BIM professional respondents from local and international companies based in the Philippines implementing the BIM technology on their design and construction projects.

Figure 1 presents the applications commonly utilized by companies with BIM technology. Among the specified applications, three applications showed the most frequency; 2D and 3D Modeling, Coordination, and Drawing Production, which are the most used applications in the Philippines.

Figure 2 indicates that most of the companies have used Autodesk Revit followed by Autodesk Navisworks. Of these nine identified software products, only 78.10% of the respondents indicated that Autodesk BIM 360 has been utilized while Autodesk AutoCAD are used by 62.50% of them. In relation to the results, this signifies that the most used software is the basic need of any BIM Firms for BIM application of 2D and 3D modeling and coordination in their projects. In this study, the researcher has considered four software: Autodesk Revit, Navisworks, BIM360, and AutoCAD in the web-based costing tool.

Table 3 shows that among the BIM dimensions, 3D modeling dominated with 96.90% respondents. Some companies utilize the use of 4D and 5D. However, few companies use 6D and 7D. The results indicate that majority of BIM companies have established 3D modeling in their BIM works. However, from the results, only few BIM companies based in the Philippines have established 4D to 7D, this may indicate the lack of financial budget and BIM professionals to work on these BIM dimensions.

Table 4 shows that LOD 300 is mostly used in design development, while 84.40% of the respondents used LOD 400 in construction stage. This implies that information

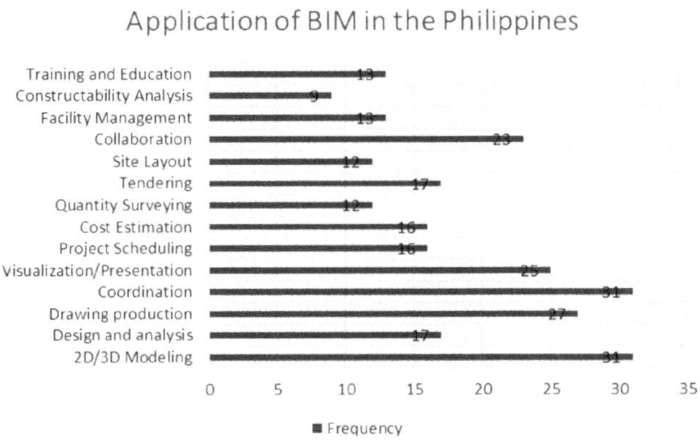

Fig. 1 Application of BIM in the Philippines

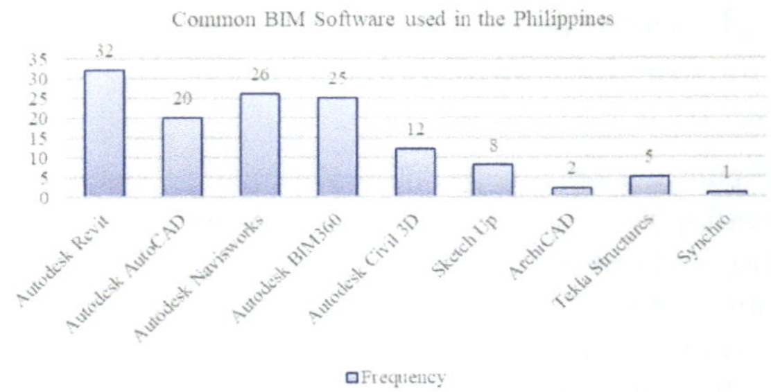

Fig. 2 Common BIM software used in the Philippines

Table 3 Current BIM dimension applied by BIM companies in the Philippines

BIM dimension	Frequency	Percentage
3D (Modeling)	31	96.90
4D (Simulation)	19	59.40
5D (Cost estimation)	18	56.30
6D (Sustainability)	4	12.50
7D (Facility management)	10	31.30

Table 4 Level of development (LOD) applied by BIM companies in the Philippines

Level of detail (LOD)	Frequency	Percentage
LOD 100 (Pre-design)	9	28.10
LOD 200 (Schematic design)	13	40.60
LOD 300 (Design development)	29	90.60
LOD 400 (Construction stage)	27	84.40
LOD 500 (As built)	17	53.10

in relation to design development and construction stage is the most important thing that needs to be place in a BIM model. For this study, the researcher considered all Level of Development (LOD 100 ~ 500) in the web-based costing tool.

Moreover, in Table 5, it illustrated the project parameter used by companies as the basis for their estimation. Areas, volumes, spaces, lengths, perimeters, and other similar measurements are all examples of parameters that can be used in building cost estimates.

Table 5 Project parameter used for cost estimation in the Philippines

Project information	Frequency	Percentage
Project area	26	81.30
Other (volumes, space, lengths, perimeter, etc.)	6	18.70

3.4 Identifying the Key Parameters to Be Considered for the Development of the Cost Estimate Tool

The following results obtained from 32 BIM experts in Philippine-based BIM firms, respectively, indicate the experiences and viewpoints when it comes to manpower and project duration for Architecture, Structure, and MEPFS in BIM applications such as 2D/3D Modeling and Coordination. Manpower organization considered in this study are BIM Modeler/Technician, Coordinator, and Manager. This further presents the difference between the BIM Level of Development based on the number of BIM employees involved and the time allotted in project that has project area of 500 square meter or less up to 10,000 sqm.

To elaborate the organization of manpower in BIM stated above, the AEC (UK) BIM Protocol describes that it is the role of the BIM Modeler/Technician to manage, model information by using BIM authoring software and producing drawings. BIM Coordinators develop and maintain BIM protocols, execution plans, documentation, and coordinate work information flows. Lastly, an individual who is responsible for developing, managing, and implementing BIM Standards is called a BIM Manager [7].

Figure 3, 4, 5 and 6 illustrates the average number of manpower and average project duration in months for floor areas ranging from 500 m^2 or less up to 10,000 m^2, based on the level of development (LOD) 100–500 for Architecture, Structure, and MEPFS disciplines.

In Fig. 3, for project area of 500 m^2 or less, the average BIM Manpower for Architecture, Structure, and MEPFS is shown. For LOD 100 ~ 200, one BIM modeler/ technician and manager are the minimum requirements. For LOD 300 ~ 500, one BIM modeler/technician, coordinator, manager are the minimum requirements. Project duration for Architecture and Structure in LOD 100 ~ 500 is about 1 month; however, it is about 2 months in project duration if the modeling scope of works is MEPFS discipline.

In Fig. 4, for project area of 500–1,000 m^2, the average BIM Manpower for Architecture and Structure for LOD 100 ~ 200 can be seen as one BIM modeler/technician and manager as the minimum requirements. However, for MEPFS discipline under the same LOD, the required BIM modeler/technician becomes two and one manager. For LOD 300 ~ 500, one BIM modeler/technician, coordinator, manager are the minimum requirements if the discipline is Architecture and Structure, and two BIM modelers are required when it comes to MEPFS discipline. Project duration for Architecture and Structure in LOD 100 ~ 500 is about 1 month; however, it is about 2 months in project duration if the modeling scope of works is MEPFS discipline.

In Fig. 5, for project area of 1,000 m^2–5,000 m^2, the average BIM Manpower for Architecture and Structure for LOD 100~2 00 can be seen as three BIM modeler/ technician and one manager as the minimum requirement for each discipline. However, for MEPFS discipline under the same LOD, the required BIM modeler/ technician becomes four and one manager. For LOD 300 for Architecture discipline, four BIM modeler/technician, one coordinator, and manager are the minimum

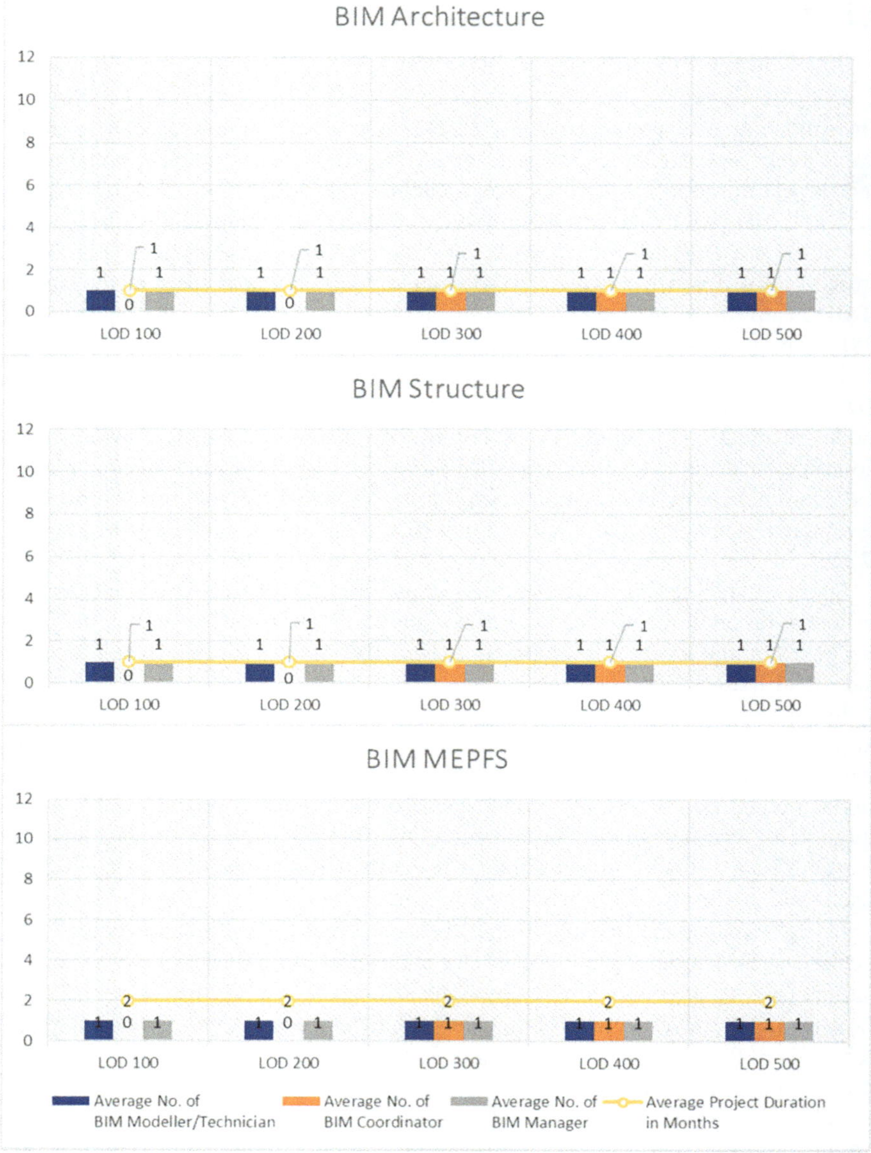

Fig. 3 BIM manpower and project duration for project area 500 m^2 or less

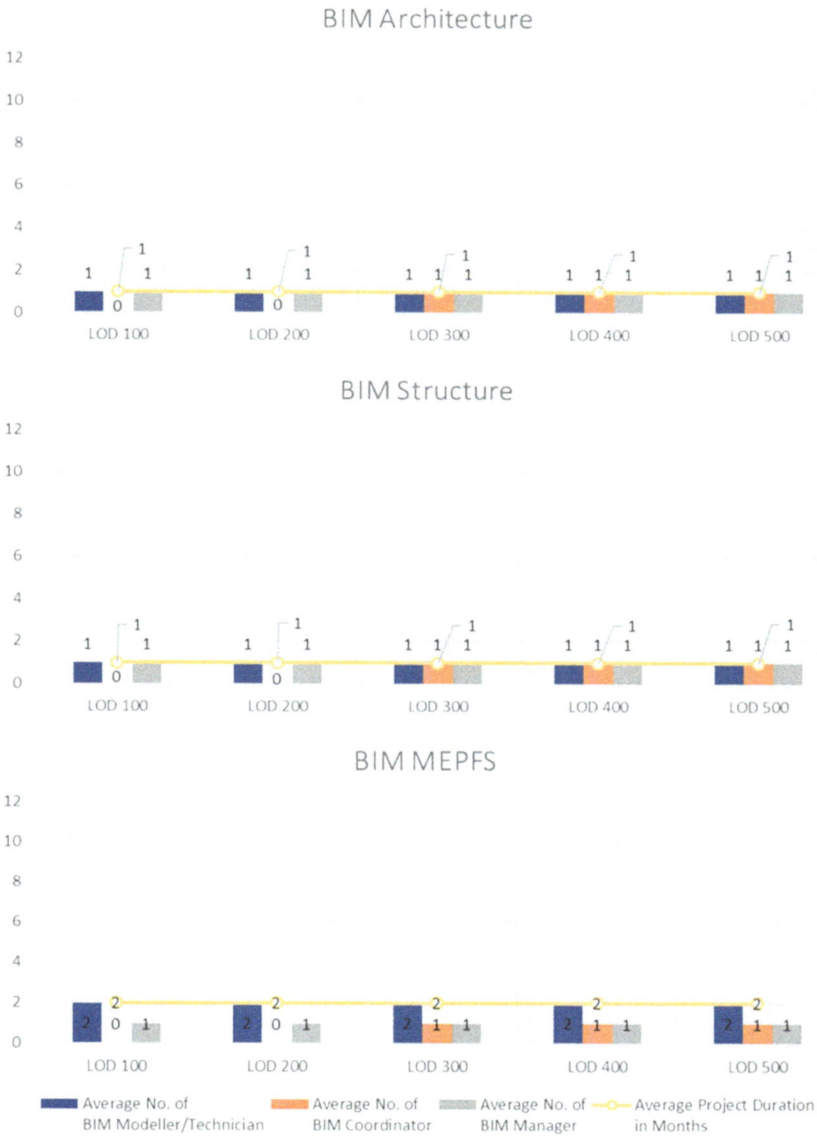

Fig. 4 BIM manpower and project duration for project area 500–1000 m^2

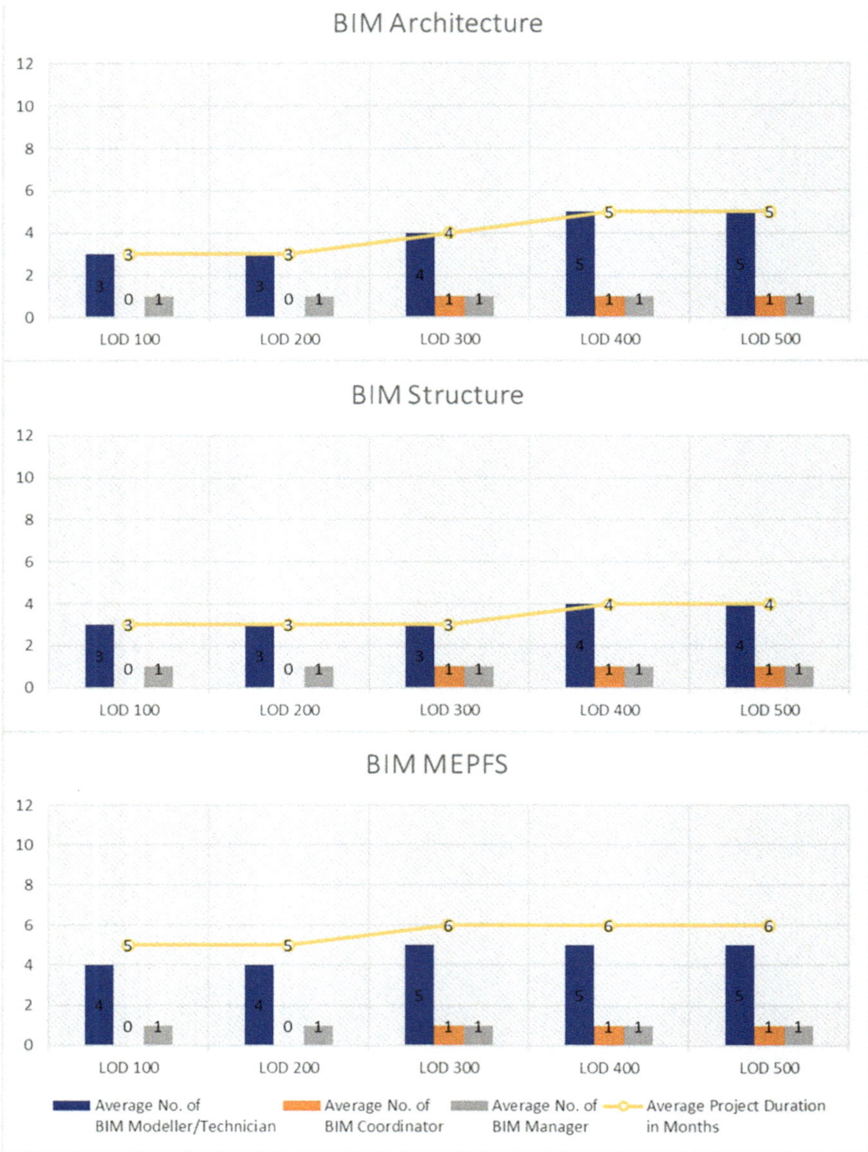

Fig. 5 BIM manpower and project duration for project area 1000–5000 m^2

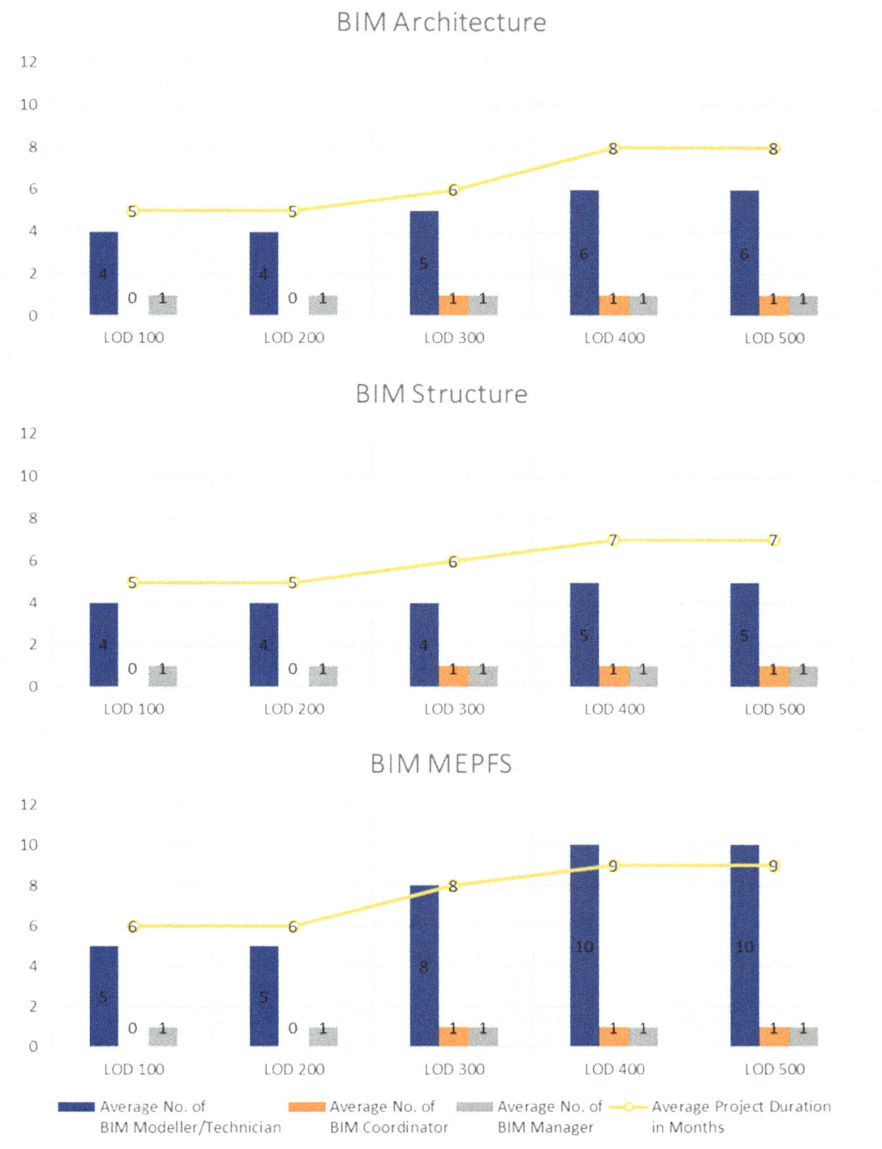

Fig. 6 BIM manpower and project duration for project area 5000–10,000 m^2

requirements while the Structure discipline needs three BIM modeler/technician and about five BIM modeler/technician for MEPFS. For LOD 400~500, both Architecture and MEPFS discipline required five BIM modeler/technician, one coordinator, and manager, compared to Structure, which requires only four BIM modeler/technician, one coordinator, and manager. Project duration for Architecture and Structure is varying from 3 months for LOD 100 up to 5 and 4 months for LOD 500 in both disciplines. Meanwhile, MEPFS project duration varies from 5 months for LOD 100 ~200 up to 6 months for LOD 300~500.

In Fig. 6, for project area of 5,000 m^2–10,000 m^2, the average BIM Manpower for Architecture and Structure for LOD 100~2 00 can be seen as four BIM modeler/ technician and one manager as the minimum requirements for each discipline. However, for MEPFS discipline under the same LOD, the required BIM modeler/ technician becomes five and one manager. For LOD 300 for Architecture discipline, five BIM modeler/technician, one coordinator, and manager are the minimum requirements while the Structure discipline needs four BIM modeler/technician and about eight BIM modeler/technician for MEPFS. For LOD 400 ~500, both Architecture and Structure discipline required six and five BIM modeler/technician, one coordinator, and manager, respectively, compared to MEPFS, which requires 10 number of BIM modeler/technician, one coordinator, and manager. Project duration for Architecture and Structure varies from 5 months for LOD100 up to 8 and 7 months for LOD 500, respectively. Meanwhile, MEPFS project duration varies from 6 months for LOD 100~200 up to 9 months for LOD 300~500.

Moreover, Fig. 7 shows the BIM work complexity of each project type. Based on the results, the complexity of BIM deliverables varies from project to project. As shown in the table, the least complex types of projects in terms of BIM works are structures like workshop, warehouses, educational facilities, and residential type of projects. On the other hand, theaters, laboratories, exhibition halls, health care facilities, mid- and high-rise buildings are the most complex type of project when dealing on BIM deliverables.

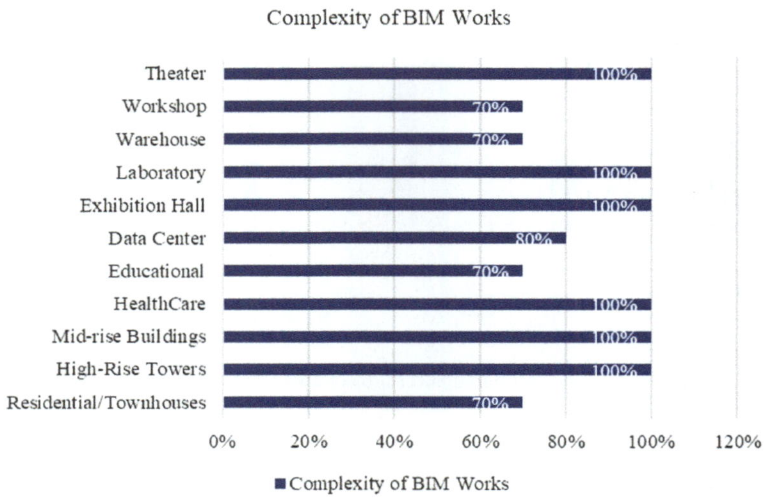

Fig. 7 BIM work complexity for each project type

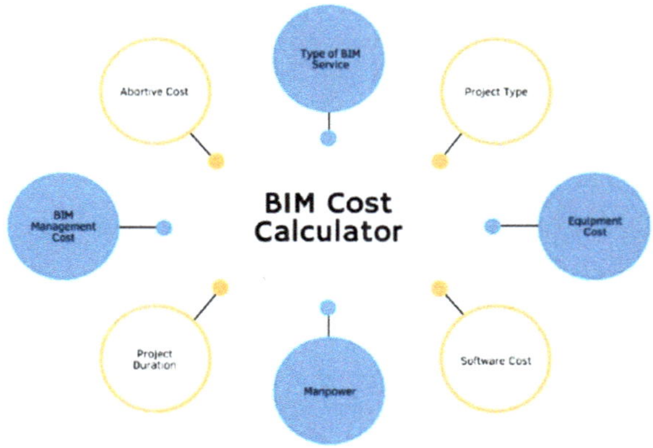

Fig. 8 Framework for the development of BIM cost calculator

3.5 *Development of BIM Services Costing Framework*

Figure 8 shows the developed BIM services costing framework, which serves as the basis and concept of the proposed web-based BIM cost calculator. The framework considered the key parameters needed to properly establish and estimate the cost of different BIM services. These include the type of BIM services, the project type which BIM deliverables complexity varies, the cost of the equipment and software needed, the manpower or human resources cost, the total project duration. Moreover, BIM management and abortive cost were included as this is important to properly manage the BIM projects and have a tolerance for any abortive work.

3.6 *Facilitating the Development of the Web-Based BIM Services Cost Estimate Tool*

From the results and interpretations of survey data and comparative analysis, the researcher has utilized the use of parameters; project floor area (m^2), project duration, project type, and manpower, and adopting parametric cost estimate strategy into the developed web-based costing tool: BIM Cost Calculator. Using angular database and firebase as the framework used and waterfall as the development methodology, BIM Cost Calculator were created.

Figure 9 shows the web-tool home page and cost summary page. The web page contains a simple instruction guide and a tab to direct the users to input parameters. BIM Cost Calculator (web-based tool) will ask the user to provide the project's information, the calculation parameters, the scope of work and client information which when completed will direct the users to the Results summary page. The summary page will show the user the breakdown of cost of BIM services for each discipline as per project requirements.

Fig. 9 Developed BIM cost calculator landing page and cost summary page. https://bim-calculator.web.app/landing

3.7 Evaluating the Performance and Precision of the Newly Built Cost Estimation Tool

After a series of test calculations on the development of web-based costing tool, the researcher has led to adjust the total BIM budget allocation based on actual calculations to verify the BIM Price for each project and is shown in Table 6.

While Table 7 shows that most of the respondents were very satisfied in terms of easily accessible, ease of use, browser compatibility, security, look and feel, overall reliability, and overall performance of BIM Cost Calculator.

Table 6 BIM budget allocation: survey result versus actual

Description	Survey result (%)	Verification (Based on actual) (%)
Technical equipment and software	10–20	66
BIM services: 2D/3D modeling and coordination	20–30	23
BIM services: drawing production	20–30	5
BIM services: management	20–30	3
Abortive costs	10–20	3

Table 7 Effectiveness and accuracy of BIM cost calculator usage

Description	N	Min.	Max.	Mean	Standard deviation	Interpretation
Easily accessible	32	3	6	5.7188	0.63421	Very effective
Ease of use	32	5	6	5.8438	0.3689	Very effective
Browser compatibility	32	5	6	5.9375	0.24593	Very effective
Security	32	3	6	5.5625	0.66901	Very effective
Look and feel	32	3	6	5.8438	0.57414	Very effective
Overall reliability	32	3	6	5.7188	0.63421	Very effective
Overall tool performance	32	5	6	5.875	0.33601	Very effective

Table 8 Significant relationship between BIM cost calculator results' reliability and usage satisfaction

Correlations			
		BIM cost calculator results reliability	BIM cost calculator usage satisfaction
How do you find the result?	Pearson correlation	1	0.258
	Sig. (two-tailed)		0.154
	N	32	32
Overall usage satisfaction	Pearson correlation	0.258	1
	Sig. (two-tailed)	0.154	
	N	32	32

3.8 Evaluating the Significant Relationship Between the BIM Cost Calculator Results' Reliability and Usage Satisfaction

In Table 8, the statistical results revealed that BIM Cost Calculator results' reliability has no significant relationship when compared to the tool usage satisfaction since $r = 0.258$, $p = 0.153$ where p-value is greater than the level of significance ($p < 0.001$).

4 Discussions

The comparison between traditional and BIM-based cost strategies was done to determine the most appropriate method to use for costing and pricing estimation in construction projects. For the traditional-based cost strategy, activity-based costing

(ABC) inputs are tracked, and outputs are attributed a monetary value. On the other hand, parametric cost estimating makes use of both past data and statistical models to pin down precise project expenditures. Parametric cost estimating, also called the cost estimating relationships method (CER's), is a strategy that, according to the study, can help with cost planning for different parts of a design by establishing a set of specifications for the factors that will have the most impact on those costs [8]. In the current 3D (BIM)-based detailed design, the technical level of the cost estimation approach is taken into consideration. Parametric cost estimation, as well as construction cost estimation using this method, involves calculations using parameters, and a crucial component of accurate construction cost estimation lies in extracting valid parameters [9].

Research is needed to identify the parameters that influence the accuracy of the estimates, develop datasets with high objectivity, and propose a BIM method that can make use of the parameters and dataset. Using the parametric method, cost estimation is performed by extracting parameters that reflect the characteristics of the subject [9]. In addition to scheduling and cost projections, Building Information Modeling (BIM) is required for 3D, 4D, and 5D modeling [10, 11]. The current BIM design situation presents the challenges when implementing these methods. As a method of data analysis, the parametric method utilizes datasets with parameters and is suitable when there is insufficient information or data for the scope of the project [12].

Based on the findings, 2D and 3D Modeling, Drawing Production, and Coordination are the most used applications in BIM firms. For the purpose of this study, we considered these applications for the development of web-based cost tools, since we know that they will support the full utilization of BIM in the Philippine construction and design industry.

Similarly, the above figure indicates that most firms have used Autodesk Revit, with 32 respondents (100%). Then Autodesk Navisworks (81.30%). 25 out of 32 respondents (78.10%) reported using Autodesk BIM 360. AutoCAD appears 20 times (62.50%). Accordingly, the most used software is essential for any BIM firm's 2D and 3D modeling, project coordination, and drawing production. In this study, the researcher has considered this software for the development of web-based costing tool. In support, it was noted that BIM refers to "Building Information Modeling," the latest innovation in construction and architectural techniques uses new software and design procedures to design and construct buildings using data-rich 3D models [13].

LOD 300, on one hand, is used most often in design development (90.60%) and construction (84.4%). This means design development and construction information must be included in a BIM model. On this study, all Level of Development (LOD) 100–500 in the web-based costing tool was considered. BIM is critical to the design and development of built environments because it allows architects, engineers, and other professionals to reduce costs and time while ensuring high quality [14].

For estimating, parameters provide the only data quantities that describe a building structure; these may be areas, volumes, spaces, lengths, or perimeters. Parametric cost estimates are calculated using major building parameters [15]. Majority of BIM respondents are using Project Area (m^2) as a parameter to describe a building structure with a frequency of 26 respondents (81.3%), the remaining 6 respondents (18.7%) uses other building description. For this study, the researcher considered Project Area (m^2) in the web-based costing tool as one of the parameters.

According to a study, manpower in any building project is essential. In the Philippines, there are a lot of construction jobs, but very little manpower is available for BIM. There have been many studies on the estimation of materials in the construction sector, but few on the costing of BIM services. However, Construction projects require more than just materials, regardless of how small or big the project is, manpower is still a necessity to do the job [16]. Following the results obtained from the survey, project size and Level of Development (LOD) influence the designation of manpower in a BIM Project. For instance, a project with an area 500 m^2 or less for BIM Architecture, Structure, and MEPFS will require at least one BIM Modeler/Technician for LOD 100~500 as majority of projects with small areas will demand less work complexity for BIM. As area gets larger and imposes a certain level of work complexity through LOD, more manpower will be required for BIM modeler/technician in LOD 100~500. Further, it will require a BIM Manager that will perform management duties. However, for cost practicality, BIM Coordinators will not be required for LOD 100~200 as a BIM Manager could cover such duties.

In a research study, project cost was used as a dependent variable, while revised duration and variation order were used as independent variables [17]. From the results obtained in this study, varying areas and LOD's impacts the course duration of any project, regardless the type or building. For instance, in BIM Architecture, structure, and MEPFS, projects with small areas require at least 1–2 months of project duration in LOD 100~500 applying in consideration the required manpower designation. In project with larger areas may take longer project duration up to 12 months or more applying considerations as well for manpower designation. This study, however, has only considered project area up to 10,000 m^2 that resulted in a duration leading up to 9 months with at least 10 BIM Modelers/technicians. However, if less personnel are assigned, BIM services could take more than 9 months. As project duration increases, increasing manpower utilization costs and the importance of predicting manpower utilization patterns [18].

In 2010, research revealed that it is crucial to identify the complexity of a project at the earliest possible stage in order to properly manage it [19]. Based on the results of the survey, that all the above-mentioned project types have an ideal 70 to 100% BIM works complexity. This significantly shows that high-rise and special-type projects have more critical BIM works that projects such as residential house which are deemed less complex.

To summarize the above investigation, the researcher developed a BIM services costing framework and has become the basis of web-based costing tool: BIM Cost Calculator. The web-tool development has considered the type of BIM Service, Project Type, Equipment Cost, Software Cost, Manpower, Project Duration, BIM Management Cost, and abortive costs which then comprises the functionalities of BIM Cost Calculator.

5 Conclusion

The findings revealed that the parametric estimating method has been considerably used for costing estimation among several BIM-based costing strategies due to its accuracy, appropriateness, and reliability. This method also can generate a lot of different measurable indicators of significance and quality, and it can also be conveniently used to assess the financial effects of changes in design, efficiency, and quality assurance.

Furthermore, the research led to the discovery of the most used services and procurements for BIM in the Philippines. This study has considered BIM applications such as 2D and 3D modeling, coordination, and drawing production in the development of the web-based costing tool as these have been utilized by most BIM Companies in the Philippines. BIM Software such as Autodesk Revit, Autodesk Navisworks, Autodesk BIM360 and Autodesk AutoCAD are most used to BIM-based projects and majority of companies has established the application of all Level of Development 100–500 into their BIM works. These are only some of the various applications that are used. So, this means that there are still businesses in the Philippines that don't use the BIM technology and application to figure out how much architectural, engineering, and construction projects will cost and how much they will cost.

Furthermore, several factors have been taken into account in the evaluation of the BIM implementation by different companies to ensure that the interface and implementation are effective and efficient. As a result, most companies are aware of how crucial and helpful it is to focus their business needs on the core components outlined above, but not all take advantage of this knowledge.

More so, data derived from a structure's building description are determined based on effectiveness and practicability through analysis, categorization of BIM works, and the use of a parametric basis. The research considered Project Area (m^2) in the web-based costing tool as it is widely used by BIM companies in the Philippines to identify the buildings' description.

On the other hand, the research has considered the three manpower assignments per BIM Project; BIM Modeler/Technician, BIM Coordinator, and BIM Manager, this generally composes the BIM organization in a company. The cost of BIM work for architectural, structural, and MEPFS 2D and 3D modeling, drawing production and coordination is based on the project area, the amount of time that is allocated for the project, and the number of personnel that need to be employed based on the areas, and the complexity of the project. Average project durations and manpower results do not have a fixed value but analyzed results are ideal based on experts' experience. At the same time, both fiscal necessity and allocation are taken into consideration. As a direct consequence of this, the organization now possesses the knowledge necessary to assist in the development of a cost estimation tool.

The developed framework of BIM services costing serves as the basis and concept of the proposed web-based BIM cost calculator. The framework considered the key parameters needed to properly establish and estimate the costing of different

BIM services. These include the type of BIM services, the project type in which BIM deliverable complexity varies, the cost of the equipment and software needed, the manpower or human resources cost, the total project duration. Moreover, BIM management and abortive cost were included as this is important to properly manage the BIM projects and have a tolerance for any abortive works.

In addition, specific parameters have been taken into consideration based on the data that have been acquired and have been used to the development of web-based costing tool. These are project floor area (m^2), project duration, project type, and manpower. It was determined that these parameters are significant contributors to the overall cost estimation. Because of this, businesses take the parameter into consideration when developing cost estimation tools for project contract prices for BIM services.

Therefore, contract costing of procurements and services for 2D and 3D Modeling and coordination in this study has relied on parametric estimating strategy and were utilized in the development of web-based costing tool: "BIM Cost Calculator". Likewise, the BIM Cost Calculator has reliable results and is effective based on the underlying indicators.

References

1. Maunula, A., Smeds, R., Hirvensalo, A.: Implementation of building information modeling (BIM)–a process perspective. APMS Innovations Netw. 379–386 (2008)
2. Taboada, J. A., Garrido-Lecca, A.: Case study on the use of BIM at the bidding stage of a building project. In Proceedings IGLC-22, pp. 1473–1482 (2014)
3. Bean, T.K., Mustapa, M., Mustapa, F.D.: A preliminary review on transcation cost components within the BIM adopted procurements. Int. J. Built Environ. Sustain. **6**, 161–167 (2019)
4. Silva, D., De Jesus, K.L., Villaverde, B., Enciso, A.I., Mecija, A.N.: Interdisciplinary framework: a building information modeling using structural equation analysis in lean construction project management. In: Tallón-Ballesteros, A.J. (ed.) Modern Management based on Big Data II and Machine Learning and Intelligent Systems III, pp. 234–240 (2021)
5. Noor, A.A.I., Idris, N.H., Ramli, H., Rooshdi, R.R.R.M., Sahamir, S.R.: The relationship between cost estimates reliability and BIM adoption: SEM analysis. In IOP Conference Series: Earth and Environmental Science, vol. 117 (2018)
6. Rodriguez, L.A.V., Bagcal, O.R., Baccay, M.A., Barbier, B.M.: Adoption of building information modeling (BIM) in the Philippines' AEC industry: prospects issues and challenges. J. Constr. Eng. Technol. Manag. (2019)
7. AEC: AEC (UK) BIM protocol implementing UK BIM standards for the architectural, engineering and construction industry version 2.0 (2012)
8. Cho, J.H., Son, B.S., Chun, J.Y.: Application of OLAP information model to parametric cost estimate and BIM. J. Asian Archit. Build. Eng. **10**, 319–326 (2011)
9. Yang, S.W., Moon, S.W., Jang, H., Choo, S., Kim, S.A.: Parametric method and building information modeling-based cost estimation model for construction cost prediction in architectural planning. Appl. Sci. **12**(19), 9553 (2022)
10. Vassen, S.A.: Impact of BIM-based quantity take off for accuracy of cost estimation. Int. J. Constr. Eng. Manag. **10**(3), 55–69 (2021)
11. Miner, R.G., Thomson, D.B.: Building information modeling-BIM: contractual risks are changing with technology. Consult. Specif. Eng. **40**(2), 54–66 (2006)

12. Hegazy, T., Ayed, A.: Neural network model for parametric cost estimation of highway projects. J. Constr. Eng. Manag. **1998**(124), 210–218 (2004)
13. Moreno, C., Olbina, S., Issa, R.R.: BIM use by architecture, engineering and construction (AEC) industry in educational facility projects. In Advances in Civil Engineering, pp. 1–19 (2019)
14. Hossain, M., Ahmend, S.: Developing an automated safety checking system using BIM: a case study in the Bangladeshi construction industry. Int. J. Constr. Manag. **22**(7), 1206–1224 (2019)
15. Eastman, C., Teicholz, P., Sacks, R., Liston, K.: BIM Handbook: A Guide to Building Information Modeling for Owners, Managers, Designers, Engineers, and Contractors. John Wiley & Sons, Inc., New Jersey (2008)
16. Gangcuangco, D. J., Villaverde, B., Silva, D.: Compressed construction work schedule: modelling of man-hour estimation of engineering activities for housing projects. In ACM International Conference Proceeding Series, pp. 17–21 (2020)
17. Gante, D., Silva, D., Leopoldo, M.: Forecasting construction cost using artificial neural network for road projects in the department of public works and highways region XI. In: Tallón-Ballesteros, A.J. (ed.) Frontiers in Artificial Intelligence and Applications, pp. 64–71 (2022)
18. Wirth, I.: Manpower utilization patterns in project management. Project Manag. Q. **14**(1), 56–63 (1983)
19. Wood, H.L., Ashton, P.: The factors of project complexity. In 18th CIB World Building Congress, pp. 69–80 (2010)